Stephen Lynch

Dynamical Systems with Applications using MATLAB®

Birkhäuser
Boston • Basel • Berlin

Stephen Lynch
Department of Computing and Mathematics
Manchester Metropolitan University
Manchester M1 5GD
United Kingdom
S.Lynch@mmu.ac.uk

MATLAB® and Simulink® are registered trademarks of The MathWorks, Inc.

Mathematics Subject Classification (2000): 34Axx, 34Cxx, 34Dxx, 37Exx, 37Gxx, 58F10, 58F14, 58F21, 78A25, 78A60, 78A97, 92Bxx, 92Exx, 93Bxx, 93Cxx, 93Dxx

Library of Congress Cataloging-in-Publication Data
Lynch, Stephen, 1964-
 Dynamical systems with applications using MATLAB / Stephen Lynch.
  p. cm.
 Includes bibliographical references and indexes.
 ISBN 0-8176-4321-4 (alk. paper)
  1. Differentiable dynamical systems—Data processing. 2. MATLAB. I. Title.

QA614.8.L97 2003
514'.74–dc22   2003062953
                                                            CIP

ISBN-10 0-8176-4321-4
ISBN-13 978-0-8176-4321-8

Printed on acid-free paper.

©2004 Birkhäuser Boston   *Birkhäuser*

All rights reserved. This work may not be translated or copied in whole or in part without the written permission of the publisher (Birkhäuser Boston, c/o Springer Science+Business Media Inc., 233 Spring Street, New York, NY 10013, USA), except for brief excerpts in connection with reviews or scholarly analysis. Use in connection with any form of information storage and retrieval, electronic adaptation, computer software, or by similar or dissimilar methodology now known or hereafter developed is forbidden.
The use in this publication of trade names, trademarks, service marks and similar terms, even if they are not identified as such, is not to be taken as an expression of opinion as to whether or not they are subject to proprietary rights.

Printed in the United States of America.   (JLS/SB)

9 8 7 6 5 4 3 2

www.birkhauser.com

Contents

Preface   **xi**

0 A Tutorial Introduction to MATLAB and the Symbolic Math Toolbox   **1**
    0.1 Tutorial One: The Basics and the Symbolic Math Toolbox (One Hour)   2
    0.2 Tutorial Two: Plots and Differential Equations (One Hour) . . .   4
    0.3 MATLAB Program Files, or M-Files   7
    0.4 Hints for Programming   10
    0.5 MATLAB Exercises   11

1 Linear Discrete Dynamical Systems   **15**
    1.1 Recurrence Relations   16
    1.2 The Leslie Model   21
    1.3 Harvesting and Culling Policies   25
    1.4 MATLAB Commands   30
    1.5 Exercises   30

2 Nonlinear Discrete Dynamical Systems   **35**
    2.1 The Tent Map and Graphical Iterations   36
    2.2 Fixed Points and Periodic Orbits   40
    2.3 The Logistic Map, Bifurcation Diagram, and Feigenbaum Number   48
    2.4 Gaussian and Hénon Maps   55

2.5	Applications	59
2.6	MATLAB Commands	62
2.7	Exercises	66

3 Complex Iterative Maps — 69
3.1	Julia Sets and the Mandelbrot Set	70
3.2	Boundaries of Periodic Orbits	74
3.3	MATLAB Commands	76
3.4	Exercises	79

4 Electromagnetic Waves and Optical Resonators — 81
4.1	Maxwell's Equations and Electromagnetic Waves	82
4.2	Historical Background	84
4.3	The Nonlinear SFR Resonator	89
4.4	Chaotic Attractors and Bistability	91
4.5	Linear Stability Analysis	94
4.6	Instabilities and Bistability	97
4.7	MATLAB Commands	101
4.8	Exercises	104

5 Fractals and Multifractals — 109
5.1	Construction of Simple Examples	110
5.2	Calculating Fractal Dimensions	116
5.3	A Multifractal Formalism	121
5.4	Multifractals in the Real World and Some Simple Examples	126
5.5	MATLAB Commands	133
5.6	Exercises	137

6 Controlling Chaos — 143
6.1	Historical Background	144
6.2	Controlling Chaos in the Logistic Map	148
6.3	Controlling Chaos in the Hénon Map	151
6.4	MATLAB Commands	155
6.5	Exercises	157

7 Differential Equations — 161
7.1	Simple Differential Equations and Applications	162
7.2	Applications to Chemical Kinetics	169
7.3	Applications to Electric Circuits	172
7.4	Existence and Uniqueness Theorem	176
7.5	MATLAB Commands	179
7.6	Exercises	180

Contents

8 Planar Systems **185**
  8.1 Canonical Forms 186
  8.2 Eigenvectors Defining Stable and Unstable Manifolds 190
  8.3 Phase Portraits of Linear Systems in the Plane 193
  8.4 Linearization and Hartman's Theorem 197
  8.5 Constructing Phase Plane Diagrams 199
  8.6 MATLAB Commands 208
  8.7 Exercises 210

9 Interacting Species **215**
  9.1 Competing Species 215
  9.2 Predator–Prey Models 218
  9.3 Other Characteristics Affecting Interacting Species 224
  9.4 MATLAB Commands 225
  9.5 Exercises 226

10 Limit Cycles **229**
  10.1 Historical Background 229
  10.2 Existence and Uniqueness of Limit Cycles in the Plane 232
  10.3 Nonexistence of Limit Cycles in the Plane 237
  10.4 Exercises 241

11 Hamiltonian Systems, Lyapunov Functions, and Stability **243**
  11.1 Hamiltonian Systems in the Plane 244
  11.2 Lyapunov Functions and Stability 249
  11.3 Exercises 253

12 Bifurcation Theory **257**
  12.1 Bifurcations of Nonlinear Systems in the Plane 258
  12.2 Multistability and Bistability 264
  12.3 MATLAB Commands 267
  12.4 Exercises 268

13 Three-Dimensional Autonomous Systems and Chaos **271**
  13.1 Linear Systems and Canonical Forms 272
  13.2 Nonlinear Systems and Stability 276
  13.3 The Rössler System and Chaos 280
  13.4 The Lorenz Equations, Chua's Circuit, and the
        Belousov–Zhabotinski Reaction 284
  13.5 MATLAB Commands 291
  13.6 Exercises 292

14 Poincaré Maps and Nonautonomous Systems in the Plane **297**
  14.1 Poincaré Maps 298

	14.2	Hamiltonian Systems with Two Degrees of Freedom	304
	14.3	Nonautonomous Systems in the Plane	307
	14.4	MATLAB Commands	316
	14.5	Exercises	320

15 Local and Global Bifurcations — 323
- 15.1 Small-Amplitude Limit Cycle Bifurcations 324
- 15.2 Melnikov Integrals and Bifurcating Limit Cycles from a Center . 328
- 15.3 Homoclinic Bifurcations 330
- 15.4 MATLAB Commands 332
- 15.5 Exercises 332

16 The Second Part of Hilbert's Sixteenth Problem — 335
- 16.1 Statement of Problem and Main Results 336
- 16.2 Poincaré Compactification 338
- 16.3 Global Results for Liénard Systems 345
- 16.4 Local Results for Liénard Systems 352
- 16.5 Exercises 354

17 Neural Networks — 359
- 17.1 Introduction 360
- 17.2 The Delta Learning Rule and Backpropagation 366
- 17.3 The Hopfield Network and Lyapunov Stability 370
- 17.4 Neurodynamics 379
- 17.5 MATLAB Commands 383
- 17.6 Exercises 391

18 Simulink — 397
- 18.1 Introduction 398
- 18.2 Electric Circuits 401
- 18.3 A Mechanical System 402
- 18.4 Nonlinear Optics 403
- 18.5 The Lorenz Equations and Chaos Synchronization 405
- 18.6 Exercises 405

19 Solutions to Exercises — 409
- 19.0 Chapter 0 409
- 19.1 Chapter 1 411
- 19.2 Chapter 2 412
- 19.3 Chapter 3 413
- 19.4 Chapter 4 414
- 19.5 Chapter 5 415
- 19.6 Chapter 6 416
- 19.7 Chapter 7 416

Contents

19.8	Chapter 8	417
19.9	Chapter 9	419
19.10	Chapter 10	420
19.11	Chapter 11	422
19.12	Chapter 12	422
19.13	Chapter 13	424
19.14	Chapter 14	425
19.15	Chapter 15	426
19.16	Chapter 16	426
19.17	Chapter 17	427
19.18	Chapter 18	428

References   **429**
   Textbooks 429
   Research Papers 434

MATLAB Program File Index   **443**

Simulink Model File Index   **447**

Index   **449**

Preface

This book provides an introduction to the theory of dynamical systems with the aid of MATLAB®,[1] Simulink®, and the Symbolic Math Toolbox. It is written for final-year undergraduates, postgraduate students, and working scientists in both industry and academia. The author has gone for breadth of coverage rather than fine detail and theorems with proof have been kept at a minimum. The material is not clouded by functional analytic and group theoretical definitions, and so is intelligible to readers with a general mathematical background. The book has a hands-on approach, guiding the student through the theory via example and the beautiful graphical interface of MATLAB. The reader is taken from basic theory right through to recently published research articles. It has been found that some of the more advanced chapters (Chapters 4–6 and 14–18) are especially useful as reference material for senior undergraduate project work.

MATLAB is extremely popular with a wide range of researchers from all sorts of disciplines. It has a very user-friendly interface and extensive visualization and numerical computation capabilities. It is an ideal package to adopt for the study of nonlinear dynamical systems; the numerical algorithms work very quickly, and complex pictures can be plotted within seconds. The Simulink accessory to MATLAB is used for simulating dynamical processes. It is as close as one can get to building apparatus and investigating the output for a given input without the

---

[1] MATLAB and Simulink are registered trademarks of The MathWorks, Inc. For MATLAB product information, please contact The MathWorks, Inc., 3 Apple Hill Drive, Natick, MA 01760-2098 USA, 508-647-7000, Fax: 508-647-7101, info@mathworks.com, www.mathworks.com/.

xii **Preface**

need for an actual physical model. For this reason, Simulink is popular in the field of engineering. The student version of MATLAB comes with a Symbolic Math Toolbox and Simulink built in.

The first chapter of this book provides an efficient tutorial introduction to MATLAB. New users will find that the tutorials enable them to become familiar with MATLAB within a few hours. Both engineering and mathematics students appreciate this method of teaching, and the author has found that it generally works well with a ratio of one staff member to about 20 students in a computer laboratory. Those moderately familiar with the package and even the expert users will find Chapter 0 to be a useful source of reference. The MATLAB program files are listed at the end of each chapter to avoid unnecessary cluttering in the text. The MATLAB programs have been kept as simple as possible and should run under later versions of the package. The MATLAB program files and Simulink model files (including updates) can be downloaded from the Web at

http://www.mathworks.com/matlabcentral/fileexchange/

under the links "Companion Software for Books" and "Mathematics." Readers will find that they can reproduce the figures given in the text, and then it is not too difficult to change parameters or equations to investigate other systems.

Chapters 1–6 introduce the theory of discrete dynamical systems. Starting with discrete systems gives the reader a more gentle introduction to the MATLAB package. Chapter 1 starts with a general introduction to iteration and linear recurrence (or difference) equations. The bulk of the chapter is concerned with the Leslie model used to investigate the population of a single species split into different age classes. Harvesting and culling policies are then investigated and optimal solutions are sought. Nonlinear discrete dynamical systems are dealt with in Chapter 2. Bifurcation diagrams, chaos, intermittency, Lyapunov exponents, periodicity, quasiperiodicity, and universality are some of the topics introduced. The theory is then applied to real-world problems from a broad range of disciplines including population dynamics, biology, economics, nonlinear optics, and neural networks. The next chapter is concerned with complex iterative maps; Julia sets and the now famous Mandelbrot set are plotted. Basins of attraction are investigated for the first time in this text. As a simple introduction to optics, electromagnetic waves and Maxwell's equations are studied at the beginning of Chapter 4. Complex iterative equations are used to model the propagation of light waves through nonlinear optical fibers. A brief history of nonlinear bistable optical resonators is discussed and the simple fiber ring resonator is analyzed in particular. Chapter 4 is devoted to the study of these optical resonators and phenomena such as bistability, chaotic attractors, feedback, hysteresis, instability, linear stability analysis, multistability, nonlinearity, and steady states are discussed. Some simple fractals may be constructed using pencil and paper in Chapter 5, and the concept of fractal dimension is introduced. Fractals may be thought of as identical motifs repeated on ever-reduced scales. Unfortunately, most of the fractals appearing in nature are not homogeneous but are more heterogeneous, hence the need for the multifractal theory given later in the chapter. It has been found that the distribution of stars and galaxies in our universe is multifractal, and there is even evidence of multifractals in the stock markets and heartbeat rhythms. Applications in materials science, geoscience, and image processing are briefly discussed. The next chapter is devoted to the new and exciting theory behind chaos control. For most systems, the maxim used by engineers in the past has been "stability good, chaos bad," but more often nowadays this is being replaced with "stability good, chaos better." There are exciting and novel applications in cardiology, engineering, laser technology, and space research, for example.

The next ten chapters are concerned with the theory of continuous dynamical systems. These chapters cover some theory of ordinary differential equations and applications to models in the real world. The Symbolic Math Toolbox can help when finding analytical solutions. A variety of numerical procedures are available when solving stiff and nonstiff systems when an analytic solution does not exist or is extremely difficult to find. The theory of differential equations applied to chemical kinetics and electric circuits is discussed in some detail in Chapter 7, which ends with the existence and uniqueness theorem for the solutions of certain types of differential equations. The theory behind the construction of phase plane portraits for two-dimensional systems is dealt with in Chapter 8. Applications are taken from chemical kinetics, economics, electronics, epidemiology, mechanics, population dynamics; and modeling the populations of interacting species is discussed in some detail in Chapter 9. Limit cycles, or isolated periodic solutions, are investigated in Chapter 10. Since we live in a periodic world, these are the most common form of solution found when modeling both the technological and natural sciences. Hamiltonian, or conservative, systems and stability are discussed in Chapter 11, and Chapter 12 is concerned with how planar systems vary depending upon a parameter. Bifurcation, bistability, and multistabilty are discussed.

Three-dimensional systems are introduced in Chapter 13. These higher-dimensional systems can exhibit strange attractors and chaotic dynamics. One can rotate the three-dimensional objects in MATLAB and plot time series plots to get a better understanding of the dynamics involved. Once again the theory can be applied to chemical kinetics (including stiff systems), electric circuits, and epidemiology; a simplified model for the weather is also briefly discussed. The next chapter deals with Poincaré first return maps that can be used to untangle complicated interlacing trajectories in higher-dimensional spaces. A periodically driven nonlinear pendulum is also investigated by means of a nonautonomous differential equation. Both local and global bifurcations are investigated in Chapter 15. The main results and statement of the famous second part of Hilbert's sixteenth problem are listed in Chapter 16. To understand these results, Poincaré compactification is introduced. The study of continuous systems ends with one of the author's specialities—limit cycles of Liénard systems. There is some detail on Liénard sysxiv   **Preface**

tems in particular in this part of the book, but they do have a ubiquity for systems in the plane.

A brief introduction to the enticing field of neural networks is presented in Chapter 17. Imagine trying to make a computer mimic the human brain. One could ask the question: In the future will it be possible for computers to think and be conscious? The human brain will always be more powerful than traditional, sequential, logic-based digital computers and scientists are trying to incorporate some features of the brain into modern computing. Neural networks perform through learning and no underlying equations are required. Mathematicians and computer scientists are attempting to mimic the way neurons work together via synapses; indeed, a neural network can be thought of as a crude multidimensional model of the human brain. The potential for this theory is still largely unexplored, but the expectations are high for future applications in a broad range of disciplines. Neural networks are already being used in pattern recognition (credit card fraud, prediction and forecasting, disease recognition, facial and speech recognition), psychological profiling, and control problems, for example. They also provide a parallel architecture allowing for very fast computational and response times. In recent years, the disciplines of neural networks and nonlinear dynamics have increasingly coalesced and a new branch of science called neurodynamics is emerging. Lyapunov functions can be used to determine the stability of certain types of neural networks. There is also evidence of chaos, feedback, nonlinearity, periodicity, and chaos synchronization in the brain.

Examples of Simulink models, referred to in earlier chapters in the book, are presented in Chapter 18. It is possible to change the type of input into the system, or parameter values, and investigate the output very quickly. The chapter ends with an example of chaos synchronization in two Lorenz systems. This is as close as one can get to experimentation without the need for expensive equipment.

Both textbooks and research papers are presented in the list of references. The textbooks can be used to gain more background material, and the research papers have been given to encourage further reading and independent study.

This book is informed by the research interests of the author, which are currently nonlinear ordinary differential equations, nonlinear optics, multifractals, and neural networks. Some references include recently published research articles by the author.

The prerequisites for studying dynamical systems using this book are undergraduate courses in linear algebra, real and complex analysis, calculus and ordinary differential equations; a knowledge of a computer language such as C or Fortran would be beneficial but not essential.

I would like to express my sincere thanks to MathWorks for supplying me with MATLAB, Simulink, and all their add-ons. My thanks to Min Poh for her technical help with MATLAB. Thanks also go to the four reviewers from MathWorks for their meticulous reading of the first draft. Special thanks go to Tom Grasso, Seth Barnes, and Ann Kostant (Executive Editor, Mathematics and Physics) at

Birkhäuser; to Tricia Manning for her careful and thorough copy edit; to Elizabeth Loew (TeXniques) for all her help with the book's cover, production, and manufacturing; and to John Spiegelman (John L. Spiegelman Publication Services) for typesetting my LaTeX files. I would also like to acknowledge the following individuals for their help with certain topics in the book: Zuhair Bandar (MMU) and Martin Brown (UMIST), Alan Steele (Carleton, Canada), Lida Nejad (MMU), and my Ph.D. supervisor, Noel Lloyd (University College of Wales, Aberystwyth). Finally, thanks to my family and especially my wife Gaynor, and our new baby son Sebastian, for all their love, inspiration, and support.

Stephen Lynch

*Dynamical Systems
with Applications
using MATLAB®*

0
A Tutorial Introduction to MATLAB and the Symbolic Math Toolbox

Aims and Objectives

- To provide a tutorial guide to MATLAB.
- To give practical experience in using the package.
- To promote self-help using the online help facilities.
- To provide a concise reference source for experienced users.

On completion of this chapter the reader should be able to

- use MATLAB and the Symbolic Math Toolbox;
- produce simple MATLAB programs;
- access some MATLAB commands and programs over the Web.

It is assumed that the reader is familiar with either the Windows or UNIX platform. This book was prepared using MATLAB (Version 6[1]) but most programs should work under earlier and later versions of the package.

The commands and programs listed in this chapter have been chosen to allow the reader to become familiar with MATLAB and the Symbolic Math Toolbox

---

[1] As this book was in production, The MathWorks, Inc. was preparing to release MATLAB 7. Please note that the program files presented in this book will work with the updated version of MATLAB.

within a few hours. They provide a concise summary of the type of commands that will be used throughout the text. New users should be able to start on their own problems after completing the chapter, and experienced users should find this chapter an excellent source of reference. Of course, there are many MATLAB textbooks on the market for those who require further applications or more detail.

If you experience any problems there are several options. There is an excellent index within MATLAB, and MATLAB program files (or M-files) can be downloaded from the MATLAB Central File Exchange at

http://www.mathworks.com/matlabcentral/.

The M-files can be found at the link "Companion Software for Books." Download the zipped M-files and **Extract** the relevant M-files from the archive onto your computer.

You can either run the tutorials interactively in the Command Window or save the commands onto an M-file. When in the Command Window, command lines may be returned to and edited using the up and down arrow keys. M-files will be discussed in Section 0.3.

0.1 Tutorial One: The Basics and the Symbolic Math Toolbox (One Hour)

There is no need to copy the comments; they are there to help you. Click on the MATLAB icon and copy the command after the >> prompt. Press **Return** at the end of a line to see the answer or use a semicolon to suppress the output. You can interrupt a computation at any time by holding **CTRL+C**. This tutorial should be run in the Command Window.

MATLAB Commands	Comments
`>> % This is a comment.`	`% Helps when writing` `% programs.`
`>> 3^2*4-3*2^5*(4-2)`	`% Simple arithmetic.`
`>> sqrt(16)`	`% The square root.`
`>> % Vectors and Matrices.`	
`>> u=1:2:9`	`% A vector u=(1,3,5,7,9).`
`>> v=u.^2`	`% A vector v=(1,9,25,49,81).`
`>> A=[1,2;3,4]`	`% A is a 2 by 2 matrix.`
`>> A'`	`% The transpose of matrix A.`
`>> det(A)`	`% The determinant of a matrix.`
`>> B=[0,3,1;0.3,0,0;0,0.5,0]`	`% B is a 3 by 3 matrix.`

0.1. Tutorial One: The Basics and the Symbolic Math Toolbox (One Hour)   3

```
>> eig(B)                        % The eigenvalues of B.
>> % List the eigenvectors and eigenvalues of B.
>> [Vects,Vals]=eig(B)
>> C=[100;200;300]               % A 3 by 1 matrix.
>> D=B*C                         % Matrix multiplication.
>> E=B^4                         % Powers of a matrix.
>> % Complex numbers.
>> % The semicolons suppress the output.
>> z1=1+i;z2=1-i;z3=2+i;
>> z4=2*z1-z2*z3                 % Simple arithmetic.
>> abs(z1)                       % The modulus.
>> real(z1)                      % The real part.
>> imag(z1)                      % The imaginary part.
>> exp(i*z1)                     % Returns a complex number.
>> % The Symbolic Math Toolbox.
>> sym(1/2)+sym(3/4)             % Symbolic evaluation.
>> % Double precision floating point arithmetic.
>> 1/2+3/4
>> % Variable precision arithmetic (50 d.p's).
>> vpa('pi',50)
>> syms x y z                    % Symbolic objects.
>> z=x^3-y^3
>> factor(z)                     % Factorization.
>> % Symbolic expansion of the last answer.
>> expand(ans)
>> simplify(z/(x-y))             % Symbolic simplification.
>> % Symbolic solution of algebraic equations.
>> [x,y]=solve('tau*x*y','tau*x*y-y')
>> f='mu*x*(1-x)'                % Define a function.
>> subs(f,x,1/2)                 % Substitute for x, gives
                                 % f(1/2).
```

```
>> fof=subs(f,x,f)              % A function of a function.
>> syms x
>> limit(x/sin(x),x,0)          % Limits.
>> diff(f,x)                    % Differentiation.
>> % Partial differentiation (w.r.t. y twice).
>> diff('x^2+3*x*y-2*y^2','y',2)
>> int('sin(x)*cos(x)',x,0,pi/2)
                                % Integration.
>> int('1/x',x,0,inf)           % Improper integration.
>> syms n
>> s1=symsum(1/n^2,1,inf)       % Symbolic summation.
>> syms x
>> g=exp(x)
>> taylor(g,10)                 % Taylor series expansion.
>> syms x s
>> laplace(x^3)                 % Laplace transform.
>> ilaplace(6/s^4)              % Inverse Laplace transform.
>> syms x w
>> fourier(-2*pi*abs(x))        % Fourier transform.
>> ifourier(4*pi/w^2)           % Inverse Fourier transform.
>> quit                         % Exit MATLAB.
>> % End of Tutorial 1.
```

0.2 Tutorial Two: Plots and Differential Equations (One Hour)

More involved plots will be discussed in Section 0.3. You will need the Symbolic Math Toolbox for most of the commands in this section.

MATLAB Commands

```
>> % Plotting graphs.

>> % Set up the domain and plot a simple function.
>> x=-2:.01:2;
>> plot(x,x.^2)
```

0.2. Tutorial Two: Plots and Differential Equations (One Hour)

```
>> % Plot two curves on one graph and label them.
>> t=0:.1:100;
>> y1=exp(-.1.*t).*cos(t);y2=cos(t);
>> plot(t,y1,t,y2),legend('y1','y2')

>> % Plotting with the Symbolic Math Toolbox.
>> ezplot('x^2',[-2,2])

>> % Plot with labels and title.
>> ezplot('exp(-t)*sin(t)'),xlabel('time'),ylabel('current'),
   title('decay')

>> % Contour plot on a grid of 50 by 50.
>> ezcontour('y^2/2-x^2/2+x^4/4',[-2,2],50)

>> % Surface plot. Rotate the plot using the mouse.
>> ezsurf('y^2/2-x^2/2+x^4/4',[-2,2],50)

>> % Combined contour and surface plot.
>> ezsurfc('y^2/2-x^2/2+x^4/4',[-2,2])

>> % A parametric plot.
>> ezplot('t^3-4*t','t^2',[-3,3])

>> % 3D parametric curve.
>> ezplot3('sin(t)','cos(t)','t',[-10,10])

>> % Implicit plot.
>> ezplot('2*x^2+3*y^2-12')
```

Symbolic solutions to ordinary differential equations (ODEs). If an analytic solution cannot be found the message "Warning: explicit solution could not be found" is shown.

```
>> % The D denotes differentiation w.r.t. t.
>> dsolve('Dx=-x/t')

>> % An initial value problem.
>> dsolve('Dx=-t/x','x(0)=1')

>> % A second-order ODE
>> dsolve('D2I+5*DI+6*I=10*sin(t)','I(0)=0','DI(0)=0')

>> % A linear system of ODEs.
>> [x,y]=dsolve('Dx=3*x+4*y','Dy=-4*x+3*y')

>> % A system with initial conditions.
```

Table 0.1: ODE function summary. There are other solvers, but these should suffice for the material covered in this book.

Solver	ODE	Method	Summary
ode45	Nonstiff	Runge–Kutta	Try this first
ode23	Nonstiff	Runge–Kutta	Crude tolerances
ode113	Nonstiff	Adams	Multistep solver
ode15s	Stiff	Gears	If you suspect stiffness
ode23s	Stiff	Rosenbrock	One-step solver

```
>> [x,y]=dsolve('Dx=x^2','Dy=y^2','x(0)=1,y(0)=1')

>> % A 3D linear system.
>> [x,y,z]=dsolve('Dx=x','Dy=y','Dz=-z')
```

Numerical Solutions to ODEs. These methods are usually used when an analytic solution cannot be found—even though one may exist! Table 0.1 lists some initial value problem solvers in MATLAB. See the online MATLAB index for more details or type >> help *topic*.

```
>> % Set up a 1D system with initial condition 50,
>> % and run for 0<t<100.
>> deq1=inline('x(1)*(.1-.01*x(1))','t','x');
>> [t,xa]=ode45(deq1,[0 100],50);
>> plot(t,xa(:,1))

>> % A 2D system with initial conditions [.01,0],
>> % and run for 0<t<50.
>> deq2=inline('[.1*x(1)+x(2);-x(1)+.1*x(2)]','t','x');
>> [t,xb]=ode45(deq2,[0 50],[.01,0]);
>> plot(xb(:,1),xb(:,2))

>> % A 3D system with initial conditions.
>> deq3=inline('[x(3)-x(1);-x(2);x(3)-17*x(1)+16]','t','x');
>> [t,xc]=ode45(deq3,[0 20],[.8,.8,.8]);
>> plot3(xc(:,1),xc(:,2),xc(:,3))

>> % A phase plane portrait of a stiff Van der Pol system.
>> deq4=inline('[x(2);1000*(1-(x(1))^2)*x(2)-x(1)]','t','x');
>> [t,xd]=ode23s(deq4,[0 3000],[.01,0]);
>> plot(xd(:,1),xd(:,2))

>> % Displacement against time.
>> plot(t,xd(:,1))

>> % End of Tutorial 2.
```

0.3 MATLAB Program Files, or M-Files

Sections 0.1 and 0.2 illustrate the use of the Command Window and its interactive nature. More involved tasks will require more lines of code and it is better to use MATLAB program files, or M-files. Each program is displayed between horizontal lines and kept short to aid in understanding; the output is also included.

Function M-files. The following M-file defines the logistic function.

```
% Program 1 - The logistic function.
% Save the M-file as fmu.m.
function y=fmu(mu,x)
y=mu*x*(1-x);
% End of Program 1.

>> fmu(2,.1)

ans = 0.1800
```

The for... end loop. For repetitive tasks.

```
% Program 2 - The Fibonacci sequence.
% Save the M-file as Fibonacci.m.
Nmax=input('Input the number of terms of the Fibonacci
            sequence:')
F=zeros(1,Nmax);
F(1)=1;F(2)=1;
for i=3:Nmax
    F(i)=F(i-1)+F(i-2);
end
F
% End of Program 2.

>>Fibonacci
Input the number of terms of the Fibonacci sequence:10

Nmax=10

F = 1  1  2  3  5  8  13  21  34  55
```

Conditional statements. If... then... elseif... else... end, etc.

```
% Program 3 - The use of logical expressions.
% Save this M-file as test1.m.
integer=input('Which integer do you wish to test?')
if integer == 0
    disp(['The integer is zero.'])
elseif integer < 0
```

0. A Tutorial Introduction to MATLAB and the Symbolic Math Toolbox

```
    disp(['The integer is negative.'])
else
    disp(['The integer is positive.'])
end
% End of Program 3.

>> test1
Which integer do you wish to test?-100

integer=-100

The integer is negative.
```

Graphic programs. Multiple plots with text.

```
% Program 4 - Two curves on one graph (Figure 0.1).
% Save the M-file as plot1.m.
% Tex symbols can be plotted within a figure window.
clear all
fsize=15;
x=0:.1:7;
figure;
plot(x,2*x./5,x,-x.^2/6+x+7/6)
text(1,2,'\leftarrow curve 1','Fontsize',fsize)
text(2.6,1,'\leftarrow line 1','Fontsize',fsize)
xlabel('x','Fontsize',fsize);
ylabel('y','Fontsize',fsize)
title('Two curves on one plot','Fontsize',fsize)
% End of Program 4.

>> plot1
```

Labeling solution curves. Solutions to the pendulum problem.

```
% Program 5 - Solution curves to different ODEs (Figure 0.2).
% Save M-file as deplot1.m.
clear;
figure;
hold on
deqn=inline('[x(2);-25*x(1)]','t','x');
[t,xa]=ode45(deqn,[0 10],[1 0]);
plot(t,xa(:,1),'r');
clear
deqn=inline('[x(2);-2*x(2)-25*x(1)]','u','x');
[t,xb]=ode45(deqn,[0 10],[1 0]);
plot(t,xb(:,1),'b');
legend('harmonic motion','damped motion')
```

0.3. MATLAB Program Files, or M-Files 9

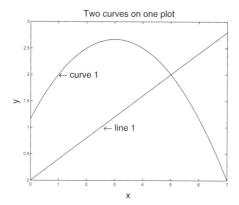

Figure 0.1: A multiple plot with text.

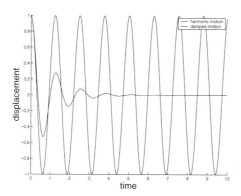

Figure 0.2: Harmonic and damped motion of a pendulum.

```
xlabel('time','Fontsize',15);
ylabel('displacement','Fontsize',15);
hold off
% End of Program 5.

>> deplot1
```

Echo commands. Useful when debugging a program.

```
% Program 6 - Use echo to display command lines.
% Save the M-file as poincare.m.
echo on
r(1)=1;
for n=2:5
```

```
    r(n)=r(n-1)/(1+2*pi*r(n-1))
end
echo off
% End of Program 6.

>> poincare
r(n)=r(n-1)/(1+2*pi*r(n-1))

r = 1.0000    0.1373    0.0737    0.0504    0.0383
```

0.4 Hints for Programming

The MATLAB language contains very powerful commands, which means that some complex programs may contain only a few lines of code. Of course, the only way to learn programming is to sit down and try it. This section has been included to point out common errors and give advice on how to troubleshoot. Remember to check the help and index pages in MATLAB and the Web if the following does not help with your particular problem.

Common Typing Errors. The author strongly advises new users to run Tutorials 0.1 and 0.2 in the Command Window environment, which should reduce typing errors.

- Be careful to include all mathematics operators.
- When dealing with vectors put a full stop (.) in front of the mathematical operator.
- Make sure brackets, parentheses, etc., match up in correct pairs.
- Remember MATLAB is case sensitive.
- Check the syntax, enter **help** followed by the name of the command.

Programming Tips. The reader should use the M-files in Section 0.3 to practice simple programming techniques.

- It is best to clear values at the start of a large program.
- Use the **figure** command when creating a new figure.
- Preallocate arrays to prevent MATLAB from having to continually resize arrays—for example, use the zeros command. An example is given in Program 2 above.
- Vectorize where possible. Convert for and while loops to equivalent vector or matrix operations.

0.5. MATLAB Exercises

- If a program involves a large number of iterations, for example 50,000, then run it for three iterations first and use the echo command.
- If the computer is not responding hold **CTRL+C** and try reducing the size of the problem.
- Click on **Debug**, **Save and Run** in the M-file window; MATLAB will list any error messages.
- Find a similar M-file in a book or on the Web, and edit it to meet your needs.
- Make sure you have the correct packages, for example, the symbolic commands will not work if you do not have the Symbolic Math Toolbox.
- Check which version of MATLAB you are using.

0.5 MATLAB Exercises

1. Evaluate the following:

    (a) $4 + 5 - 6$;

    (b) 3^{12};

    (c) $\sin(0.1\pi)$;

    (d) $(2 - (3 - 4(3 + 7(1 - (2(3 - 5))))))$;

    (e) $\frac{2}{5} - \frac{3}{4} \times \frac{2}{3}$.

2. Given that

$$A = \begin{pmatrix} 1 & 2 & -1 \\ 0 & 1 & 0 \\ 3 & -1 & 2 \end{pmatrix}, \quad B = \begin{pmatrix} 1 & 2 & 3 \\ 1 & 1 & 2 \\ 0 & 1 & 2 \end{pmatrix}, \quad C = \begin{pmatrix} 2 & 1 & 1 \\ 0 & 1 & -1 \\ 4 & 2 & 2 \end{pmatrix},$$

    determine the following:

    (a) $A + 4BC$;

    (b) the inverse of each matrix if it exists;

    (c) A^3;

    (d) the determinant of C;

    (e) the eigenvalues and eigenvectors of B.

3. Given that $z_1 = 1 + i$, $z_2 = -2 + i$, and $z_3 = -i$, evaluate the following:

    (a) $z_1 + z_2 - z_3$;

    (b) $\frac{z_1 z_2}{z_3}$;

(c) e^{z_1};

(d) $\ln(z_1)$;

(e) $\sin(z_3)$.

4. Evaluate the following limits if they exist:

   (a) $\lim_{x \to 0} \frac{\sin x}{x}$;

   (b) $\lim_{x \to \infty} \frac{x^3 + 3x^2 - 5}{2x^3 - 7x}$;

   (c) $\lim_{x \to \pi} \frac{\cos x + 1}{x - \pi}$;

   (d) $\lim_{x \to 0^+} \frac{1}{x}$;

   (e) $\lim_{x \to 0} \frac{2 \sinh x - 2 \sin x}{\cosh x - 1}$.

5. Find the derivatives of the following functions:

   (a) $y = 3x^3 + 2x^2 - 5$;

   (b) $y = \sqrt{1 + x^4}$;

   (c) $y = e^x \sin x \cos x$;

   (d) $y = \tanh x$;

   (e) $y = x^{\ln x}$.

6. Evaluate the following definite integrals:

   (a) $\int_{x=0}^{1} 3x^3 + 2x^2 - 5 \, dx$;

   (b) $\int_{x=1}^{\infty} \frac{1}{x^2} dx$;

   (c) $\int_{-\infty}^{\infty} e^{-x^2} dx$;

   (d) $\int_0^1 \frac{1}{\sqrt{x}} dx$;

   (e) $\int_0^{\frac{2}{\pi}} \frac{\sin(1/t)}{t^2} dt$.

7. Graph the following:

   (a) $y = 3x^3 + 2x^2 - 5$;

   (b) $y = e^{-x^2}$ for $-5 \leq x \leq 5$;

   (c) $x^2 - 2xy - y^2 = 1$;

   (d) $z = 4x^2 e^y - 2x^4 - e^{4y}$ for $-3 \leq x \leq 3$ and $-1 \leq y \leq 1$;

   (e) $x = t^2 - 3t$, $y = t^3 - 9t$ for $-4 \leq t \leq 4$.

8. Solve the following differential equations:
   (a) $\frac{dy}{dx} = \frac{x}{2y}$, given that $y(1) = 1$;
   (b) $\frac{dy}{dx} = \frac{-y}{x}$, given that $y(2) = 3$;
   (c) $\frac{dy}{dx} = \frac{x^2}{y^3}$, given that $y(0) = 1$;
   (d) $\frac{d^2x}{dt^2} + 5\frac{dx}{dt} + 6x = 0$, given that $x(0) = 1$ and $\dot{x}(0) = 0$;
   (e) $\frac{d^2x}{dt^2} + 5\frac{dx}{dt} + 6x = \sin(t)$, given that $x(0) = 1$ and $\dot{x}(0) = 0$.

9. Carry out 100 iterations on the recurrence relation
$$x_{n+1} = 4x_n(1 - x_n),$$
given that (a) $x_0 = 0.2$ and (b) $x_0 = 0.2001$. List the final 10 iterates in each case.

10. Type "help while" after the >> prompt to read the help page on the while command. Use a while loop to program Euclid's algorithm for finding the greatest common divisor of two integers. Use your program to find the greatest common divisor of 12,348 and 14,238.

Recommended Textbooks

[1] W. Palm III, *Introduction to MATLAB 7 for Engineers*, McGraw–Hill, New York, 2004.

[2] B. Daku, *Learn MATLAB Fast: The M-Tutor Interactive Tutorial*, John Wiley, New York, 2002.

[3] B. R. Hunt, R. L. Lipsman, J. Rosenberg, and K. R. Coombes, *A Guide to MATLAB*, Cambridge University Press, Cambridge, UK, 2001.

[4] R. Pratab, *Getting Started with MATLAB: Version 6: A Quick Introduction for Scientists and Engineers*, Oxford University Press, London, 2001.

[5] D. Hanselman and B. R. Littlefield, *Mastering MATLAB 6*, Prentice–Hall, Upper Saddle River, NJ, 2000.

1
Linear Discrete Dynamical Systems

Aims and Objectives

- To introduce recurrence relations for first- and second-order difference equations.
- To introduce the theory of the Leslie model.
- To apply the theory to modeling the population of a single species.

On completion of this chapter, the reader should be able to

- solve first- and second-order homogeneous linear difference equations;
- find eigenvalues and eigenvectors of matrices;
- model a single population with different age classes;
- predict the long-term rate of growth/decline of the population;
- investigate how harvesting and culling policies affect the model.

Many textbooks on dynamical systems start with the theory of discrete mathematics. The author has adopted this approach here in order to provide a gentle introduction to MATLAB. However, readers who are more familiar with the theory of ODEs might want to start with continuous systems in the second half of the book.

1. Linear Discrete Dynamical Systems

This chapter deals with linear discrete dynamical systems, where time is measured by the number of iterations carried out and the dynamics are not continuous. In applications this would imply that the solutions are observed at discrete time intervals.

Recurrence relations can be used to construct mathematical models of discrete systems. They are also used extensively to solve many differential equations that do not have an analytic solution; the differential equations are represented by recurrence relations (or difference equations) that can be solved numerically on a computer. Of course one has to be careful when considering the accuracy of the numerical solutions. ODEs are used to model continuous dynamical systems in the second half of the book.

The bulk of this chapter is concerned with a linear discrete dynamical system that can be used to model the population of a single species. As with continuous systems, in applications to the real world, linear models generally produce good results over only a limited range of time. The Leslie model introduced here is useful when establishing harvesting and culling policies. Nonlinear discrete dynamical systems will be discussed in the next chapter.

The Poincaré maps introduced in Chapter 14, for example, illustrate how discrete systems can be used to help in the understanding of how continuous systems behave.

1.1 Recurrence Relations

This section is intended to give the reader a brief introduction to *difference equations* and illustrate the theory with some simple models.

First-Order Linear Difference Equations. A *recurrence relation* can be defined by a difference equation of the form

(1.1) $$x_{n+1} = f(x_n),$$

where x_{n+1} is derived from x_n and $n = 0, 1, 2, 3, \ldots$. If one starts with an initial value, say, x_0, then *iteration* of (1.1) leads to a sequence of the form

$$\{x_i : i = 0 \text{ to } \infty\} = \{x_0, x_1, x_2, \ldots, x_n, x_{n+1}, \ldots\}.$$

In applications, one would like to know how this sequence can be interpreted in physical terms. Equations of the form (1.1) are called *first-order difference equations* because the suffices differ by one. Consider the following simple example.

Example 1. The difference equation used to model the interest in a bank account compounded once per year is given by

$$x_{n+1} = \left(1 + \frac{3}{100}\right) x_n, \quad n = 0, 1, 2, 3, \ldots.$$

Find a general solution and determine the balance in the account after five years given that the initial deposit is 10,000 dollars and the interest is compounded annually.

Solution. Using the recurrence relation

$$x_1 = \left(1 + \frac{3}{100}\right) \times 10{,}000,$$

$$x_2 = \left(1 + \frac{3}{100}\right) \times x_1 = \left(1 + \frac{3}{100}\right)^2 \times 10{,}000,$$

and, in general,

$$x_n = \left(1 + \frac{3}{100}\right)^n \times 10{,}000,$$

where $n = 0, 1, 2, 3, \ldots$. Given that $x_0 = 10{,}000$ and $n = 5$, the balance after five years will be $x_5 = 11{,}592.74$ dollars.

Theorem 1. *The general solution of the first-order linear difference equation*

(1.2) $$x_{n+1} = mx_n + c, \quad n = 0, 1, 2, 3, \ldots,$$

is given by

$$x_n = m^n x_0 + \begin{cases} \frac{m^n - 1}{m - 1} c & \text{if } m \neq 1, \\ nc & \text{if } m = 1. \end{cases}$$

Proof. Applying the recurrence relation given in (1.2)

$$x_1 = mx_0 + c,$$
$$x_2 = mx_1 + c = m^2 x_0 + mc + c,$$
$$x_3 = mx_2 + c = m^3 x_0 + m^2 c + mc + c,$$

and the pattern in general is

$$x_n = m^n x_0 + (m^{n-1} + m^{n-2} + \cdots + m + 1)c.$$

Using geometric series, $m^{n-1} + m^{n-2} + \cdots + m + 1 = \frac{m^n - 1}{m - 1}$, provided that $m \neq 1$. If $m = 1$, then the sum of the geometric sequence is n. This concludes the proof of Theorem 1. Note that if $|m| < 1$, then $x_n \to \frac{c}{1-m}$ as $n \to \infty$. □

Second-Order Linear Difference Equations. Recurrence relations involving terms whose suffices differ by two are known as *second-order linear difference equations*. The general form of these equations with constant coefficients is

(1.3) $$ax_{n+2} = bx_{n+1} + cx_n.$$

Theorem 2. *The general solution of the second-order recurrence relation (1.3) is*

$$x_n = k_1 \lambda_1^n + k_2 \lambda_2^n,$$

where k_1, k_2 are constants and $\lambda_1 \neq \lambda_2$ are the roots of the quadratic equation $a\lambda^2 - b\lambda - c = 0$. If $\lambda_1 = \lambda_2$, then the general solution is of the form

$$x_n = (k_3 + nk_4)\lambda_1^n.$$

Note that when λ_1 and λ_2 are complex, the general solution can be expressed as

$$x_n = k_1 \lambda_1^n + k_2 \lambda_2^n = k_1(re^{i\theta})^n + k_2(re^{-i\theta})^n = r^n(A\cos(n\theta) + B\sin(n\theta)),$$

where A and B are constants. When the eigenvalues are complex, the solution oscillates and is real.

Proof. The solution of system (1.2) gives us a clue where to start. Assume that $x_n = \lambda^n k$ is a solution, where λ and k are to be found. Substituting, equation (1.3) becomes

$$a\lambda^{n+2}k = b\lambda^{n+1}k + c\lambda^n k$$

or

$$\lambda^n k(a\lambda^2 - b\lambda - c) = 0.$$

Assuming that $\lambda^n k \neq 0$, this equation has solutions if

(1.4) $$a\lambda^2 - b\lambda - c = 0.$$

Equation (1.4) is called the *characteristic equation*. The difference equation (1.3) has two solutions and because the equation is linear, a solution is given by

$$x_n = k_1 \lambda_1^n + k_2 \lambda_2^n,$$

where $\lambda_1 \neq \lambda_2$ are the roots of the characteristic equation.

If $\lambda_1 = \lambda_2$, then the characteristic equation can be written as

$$a\lambda^2 - b\lambda - c = a(\lambda - \lambda_1)^2 = a\lambda^2 - 2a\lambda_1\lambda + a\lambda_1^2.$$

Therefore, $b = 2a\lambda_1$ and $c = -a\lambda_1^2$. Now assume that another solution is of the form $kn\lambda^n$. Substituting, equation (1.3) becomes

$$ax_{n+2} - bx_{n+1} - cx_n = a(n+2)k\lambda_1^{n+2} - b(n+1)k\lambda_1^{n+1} - cnk\lambda_1^n;$$

therefore,

$$ax_{n+2} - bx_{n+1} - cx_n = kn\lambda_1^n(a\lambda_1^2 - b\lambda_1 - c) + k\lambda_1(2a\lambda_1 - b),$$

which equates to zero from the above. This confirms that $kn\lambda_n$ is a solution to equation (1.3). Since the system is linear, the general solution is thus of the form

$$x_n = (k_3 + nk_4)\lambda_1^n.$$

The values of k_j can be determined if x_0 and x_1 are given. Consider the following simple examples. □

Example 2. Solve the following second-order linear difference equations:

(i) $x_{n+2} = x_{n+1} + 6x_n$, $n = 0, 1, 2, 3, \ldots$, given that $x_0 = 1$ and $x_1 = 2$;

(ii) $x_{n+2} = 4x_{n+1} - 4x_n$, $n = 0, 1, 2, 3, \ldots$, given that $x_0 = 1$ and $x_1 = 3$;

(iii) $x_{n+2} = x_{n+1} - x_n$, $n = 0, 1, 2, 3, \ldots$, given that $x_0 = 1$ and $x_1 = 2$.

Solution.

(i) The characteristic equation is

$$\lambda^2 - \lambda - 6 = 0,$$

which has roots at $\lambda_1 = 3$ and $\lambda_2 = -2$. The general solution is therefore

$$x_n = k_1 3^n + k_2(-2)^n, \quad n = 0, 1, 2, 3, \ldots.$$

The constants k_1 and k_2 can be found by setting $n = 0$ and $n = 1$. The final solution is

$$x_n = \frac{4}{5}3^n + \frac{1}{5}(-2)^n, \quad n = 0, 1, 2, 3, \ldots.$$

(ii) The characteristic equation is

$$\lambda^2 - 4\lambda + 4 = 0,$$

which has a repeated root at $\lambda_1 = 2$. The general solution is

$$x_n = (k_3 + k_4 n)2^n, \quad n = 0, 1, 2, 3, \ldots.$$

Substituting for x_0 and x_1 gives the solution

$$x_n = \left(1 + \frac{n}{2}\right)2^n, \quad n = 0, 1, 2, 3, \ldots.$$

(iii) The characteristic equation is

$$\lambda^2 - \lambda + 1 = 0,$$

which has complex roots $\lambda_1 = \frac{1}{2} + i\frac{\sqrt{3}}{2} = e^{\frac{i\pi}{3}}$ and $\lambda_2 = \frac{1}{2} - i\frac{\sqrt{3}}{2} = e^{\frac{-i\pi}{3}}$.
The general solution is

$$x_n = k_1\lambda_1^n + k_2\lambda_2^n, \quad n = 0, 1, 2, 3, \ldots.$$

Substituting for λ_1 and λ_2, the general solution becomes

$$x_n = (k_1 + k_2)\cos\left(\frac{n\pi}{3}\right) + i(k_1 - k_2)\sin\left(\frac{n\pi}{3}\right), \quad n = 0, 1, 2, 3, \ldots.$$

Substituting for x_0 and x_1, gives $k_1 = \frac{1}{2} - \frac{i}{2\sqrt{3}}$ and $k_2 = \frac{1}{2} + \frac{i}{2\sqrt{3}}$, and so

$$x_n = \cos\left(\frac{n\pi}{3}\right) + \sqrt{3}\sin\left(\frac{n\pi}{3}\right), \quad n = 0, 1, 2, 3, \ldots.$$

Example 3. Suppose that the national income of a small country in year n is given by $I_n = S_n + P_n + G_n$, where S_n, P_n, and G_n represent national spending by the populous, private investment, and government spending, respectively. If the national income increases from one year to the next, then assume that consumers will spend more the following year; in this case, suppose that consumers spend $\frac{1}{6}$ of the previous year's income, then $S_{n+1} = \frac{1}{6}I_n$. An increase in consumer spending should also lead to increased investment the following year, assume that $P_{n+1} = S_{n+1} - S_n$. Substitution for S_n then gives $P_{n+1} = \frac{1}{6}(I_n - I_{n-1})$. Finally, assume that the government spending is kept constant. Simple manipulation then leads to the following economic model

(1.5) $$I_{n+2} = \frac{5}{6}I_{n+1} - \frac{1}{6}I_n + G,$$

where I_n is the national income in year n, and G is a constant. If the initial national income is G dollars and one year later is $\frac{3}{2}G$ dollars, determine

(i) a general solution to this model;

(ii) the national income after five years; and

(iii) the long-term state of the economy.

Solution.

(i) The characteristic equation is given by

$$\lambda^2 - \frac{5}{6}\lambda + \frac{1}{6} = 0,$$

which has solutions $\lambda_1 = \frac{1}{2}$ and $\lambda_2 = \frac{1}{3}$. Equation (1.5) also has a constant term G. Assume that the solution involves a constant term also; try $I_n = k_3 G$, then from equation (1.5)

$$k_3 G = \frac{5}{6} k_3 G - \frac{1}{6} k_3 G + G,$$

and so $k_3 = \frac{1}{1 - \frac{5}{6} + \frac{1}{6}} = 3$. Therefore, a general solution is of the form

$$I_n = k_1 \lambda_1^n + k_2 \lambda_2^n + 3G.$$

(ii) Given that $I_0 = G$ and $I_1 = \frac{3}{2} G$, simple algebra gives $k_1 = -5$ and $k_2 = 3$. When $n = 5$, $I_5 = 2.856 G$, to three decimal places.

(iii) As $n \to \infty$, $I_n \to 3G$, since $|\lambda_1| < 1$ and $|\lambda_2| < 1$. Therefore, the economy stabilizes in the long term to a constant value of $3G$. This is obviously a very crude model.

A general n-dimensional linear discrete population model is discussed in the following sections using matrix algebra.

1.2 The Leslie Model

The Leslie model was developed around 1940 to describe the population dynamics of the female portion of a species. For most species the number of females is equal to the number of males and this assumption is made here. The model can be applied to human populations, insect populations, and animal and fish populations. The model is an example of a discrete dynamical system. As explained throughout the text, we live in a nonlinear world and universe; since this model is linear, one would expect the results to be inaccurate in the long term. However, the model can give some interesting results and it incorporates some features not discussed in later chapters. As in Chapter 9, the following characteristics are ignored—diseases, environmental effects, pollution, and seasonal effects.

Assumptions. The females are divided into n age classes; thus if N is the theoretical maximum age attainable by a female of the species, then each age class will span a period of $\frac{N}{n}$ equally spaced, days, weeks, months, years, etc. The population is observed at regular discrete time intervals that are each equal to the length of one age class. Thus the kth time period will be given by $t_k = \frac{kN}{n}$. Define $x_i^{(k)}$ to be the number of females in the ith age class after the kth time period. Let b_i denote the number of female offspring born to one female during the ith age class, and let c_i be the proportion of females who continue to survive from the ith to the $(i+1)$st age class.

In order for this to be a realistic model the following conditions must be satisfied:

(i) $b_i \geq 0, 1 \leq i \leq n$;

(ii) $0 < c_i \leq 1, 1 \leq i < n$.

Obviously, some b_i have to be positive in order to ensure that some births do occur and no c_i are zero, otherwise there would be no females in the $(i+1)$st age class.

Working with the female population as a whole, the following sets of linear equations can be derived. The number of females in the first age class after the kth time period is equal to the number of females born to females in all n age classes between the time t_{k-1} and t_k; thus

$$x_1^{(k)} = b_1 x_1^{(k-1)} + b_2 x_2^{(k-1)} + \cdots + b_n x_n^{(k-1)}.$$

The number of females in the $(i+1)$st age class at time t_k is equal to the number of females in the ith age class at time t_{k-1} who continue to survive to enter the $(i+1)$st age class, hence

$$x_{i+1}^{(k)} = c_i x_i^{(k-1)}.$$

Equations of the above form can be written in matrix form, and so

$$\begin{pmatrix} x_1^{(k)} \\ x_2^{(k)} \\ x_3^{(k)} \\ \vdots \\ x_n^{(k)} \end{pmatrix} = \begin{pmatrix} b_1 & b_2 & b_3 & \cdots & b_{n-1} & b_n \\ c_1 & 0 & 0 & \cdots & 0 & 0 \\ 0 & c_2 & 0 & \cdots & 0 & 0 \\ \vdots & \vdots & \vdots & \ddots & \vdots & \vdots \\ 0 & 0 & 0 & \cdots & c_{n-1} & 0 \end{pmatrix} \begin{pmatrix} x_1^{(k-1)} \\ x_2^{(k-2)} \\ x_3^{(k-1)} \\ \vdots \\ x_n^{(k-1)} \end{pmatrix},$$

or

$$X^{(k)} = L X^{(k-1)}, \quad k = 1, 2, \ldots,$$

where $X \in \Re^n$ and the matrix L is called the *Leslie matrix*.

Suppose that $X^{(0)}$ is a vector giving the initial number of females in each of the n age classes, then

$$X^{(1)} = L X^{(0)},$$
$$X^{(2)} = L X^{(1)} = L^2 X^{(0)},$$
$$\vdots$$
$$X^{(k)} = L X^{(k-1)} = L^k X^{(0)}.$$

Therefore, given the initial age distribution and the Leslie matrix L, it is possible to determine the female age distribution at any later time interval.

1.2. The Leslie Model

Example 4. Consider a species of bird that can be split into three age groupings: those aged 0–1 year, those aged 1–2 years, and those aged 2–3 years. The population is observed once a year. Given that the Leslie matrix is equal to

$$L = \begin{pmatrix} 0 & 3 & 1 \\ 0.3 & 0 & 0 \\ 0 & 0.5 & 0 \end{pmatrix},$$

and the initial population distribution of females is $x_1^{(0)} = 1000$, $x_2^{(0)} = 2000$, and $x_3^{(0)} = 3000$, compute the number of females in each age group after

(a) 10 years;

(b) 20 years;

(c) 50 years.

Solution. Using the above,

(a) $$X^{(10)} = L^{10} X^{(0)} = \begin{pmatrix} 5383 \\ 2177 \\ 712 \end{pmatrix},$$

(b) $$X^{(20)} = L^{20} X^{(0)} = \begin{pmatrix} 7740 \\ 2388 \\ 1097 \end{pmatrix},$$

(c) $$X^{(50)} = L^{50} X^{(0)} = \begin{pmatrix} 15695 \\ 4603 \\ 2249 \end{pmatrix}.$$

The numbers are rounded down to whole numbers since it is not possible to have a fraction of a living bird. Obviously, the populations cannot keep on growing indefinitely. However, the model does give useful results for some species when the time periods are relatively short.

In order to investigate the limiting behavior of the system it is necessary to consider the eigenvalues and eigenvectors of the matrix L. These can be used to determine the eventual population distribution with respect to the age classes.

Theorem 3. *Let the Leslie matrix L be as defined above and assume that*

(a) $b_i \geq 0$ *for* $1 \leq i \leq n$;

(b) *at least two succesive b_i are strictly positive; and*

(c) $0 < c_i \leq 1$ *for* $1 \leq i < n$.

Then

24   1. Linear Discrete Dynamical Systems

(i) *matrix L has a unique positive eigenvalue, say, λ_1;*

(ii) *λ_1 is simple, or has algebraic multiplicity one;*

(iii) *the eigenvector—X_1, say—corresponding to λ_1 has positive components;*

(iv) *any other eigenvalue, $\lambda_i \neq \lambda_1$, of L satisfies*

$$|\lambda_i| < \lambda_1,$$

and the positive eigenvalue λ_1 is called strictly dominant.

The reader will be asked to prove part (i) in the exercises at the end of the chapter.

If the Leslie matrix L has a unique positive strictly dominant eigenvalue, then an eigenvector corresponding to λ_1 is a nonzero vector solution of

$$LX = \lambda_1 X.$$

Assume that $x_1 = 1$, then a possible eigenvector corresponding to λ_1 is given by

$$X_1 = \begin{pmatrix} 1 \\ \frac{c_1}{\lambda_1} \\ \frac{c_1 c_2}{\lambda_1^2} \\ \vdots \\ \frac{c_1 c_2 \cdots c_{n-1}}{\lambda_1^{n-1}} \end{pmatrix}.$$

Assume that L has n linearly independent eigenvectors, say, $X_1, X_2, \ldots, X_n$. Therefore, L is diagonalizable. If the initial population distribution is given by $X^{(0)} = X_0$, then there exist constants $b_1, b_2, \ldots, b_n$ such that

$$X_0 = b_1 X_1 + b_2 X_2 + \cdots + b_n X_n.$$

Since

$$X^{(k)} = L^k X_0 \quad \text{and} \quad L^k X_i = \lambda_i^k X_i,$$

then

$$X^{(k)} = L^k(b_1 X_1 + b_2 X_2 + \cdots + b_n X_n) = b_1 \lambda_1^k X_1 + b_2 \lambda_2^k X_2 + \cdots b_n \lambda_n^k X_n.$$

Therefore,

$$X^{(k)} = \lambda_1^k \left(b_1 X_1 + b_2 \left(\frac{\lambda_2}{\lambda_1}\right)^k X_2 + \cdots + b_n \left(\frac{\lambda_n}{\lambda_1}\right)^k X_n \right).$$

Since λ_1 is dominant, $|\frac{\lambda_i}{\lambda_1}| < 1$ for $\lambda_i \neq \lambda_1$, and $(\frac{\lambda_i}{\lambda_1})^k \to 0$ as $k \to \infty$. Thus for large k,

$$X^{(k)} \approx b_1 \lambda_1^k X_1.$$

In the long run, the age distribution stabilizes and is proportional to the vector X_1. Each age group will change by a factor of λ_1 in each time period. The vector X_1 can be normalized so that its components sum to one; the normalized vector then gives the eventual proportions of females in each of the n age groupings.

Note that if $\lambda_1 > 1$, the population eventually increases; if $\lambda_1 = 1$, the population stabilizes; and if $\lambda_1 < 1$, the population eventually decreases.

Example 5. Determine the eventual distribution of the age classes for Example 4.

Solution. The characteristic equation is given by

$$\det(L - \lambda I) = \begin{vmatrix} -\lambda & 3 & 1 \\ 0.3 & -\lambda & 0 \\ 0 & 0.5 & -\lambda \end{vmatrix} = -\lambda^3 + 0.9\lambda + 0.15 = 0.$$

The roots of the characteristic equation are

$$\lambda_1 = 1.023, \qquad \lambda_2 = -0.851, \qquad \lambda_3 = -0.172$$

to three decimal places. Note that λ_1 is the dominant eigenvalue.

To find the eigenvector corresponding to λ_1, solve

$$\begin{pmatrix} -1.023 & 3 & 1 \\ 0.3 & -1.023 & 0 \\ 0 & 0.5 & -1.023 \end{pmatrix} \begin{pmatrix} x_1 \\ x_2 \\ x_3 \end{pmatrix} = \begin{pmatrix} 0 \\ 0 \\ 0 \end{pmatrix}.$$

One solution is $x_1 = 2.929$, $x_2 = 0.855$, and $x_3 = 0.420$. Divide each term by the sum to obtain the normalized eigenvector

$$\hat{X}_1 = \begin{pmatrix} 0.696 \\ 0.204 \\ 0.1 \end{pmatrix}.$$

Hence after a number of years, the population will increase by approximately 2.3% every year. The percentage of females aged 0–1 year will be 69.6%, aged 1–2 years will be 20.4%, and aged 2–3 years will be 10%.

1.3 Harvesting and Culling Policies

This section is concerned with insect and fish populations only. The model has applications when considering insect species that survive on crops, for example. An insect population can be culled each year by applying either an insecticide or a predator species. Harvesting of fish populations is particularly important nowadays;

26   1. Linear Discrete Dynamical Systems

certain policies have to be employed to avoid depletion and extinction of the fish species. Harvesting indiscriminately could cause extinction of certain species of fish from our oceans.

A harvesting or culling policy should only be used if the population is increasing.

Definition 1. A harvesting or culling policy is said to be *sustainable* if the number of fish, or insects, killed and the age distribution of the population remaining are the same after each time period.

Assume that the fish or insects are killed in short sharp bursts at the end of each time period. Let X be the population distribution vector for the species just before the harvesting or culling is applied. Suppose that a fraction of the females about to enter the $(i+1)$st class are killed, giving a matrix

$$D = \begin{pmatrix} d_1 & 0 & 0 & \cdots & 0 \\ 0 & d_2 & 0 & \cdots & 0 \\ 0 & 0 & d_3 & \cdots & 0 \\ \vdots & \vdots & \vdots & \ddots & \vdots \\ 0 & 0 & 0 & \cdots & d_n \end{pmatrix}.$$

By definition, $0 \leq d_i \leq 1$, where $1 \leq i \leq n$. The numbers killed will be given by DLX and the population distribution of those remaining will be

$$LX - DLX = (I - D)LX.$$

In order for the policy to be sustainable one must have

(1.6) $$(I - D)LX = X.$$

If the dominant eigenvalue of $(I - D)L$ is one, then X will be an eigenvector for this eigenvalue and the population will stabilize. This will impose certain conditions on the matrix D. Hence

$$I - D = \begin{pmatrix} (1-d_1) & 0 & 0 & \cdots & 0 \\ 0 & (1-d_2) & 0 & \cdots & 0 \\ 0 & 0 & (1-d_3) & \cdots & 0 \\ \vdots & \vdots & \vdots & \ddots & \vdots \\ 0 & 0 & 0 & \cdots & (1-d_n) \end{pmatrix}$$

and the matrix, say, $M = (I - D)L$, is easily computed. The matrix M is also a Leslie matrix and hence has an eigenvalue $\lambda_1 = 1$ if and only if

(1.7) $$(1-d_1)(b_1 + b_2c_1(1-d_1) + b_3c_1c_2(1-d_2)(1-d_3) \\ + \cdots + b_nc_1 \cdots c_{n-1}(1-d_1) \cdots (1-d_n)) = 1.$$

1.3. Harvesting and Culling Policies

Only values of $0 \leq d_i \leq 1$ that satisfy equation (1.7) can produce a sustainable policy.

A possible eigenvector corresponding to $\lambda_1 = 1$ is given by

$$X_1 = \begin{pmatrix} 1 \\ (1-d_2)c_1 \\ (1-d_2)(1-d_3)c_1c_2 \\ \vdots \\ (1-d_2)\cdots(1-d_n)c_1c_2\cdots c_{n-1} \end{pmatrix}.$$

The sustainable population will be $C_1 X_1$, where C_1 is a constant. Consider the following policies.

Sustainable Uniform Harvesting, or Culling. Let $d = d_1 = d_2 = \cdots = d_n$, then (1.6) becomes

$$(1-d)LX = X,$$

which means that $\lambda_1 = \frac{1}{1-d}$. Hence a possible eigenvector corresponding to λ_1 is given by

$$X_1 = \begin{pmatrix} 1 \\ \frac{c_1}{\lambda_1} \\ \frac{c_1 c_2}{\lambda_1^2} \\ \vdots \\ \frac{c_1 c_2 \cdots c_{n-1}}{\lambda_1^{n-1}} \end{pmatrix}.$$

Sustainable Harvesting, or Culling of the Youngest Class. Let $d_1 = d$ and $d_2 = d_3 = \cdots = d_n = 0$; therefore, (1.7) becomes

$$(1-d)(b_1 + b_2 c_1 + b_3 c_1 c_2 + \cdots + b_n c_1 c_2 \cdots c_{n-1}) = 1,$$

or, equivalently,

$$(1-d)R = 1,$$

where R is known as the *net reproduction rate*. Harvesting or culling is only viable if $R > 1$. The age distribution after each harvest or cull is then given by

$$X_1 = \begin{pmatrix} 1 \\ c_1 \\ c_1 c_2 \\ \vdots \\ c_1 c_2 \cdots c_{n-1} \end{pmatrix}.$$

28   1. Linear Discrete Dynamical Systems

Definition 2. An *optimal sustainable* harvesting or culling policy is one in which either one or two age classes are killed. If two classes are killed, then the older age class is completely killed.

Example 6. A certain species of fish can be divided into three six-month age classes and has Leslie matrix

$$L = \begin{pmatrix} 0 & 4 & 3 \\ 0.5 & 0 & 0 \\ 0 & 0.25 & 0 \end{pmatrix}.$$

The species of fish is to be harvested by fishermen using one of four different policies, which are uniform harvesting or harvesting one of the three age classes, respectively. Which of these four policies are sustainable? Decide which of the sustainable policies the fishermen should use.

Solution. The characteristic equation is given by

$$\det(L - \lambda I) = \begin{vmatrix} -\lambda & 4 & 3 \\ 0.5 & -\lambda & 0 \\ 0 & 0.25 & -\lambda \end{vmatrix} = -\lambda^3 + 2\lambda + 0.375 = 0.$$

The eigenvalues are given by $\lambda_1 = 1.5$, $\lambda_2 = -0.191$, and $\lambda_3 = -1.309$ to three decimal places. The eigenvalue λ_1 is dominant and the population will eventually increase by 50% every six months. The normalized eigenvector corresponding to λ_1 is given by

$$\hat{X}_1 = \begin{pmatrix} 0.529 \\ 0.177 \\ 0.294 \end{pmatrix}.$$

So after a number of years there will be 52.9% of females aged 0–6 months; 17.7% of females aged 6–12 months, and 29.4% of females aged 12–18 months.

If the harvesting policy is to be sustainable, then (1.7) becomes

$$(1 - d_1)(b_1 + b_2 c_1(1 - d_2) + b_3 c_1 c_2(1 - d_2)(1 - d_3)) = 1.$$

Suppose that $h_i = (1 - d_i)$, then

(1.8) $$h_1 h_2 (2 + 0.375 h_3) = 1.$$

Consider the four policies separately.

(i) *Uniform harvesting*: Let $\mathbf{h} = (h, h, h)$. Equation (1.8) becomes

$$h^2(2 + 0.375h) = 1,$$

1.3. Harvesting and Culling Policies

which has solutions $h = 0.667$ and $d = 0.333$. The normalized eigenvector is given by

$$\hat{X}_U = \begin{pmatrix} 0.720 \\ 0.240 \\ 0.040 \end{pmatrix}.$$

(ii) *Harvesting the youngest age class*: Let $\mathbf{h} = (h_1, 1, 1)$. Equation (1.8) becomes

$$h_1(2 + 0.375) = 1,$$

which has solutions $h_1 = 0.421$ and $d_1 = 0.579$. The normalized eigenvector is given by

$$\hat{X}_{A_1} = \begin{pmatrix} 0.615 \\ 0.308 \\ 0.077 \end{pmatrix}.$$

(iii) *Harvesting the middle age class*: Let $\mathbf{h} = (1, h_2, 1)$. Equation (1.8) becomes

$$h_2(2 + 0.375) = 1,$$

which has solutions $h_2 = 0.421$ and $d_2 = 0.579$. The normalized eigenvector is given by

$$\hat{X}_{A_2} = \begin{pmatrix} 0.791 \\ 0.167 \\ 0.042 \end{pmatrix}.$$

(iv) *Harvesting the oldest age class*: Let $\mathbf{h} = (1, 1, h_3)$. Equation (13.3) becomes

$$1(2 + 0.375 h_3) = 1,$$

which has no solutions if $0 \leq h_3 \leq 1$.

Therefore, harvesting policies (i)–(iii) are sustainable and policy (iv) is not.

The long-term distributions of the populations of fish are determined by the normalized eigenvectors $\hat{X}_U$, $\hat{X}_{A_1}$, and $\hat{X}_{A_2}$ given above. If, for example, the fishermen wanted to leave as many fish as possible in the youngest age class, then the policy that should be adopted is the second age class harvesting. Then 79.1% of the females would be in the youngest age class after a number of years.

1.4 MATLAB Commands

```
% Program_1a - Solving recurrence relations.

% Solving a first-order recurrence relation (Example 1).
% Call a Maple command from the Symbolic Math Toolbox.
xn=maple('rsolve({x(n+1)-(1+(3/(100)))*x(n),x(0)=10000},x)')
n=5
savings=vpa(eval(xn),7)

% Solving a second-order recurrence relation (Example 2(i)).
clear
xn=maple('rsolve({x(n+2)-x(n+1)-6*x(n),x(0)=1,x(1)=2},x)')

% Solving a characteristic equation (Example 2(iii)).
syms lambda
CE=lambda^2-lambda+1
lambda=solve(CE)
% End of Program_1a.
```

```
% Program_1b - The Leslie matrix.

% Define a 3x3 Leslie Matrix (Example 4).
L=[0 3 1; 0.3 0 0; 0 0.5 0]

% Set initial conditions.
X0=[1000;2000;3000]

% After 10 years the population distribution will be:
X10=L^10*X0

% Find the eigenvectors and eigenvalues of L (Example 5).
[v,d]=eig(L)
% End of Program_1b.
```

1.5 Exercises

1. The difference equation used to model the length of a carpet, say, l_n, rolled n times is given by

$$l_{n+1} = l_n + \pi(4 + 2cn), \quad n = 0, 1, 2, 3, \ldots,$$

where c is the thickness of the carpet. Solve this recurrence relation.

2. Solve the following second-order linear difference equations:

(a) $x_{n+2} = 5x_{n+1} - 6x_n$, $n = 0, 1, 2, 3, \ldots$, if $x_0 = 1$, $x_1 = 4$;
(b) $x_{n+2} = x_{n+1} - \frac{1}{4}x_n$, $n = 0, 1, 2, 3, \ldots$, if $x_0 = 1$, $x_1 = 2$;
(c) $x_{n+2} = 2x_{n+1} - 2x_n$, $n = 0, 1, 2, 3, \ldots$, if $x_0 = 1$, $x_1 = 2$;
(d) $F_{n+2} = F_{n+1} + F_n$, $n = 0, 1, 2, 3, \ldots$, if $F_1 = 1$ and $F_2 = 1$ (the sequence of numbers is known as the Fibonacci sequence);
(e) $x_{n+2} = x_{n+1} + 2x_n - f(n)$, $n = 0, 1, 2, \ldots$, given that $x_0 = 2$ and $x_1 = 3$, when (i) $f(n) = 2$, (ii) $f(n) = 2n$, and (iii) $f(n) = e^n$ (use MATLAB for part (iii) only).

3. Consider a human population that is divided into three age classes: those aged 0–15 years, those aged 15–30 years, and those aged 30–45 years. The Leslie matrix for the female population is given by

$$L = \begin{pmatrix} 0 & 1 & 0.5 \\ 0.9 & 0 & 0 \\ 0 & 0.8 & 0 \end{pmatrix}.$$

Given that the initial population distribution of females is $x_1^{(0)} = 10000$, $x_2^{(0)} = 15000$, and $x_3^{(0)} = 8000$, compute the number of females in each of these groupings after

(a) 225 years;
(b) 750 years;
(c) 1500 years.

4. Consider the following Leslie matrix used to model the female portion of a species

$$L = \begin{pmatrix} 0 & 0 & 6 \\ \frac{1}{2} & 0 & 0 \\ 0 & \frac{1}{3} & 0 \end{pmatrix}.$$

Determine the eigenvalues and eigenvectors of L. Show that there is no dominant eigenvalue and describe how the population would develop in the long term.

5. Consider a human population that is divided into five age classes: those aged 0–15 years, those aged 15–30 years, those aged 30–45 years, those aged 45–60 years, and those aged 60–75 years. The Leslie matrix for the female population is given by

$$L = \begin{pmatrix} 0 & 1 & 1.5 & 0 & 0 \\ 0.9 & 0 & 0 & 0 & 0 \\ 0 & 0.8 & 0 & 0 & 0 \\ 0 & 0 & 0.7 & 0 & 0 \\ 0 & 0 & 0 & 0.5 & 0 \end{pmatrix}.$$

1. Linear Discrete Dynamical Systems

Determine the eigenvalues and eigenvectors of L and describe how the population distribution develops.

6. Given that

$$L = \begin{pmatrix} b_1 & b_2 & b_3 & \cdots & b_{n-1} & b_n \\ c_1 & 0 & 0 & \cdots & 0 & 0 \\ 0 & c_2 & 0 & \cdots & 0 & 0 \\ \vdots & \vdots & \vdots & \ddots & \vdots & \vdots \\ 0 & 0 & 0 & \cdots & c_{n-1} & 0 \end{pmatrix},$$

where $b_i \geq 0, 0 < c_i \leq 1$, and at least two successive b_i are strictly positive, prove that $p(\lambda) = 1$, if λ is an eigenvalue of L, where

$$p(\lambda) = \frac{b_1}{\lambda} + \frac{b_2 c_1}{\lambda^2} + \cdots + \frac{b_n c_1 c_2 \cdots c_{n-1}}{\lambda^n}.$$

Show the following:

(i) $p(\lambda)$ is strictly decreasing.

(ii) $p(\lambda)$ has a vertical asymptote at $\lambda = 0$.

(iii) $p(\lambda) \to 0$ as $\lambda \to \infty$.

Prove that a general Leslie matrix has a unique positive eigenvalue.

7. A certain species of insect can be divided into three age classes: 0–6 months, 6–12 months, and 12–18 months. A Leslie matrix for the female population is given by

$$L = \begin{pmatrix} 0 & 4 & 10 \\ 0.4 & 0 & 0 \\ 0 & 0.2 & 0 \end{pmatrix}.$$

Determine the long-term distribution of the insect population. An insecticide is applied which kills off 50% of the youngest age class. Determine the long-term distribution if the insecticide is applied every six months.

8. Assuming the same model for the insects as in Exercise 7, determine the long-term distribution if an insecticide is applied every six months which kills 10% of the youngest age class, 40% of the middle age class, and 60% of the oldest age class.

9. In a fishery, a certain species of fish can be divided into three age groups each one year long. The Leslie matrix for the female portion of the population is given by

$$L = \begin{pmatrix} 0 & 3 & 36 \\ \frac{1}{3} & 0 & 0 \\ 0 & \frac{1}{2} & 0 \end{pmatrix}.$$

Show that, without harvesting, the fish population would double each year. Describe the long-term behavior of the system if the following policies are applied:

(i) harvest 50% from each age class;

(ii) harvest the youngest fish only, using a sustainable policy;

(iii) harvest 50% of the youngest fish;

(iv) harvest 50% of the whole population from the youngest class only;

(v) harvest 50% of the oldest fish.

10. Determine an optimal sustainable harvesting policy for the system given in Exercise 9 if the youngest age class is left untouched.

Recommended Reading

[1] M. R. S. Kulenovic and O. Merino, *Discrete Dynamical Systems and Difference Equations with Mathematica*, Chapman and Hall, London, 2002.

[2] S. Barnet, *Discrete Mathematics*, Addison–Wesley, Reading, MA, 1998.

[3] S. Jal and T. G. Hallam, Effects of delay, truncations and density dependence in reproduction schedules on stability of nonlinear Leslie matrix models, *J. Math. Biology*, **31**-4 (1993), 367–395.

[4] S. I. Grossman, *Applications for Elementary Linear Algebra*, Wadsworth, Belmont, CA, 1980.

2
Nonlinear Discrete Dynamical Systems

Aims and Objectives

- To introduce nonlinear one- and two-dimensional iterated maps.
- To investigate period-doubling bifurcations to chaos.
- To introduce the notion of universality.

On completion of this chapter, the reader should be able to

- produce graphical iterations of one-dimensional iterated maps;
- test whether or not certain systems are chaotic;
- plot bifurcation diagrams;
- apply some of the theory to model simple problems from biology, economics, neural networks, nonlinear optics, and population dynamics.

Most of the dynamics displayed by highly complicated nonlinear systems also appear for simple nonlinear systems. The reader is first introduced to the tent function, which is composed of two straight lines. The graphical method of iteration is introduced using this simple function since the constructions may be easily carried out with graph paper, rule, and pencil. The reader is also shown how to graph composite functions. The system can display periodicity, mixing, and sensitivity to initial conditions, the essential ingredients for chaos.

2. Nonlinear Discrete Dynamical Systems

The logistic map is used as a simple model for the population growth of an insect species. Bifurcation diagrams are plotted and period-doubling bifurcations to chaos are displayed.

Bifurcation diagrams are plotted for the Gaussian map. Two-dimensional Hénon maps are investigated, periodic points are found, and chaotic (or strange) attractors are produced.

2.1 The Tent Map and Graphical Iterations

As a simple introduction to one-dimensional nonlinear discrete dynamical systems, consider the *tent map*, $T : [0, 1] \to [0, 1]$ defined by

$$T(x) = \begin{cases} \mu x, & 0 \le x < \frac{1}{2}, \\ \mu(1 - x), & \frac{1}{2} \le x \le 1, \end{cases}$$

where $0 \le \mu \le 2$. The tent map is constructed from two straight lines, which makes the analysis simpler than for truly nonlinear systems. The graph of the T function may be plotted by hand and is given in Figure 2.1.

Figure 2.1: A graph of the tent function.

Define an iterative map by

(2.1) $$x_{n+1} = T(x_n),$$

where $x_n \in [0, 1]$. Although the form of the tent map is simple and the equations involved are linear, for certain parameter values, this system can display highly complex behavior and even chaotic phenomena. In fact, most of the features discussed in later chapters of this text are displayed by this relatively simple system. For certain parameter values, the mapping undergoes stretching and folding transformations and displays sensitivity to initial conditions and periodicity. Fortunately, it is not difficult to carry out simple iterations for system (2.1) as the following examples demonstrate.

2.1. The Tent Map and Graphical Iterations

Example 1. Iterate the tent function numerically for the following μ and x_0 values:

(I) $\mu = \frac{1}{2}$:

(i) $x_0 = \frac{1}{4}$, (ii) $x_0 = \frac{1}{2}$, (iii) $x_0 = \frac{3}{4}$;

(II) $\mu = 1$:

(i) $x_0 = \frac{1}{3}$, (ii) $x_0 = \frac{2}{3}$;

(III) $\mu = \frac{3}{2}$:

(i) $x_0 = \frac{3}{5}$, (ii) $x_0 = \frac{6}{13}$, (iii) $x_0 = \frac{1}{3}$;

(IV) $\mu = 2$:

(i) $x_0 = \frac{1}{3}$, (ii) $x_0 = \frac{1}{5}$, (iii) $x_0 = \frac{1}{7}$, (iv) $x_0 = \frac{1}{11}$.

Solution. A calculator or computer is not needed here. It is very easy to carry out the iterations by hand. For the sake of simplicity, the iterates will be listed as $\{x_0, x_1, x_2, \ldots, x_n, \ldots\}$. The solutions are as follows:

(I) $\mu = \frac{1}{2}$:

(i) $\{\frac{1}{4}, \frac{1}{8}, \frac{1}{16}, \ldots, \frac{1}{4 \times 2^n}, \ldots\}$;

(ii) $\{\frac{1}{2}, \frac{1}{4}, \frac{1}{8}, \ldots, \frac{1}{2^{n+1}}, \ldots\}$;

(iii) $\{\frac{3}{4}, \frac{3}{8}, \frac{3}{16}, \ldots, \frac{3}{4 \times 2^n}, \ldots\}$.

In each case, $x_n \to 0$ as $n \to \infty$.

(II) $\mu = 1$:

(i) $\{\frac{1}{3}, \frac{1}{3}, \frac{1}{3}, \ldots, \frac{1}{3}, \ldots\}$;

(ii) $\{\frac{2}{3}, \frac{1}{3}, \frac{1}{3}, \ldots, \frac{1}{3}, \ldots\}$.

The orbits tend to points of period one in the range $[0, \frac{1}{2}]$.

(III) $\mu = \frac{3}{2}$:

(i) $\{\frac{3}{5}, \frac{3}{5}, \frac{3}{5}, \ldots, \frac{3}{5}, \ldots\}$;

(ii) $\{\frac{6}{13}, \frac{9}{13}, \frac{6}{13}, \frac{9}{13}, \ldots, \frac{6}{13}, \frac{9}{13}, \ldots\}$;

(iii) $\{\frac{1}{3}, \frac{1}{2}, \frac{3}{4}, \frac{3}{8}, \frac{9}{16}, \frac{21}{32}, \frac{33}{64}, \frac{93}{128}, \frac{105}{256}, \frac{315}{512}, \frac{591}{1024}, \ldots\}$.

In case (i), the iterate x_{n+1} is equal to x_n for all n. This type of sequence displays *period-one* behavior. In case (ii), the iterate x_{n+2} is equal to x_n for all n, and the result is *period-two* behavior. In case (iii), the first 11 iterates are listed but other methods need to be used in order to establish the long-term behavior of the sequence.

(IV) $\mu = 2$:

(i) $\{\frac{1}{3}, \frac{2}{3}, \frac{2}{3}, \ldots, \frac{2}{3}, \ldots\}$;

(ii) $\{\frac{1}{5}, \frac{2}{5}, \frac{4}{5}, \frac{2}{5}, \frac{4}{5}, \ldots, \frac{2}{5}, \frac{4}{5}, \ldots\}$;

(iii) $\{\frac{1}{7}, \frac{2}{7}, \frac{4}{7}, \frac{6}{7}, \frac{2}{7}, \frac{4}{7}, \frac{6}{7}, \ldots, \frac{2}{7}, \frac{4}{7}, \frac{6}{7}, \ldots\}$;

(iv) $\{\frac{1}{11}, \frac{2}{11}, \frac{4}{11}, \frac{8}{11}, \frac{6}{11}, \frac{10}{11}, \frac{2}{11}, \ldots, \frac{2}{11}, \frac{4}{11}, \frac{8}{11}, \frac{6}{11}, \frac{10}{11}, \ldots\}$.

The sequences behave as follows: (i) there is period-one behavior, (ii) there is period-two behavior, (iii) there is a period-three sequence, and (iv) there is a period-five sequence.

Example 2. Using the tent map defined by equation (2.1) when $\mu = 2$, compute the first 20 iterates for the following two initial conditions:

(i) $x_0 = 0.2$;

(ii) $x_0 = 0.2001 = 0.2 + \epsilon$.

Solution. The iterates may be computed using the MATLAB package. Use sym(2001/10000) as an initial condition in case (ii) and the double command in the MATLAB program (see Section 2.6). The first 20 iterates for both initial conditions are listed side by side in Table 2.1.

The system clearly shows sensitivity to initial conditions for the parameter value $\mu = 2$. Comparing the numerical values in the second and third columns, it is not difficult to see that the sequences diverge markedly when $n > 9$. This test for sensitivity to initial conditions gives researchers a simple tool to determine whether or not a system is chaotic. A more in-depth description of chaos is given in Section 13.3.II.

The results of Examples 1 and 2 show that there is a rich variety of dynamics that system (2.1) can display. Indeed, a now-famous result by Li and Yorke [9], states that if a system displays period-three behavior, then the system can display periodic behavior of any period, and they go on to prove that the system can display chaotic phenomena. Hence when $\mu = 2$, system (2.1) is chaotic since it has a period-three sequence (Example 1(IV)(iii)).

Unfortunately, numerical iterations do not always give a clear insight into how the sequences develop as n gets large. Another popular method used to display the sequence of iterations more clearly is the so-called *graphical method*.

The Graphical Method. From an initial point x_0, draw a vertical line up to the function, in this case, $T(x)$. From this point, draw a horizontal line either left or right to join the diagonal $y = x$. The x-ordinate corresponds to the iterate $x_1 = T(x_0)$. From the point $(x_1, T(x_0))$, draw a vertical line up or down to join the function $T(x)$. Draw a horizontal line from this point to the diagonal at the point $(x_2, T(x_1))$. The first two iterates are shown in Figure 2.2.

2.1. The Tent Map and Graphical Iterations

Table 2.1: The first 20 iterates for both initial conditions in Example 2.

n	x_n	x_n
0	$x_0 = 0.2$	$x_0 = 0.2001$
1	0.4	0.4002
2	0.8	0.8004
3	0.4	0.3992
4	0.8	0.7984
5	0.4	0.4032
6	0.8	0.8064
7	0.4	0.3872
8	0.8	0.7744
9	0.4	0.4512
10	0.8	0.9024
11	0.4	0.1952
12	0.8	0.3904
13	0.4	0.7808
14	0.8	0.4384
15	0.4	0.8768
16	0.8	0.2464
17	0.4	0.4928
18	0.8	0.9856
19	0.4	0.0288
20	0.8	0.0576

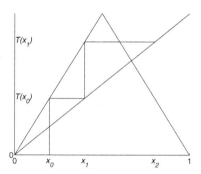

Figure 2.2: A possible graphical iteration when $n = 2$.

The iterative procedure may be summarized as a simple repeated two-step algorithm.

1. Draw a vertical line to the function (evaluation).

40   2. Nonlinear Discrete Dynamical Systems

2. Draw a horizontal line to the diagonal (feedback); go back to 1.

The algorithm generates successive iterates along the x-axis, corresponding to the sequence of points $\{x_0, x_1, x_2, \ldots, x_n, \ldots\}$.

To demonstrate the method, the numerical iterations carried out in Examples 1 and 2 will now be repeated using the graphical method.

Example 3. Iterate the tent function graphically for the following μ and x_0 values:

(I) $\mu = \frac{1}{2}$:

  (i) $x_0 = \frac{1}{4}$, (ii) $x_0 = \frac{1}{2}$, (iii) $x_0 = \frac{3}{4}$;

(II) $\mu = 1$:

  (i) $x_0 = \frac{1}{3}$, (ii) $x_0 = \frac{2}{3}$;

(III) $\mu = \frac{3}{2}$:

  (i) $x_0 = \frac{3}{5}$, (ii) $x_0 = \frac{6}{13}$, (iii) $x_0 = \frac{1}{3}$;

(IV) $\mu = 2$:

  (i) $x_0 = \frac{1}{3}$, (ii) $x_0 = \frac{1}{5}$, (iii) $x_0 = \frac{1}{7}$, (iv) $x_0 = \frac{1}{11}$.

(V) $\mu = 2$:

  (i) $x_0 = 0.2$, (ii) $x_0 = 0.2001$.

Solution. Each of the diagrams (Figures 2.3–2.7) were produced using the MAT-LAB package. Most of the graphical iterations are self-explanatory; however, Figures 2.5(c) and 2.7 warrant further explanation. When $\mu = \frac{3}{2}$, the tent map displays sensitivity to initial conditions and can be described as being chaotic. The iterative path plotted in Figure 2.5(c) appears to wander randomly. It is still not clear whether the path is chaotic or whether the path is periodic of a very high period. Figure 2.7 clearly shows the sensitivity to initial conditions. Again, it is not clear in case (ii) whether the path is chaotic or of a very high period.

What is clear from the diagrams is that the three basic properties of chaos—mixing, periodicity, and sensitivity to initial conditions—are all exhibited for certain parameter values.

2.2 Fixed Points and Periodic Orbits

Consider the general map

(2.2) $$x_{n+1} = f(x_n).$$

Definition 1. A *fixed point*, or point of period one, of system (2.2) is a point at which $x_{n+1} = f(x_n) = x_n$, for all n.

2.2. Fixed Points and Periodic Orbits   41

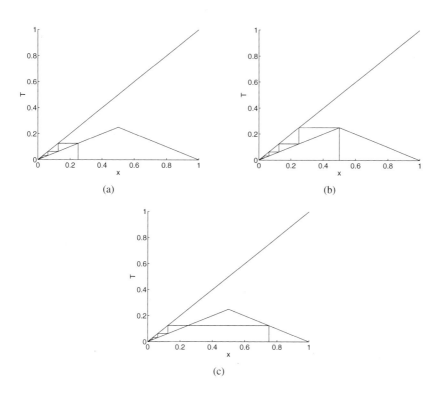

Figure 2.3: Graphical iterations when $\mu = \frac{1}{2}$: (a) $x_0 = \frac{1}{4}$, (b) $x_0 = \frac{1}{2}$, and (c) $x_0 = \frac{3}{4}$.

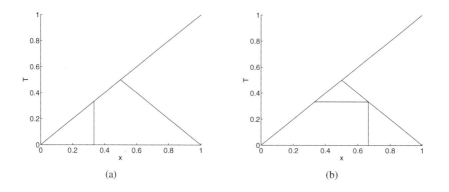

Figure 2.4: Graphical iterations when $\mu = 1$: (a) $x_0 = \frac{1}{3}$ and (b) $x_0 = \frac{2}{3}$.

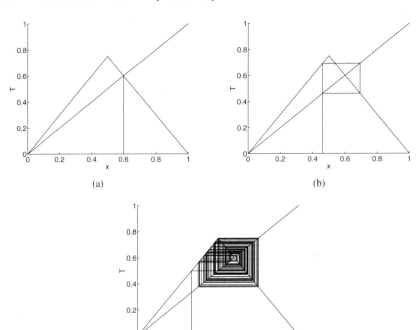

Figure 2.5: Graphical iterations when $\mu = \frac{3}{2}$: (a) $x_0 = \frac{3}{5}$, (b) $x_0 = \frac{6}{13}$, and (c) $x_0 = \frac{1}{3}$ for 200 iterations.

For the tent map, this implies that $T(x_n) = x_n$, for all n. Graphically, the fixed points can be found by identifying intersections of the function $T(x)$ with the diagonal.

As with other dynamical systems, the fixed points of period one can be attracting, repelling, or indifferent. The type of fixed point is determined from the gradient of the tangent to the function, $T(x)$ in this case, at the fixed point. For straight line segments with equation $y = mx + c$, it can be easily shown that if

- $m < -1$, the iterative path is repelled and spirals away from the fixed point;

- $-1 < m < 0$, the iterative path is attracted and spirals in to the fixed point;

- $0 < m < 1$, the iterative path is attracted and staircases in to the fixed point;

- $m > 1$, the iterative path is repelled and staircases away from the critical point.

2.2. Fixed Points and Periodic Orbits 43

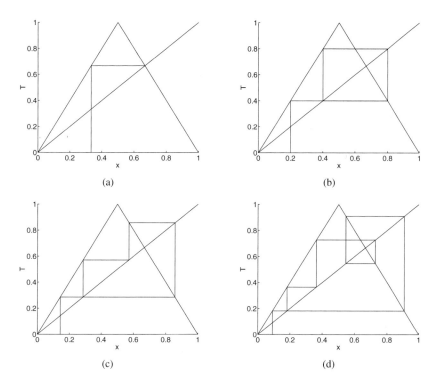

Figure 2.6: Graphical iterations when $\mu = 2$: (a) $x_0 = \frac{1}{3}$, (b) $x_0 = \frac{1}{5}$, (c) $x_0 = \frac{1}{7}$, and (d) $x_0 = \frac{1}{11}$.

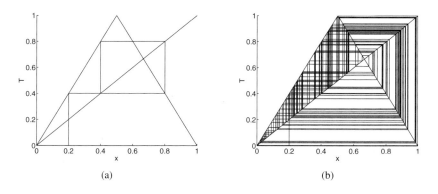

Figure 2.7: Graphical iterations when $\mu = 2$: (a) $x_0 = 0.2$ and (b) $x_0 = 0.2001$ for 200 iterations.

44   2. Nonlinear Discrete Dynamical Systems

When $|m| = 1$, the fixed point is neither repelling nor attracting and $m = 0$ is a trivial case. A test for stability of fixed points for nonlinear iterative maps will be given in Section 2.3.

Using Definition 1, it is possible to determine the fixed points of period one for the tent map (2.1). If $0 < \mu < 1$, the only fixed point is at $x = 0$ (see Figure 2.8) and since the gradient at $x = 0$ is less than one, the fixed point is stable. Note that the origin is called the *trivial fixed point*.

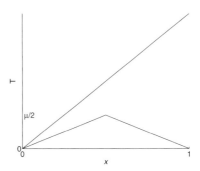

Figure 2.8: The intersection $T(x) = x$ when $0 < \mu < 1$.

When $\mu = 1$, the branch μx of $T(x)$ coincides with the diagonal and all points lying in the interval $0 \leq x \leq \frac{1}{2}$ are of period one. Once the tent function crosses the diagonal the origin becomes unstable since the gradient of the tent map at this point now exceeds one.

When $1 < \mu \leq 2$, there are two fixed points of period one, $x_{1,1} = 0$ and $x_{1,2} = \frac{\mu}{1+\mu}$ (see Figure 2.9).

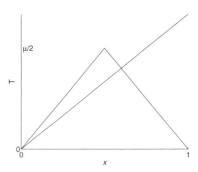

Figure 2.9: The intersections $T(x) = x$ when $1 < \mu \leq 2$. There are two intersections.

2.2. Fixed Points and Periodic Orbits

Notation. Throughout this chapter, the periodic point given by $x_{i,j}$ will denote the jth point of period i. This notation is useful when determining the number of points of period i. For example, $x_{1,1}$ and $x_{1,2}$ above are the two fixed points of period one.

The gradient of the function $T(x)$ is greater than one at $x_{1,1}$, so this point is unstable; the gradient of $T(x)$ at the point $x_{1,2}$ is less than -1. Therefore, this point is also unstable.

In summary, then, when $0 \leq \mu < 1$, there is one stable period-one point at $x = 0$; when $\mu = 1$, there are an infinite number of period-one points in the interval $0 \leq x \leq \frac{1}{2}$; and when $1 < \mu \leq 2$, there are two unstable period-one points at $x_{1,1}$ and $x_{1,2}$. The obvious question then is, where do the paths go if not to these two points of period one? The answer to this question will be given later (see Exercise 3 in Section 2.7).

Definition 2. For system (2.2), a *fixed point of period N* is a point at which $x_{n+N} = f^N(x_n) = x_n$ for all n.

In order to determine the fixed points of period two for the tent map it is necessary to find the points of intersection of $T^2(x)$ with the diagonal. Consider the case where $\mu = 2$, the methods below can be applied for any value of μ in the interval $[0, 2]$.

The function of the function $T(T(x)) = T^2(x)$ is determined by replacing x with $T(x)$ in the mapping

$$T(x) = \begin{cases} 2x, & 0 \leq x < \frac{1}{2}, \\ 2(1-x), & \frac{1}{2} \leq x \leq 1. \end{cases}$$

Hence

$$T^2(x) = \begin{cases} 2T(x), & 0 \leq T(x) < \frac{1}{2}, \\ 2(1-T(x)), & \frac{1}{2} \leq T(x) \leq 1. \end{cases}$$

The interval $0 \leq T(x) < \frac{1}{2}$ on the vertical axis corresponds to two intervals, namely, $0 \leq x < T^{-1}(\frac{1}{2})$ and $T^{-1}(\frac{1}{2}) \leq x \leq 1$ on the horizontal axis. When $\mu = 2$, it is not difficult to show that $T^{-1}(\frac{1}{2}) = \frac{1}{4}$ or $\frac{3}{4}$, depending on the branch of $T(x)$. The process may be repeated for $T(x)$ lying in the interval $[\frac{1}{2}, 1]$. Therefore, $T^2(x)$ becomes

$$T^2(x) = \begin{cases} 4x, & 0 \leq x < \frac{1}{4}, \\ 2 - 4x, & \frac{1}{4} \leq x < \frac{1}{2}, \\ 4x - 2, & \frac{1}{2} \leq x < \frac{3}{4}, \\ 4 - 4x, & \frac{3}{4} \leq x \leq 1. \end{cases}$$

This function intersects the diagonal at four points corresponding to $x = 0, \frac{2}{5}, \frac{2}{3}$, and $\frac{4}{5}$, as shown in Figure 2.10.

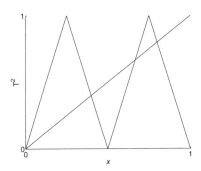

Figure 2.10: The graphs of $T^2(x)$ and $y = x$ when $\mu = 2$.

The fixed points at $x = 0$ and $x = \frac{2}{3}$ are of period one; therefore, there are two points of period two given by $x_{2,1} = \frac{2}{5}$ and $x_{2,2} = \frac{4}{5}$. Since the gradient of $|T^2(x)|$ is greater than one at these points, $x_{2,1}$ and $x_{2,2}$ are unstable.

It is not difficult to show that there are no period-two points for $0 \leq \mu \leq 1$ and there are two points of period two for $1 < \mu \leq 2$.

To determine the fixed points of period three, it is necessary to find the points of intersection of $T^3(x)$ with the diagonal. Consider the case where $\mu = 2$. The methods below can be applied for any value of μ in the interval $[0, 2]$.

The function $T(T(T(x))) = T^3(x)$ is determined by replacing x with $T(x)$ in the mapping for $T^2(x)$. Hence

$$T^3(x) = \begin{cases} 4T(x), & 0 \leq T(x) < \frac{1}{4}, \\ 2 - 4T(x), & \frac{1}{4} \leq T(x) < \frac{1}{2}, \\ 4T(x) - 2, & \frac{1}{2} \leq T(x) < \frac{3}{4}, \\ 4 - 4T(x), & \frac{3}{4} \leq T(x) \leq 1. \end{cases}$$

The interval $0 \leq T(x) < \frac{1}{4}$ on the vertical axis corresponds to two intervals, namely $0 \leq x < T^{-1}(\frac{1}{4})$ and $T^{-1}(\frac{1}{4}) \leq x \leq 1$ on the horizontal axis. When $\mu = 2$, it is not difficult to show that $T^{-1}(\frac{1}{4}) = \frac{1}{8}$ or $\frac{7}{8}$, depending on the branch of $T(x)$. The process may be repeated for $T(x)$ lying in the other intervals. Therefore, $T^3(x)$ becomes

2.2. Fixed Points and Periodic Orbits 47

$$T^3(x) = \begin{cases} 8x, & 0 \leq x < \frac{1}{8} \\ 2 - 8x, & \frac{1}{8} \leq x < \frac{1}{4}, \\ 8x - 2, & \frac{1}{4} \leq x < \frac{3}{8}, \\ 4 - 8x, & \frac{3}{8} \leq x < \frac{1}{2}, \\ 8x - 4, & \frac{1}{2} \leq x < \frac{5}{8}, \\ 6 - 8x, & \frac{5}{8} \leq x < \frac{3}{4}, \\ 8x - 6, & \frac{3}{4} \leq x < \frac{7}{8}, \\ 8 - 8x, & \frac{7}{8} \leq x \leq 1. \end{cases}$$

This function intersects the diagonal at eight points corresponding to $x = 0$, $\frac{2}{9}, \frac{2}{7}, \frac{4}{9}, \frac{4}{7}, \frac{2}{3}, \frac{6}{7}$, and $\frac{8}{9}$ as shown in Figure 2.11. Note that points of period two do not repeat on every third cycle and hence do not appear here.

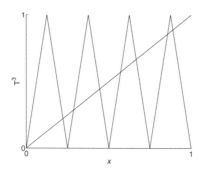

Figure 2.11: The graphs of $T^3(x)$ and $y = x$ when $\mu = 2$.

The fixed points at $x = 0$ and $x = \frac{2}{3}$ are of period one; therefore, there are six points of period three given by $x_{3,1} = \frac{2}{9}, x_{3,2} = \frac{4}{9}, x_{3,3} = \frac{8}{9}, x_{3,4} = \frac{2}{7}, x_{3,5} = \frac{4}{7}$, and $x_{3,6} = \frac{6}{7}$. Since the gradient of $|T^3(x)|$ is greater than one at these points, all six points are unstable. Thus an initial point close to the periodic orbit, but not on it, will move away and the orbits will diverge.

This process may be repeated to determine points of any period for the tent map. Recall that the results by Li and Yorke imply that the map contains periodic points of all periods. It is therefore possible to find points of period 10, 10^6, or even 10^{100}, for example. There are also aperiodic (or nonperiodic) orbits and the system is sensitive to initial conditions. Similar phenomena are observed for three-dimensional autonomous systems in Chapter 13, in fact, most of the dynamics exhibited there appear for this much simpler system.

2.3 The Logistic Map, Bifurcation Diagram, and Feigenbaum Number

In the early 1970s, May [10] and others began to investigate the equations used by fish biologists and entomologists to model the fluctuations in certain species. Simple population models have been discussed in other chapters using continuous dynamical models but the analysis here will be restricted to simple nonlinear discrete systems. Perhaps the most famous system used to model a single species is that known as the *logistic map* given by

(2.3) $$x_{n+1} = f_\mu(x_n) = \mu x_n(1 - x_n),$$

where μ is a parameter and $0 \leq x_n \leq 1$ represents the scaled population size. Consider the case where μ is related to the reproduction rate and x_n represents the population of blowflies at time n, which can be measured in hours, days, weeks, months, etc. Blowflies have a relatively short lifespan and are easy to monitor in the laboratory. Note that this model is extremely simple but as with the tent map a rich variety of behavior is displayed as the parameter μ is varied. We note that scientists would find it difficult to change reproduction rates of individual flies directly; however, for many species the reproduction rate depends on other factors such as temperature. Hence imagine a tank containing a large number of blowflies. Experimentally, we would like to observe how the population fluctuates, if at all, at different temperatures. A population of zero would imply that the tank is empty and a scaled population of one would indicate that the tank is full. The numbers produced in this model would be rounded down to guarantee that fractions would be ignored as in the continuous case.

It must be pointed out that this model does not take into account many features that would influence a population in real applications. For example, age classes, diseases, other species interactions, and environmental effects are all ignored. Even though many factors are left out of the equation, the results show a wide range of dynamical behavior that has been observed both experimentally and in the field.

Consider the logistic map $f_\mu : [0, 1] \to [0, 1]$ given by

$$x_{n+1} = f_\mu(x_n),$$

where $f_\mu(x) = \mu x(1 - x)$. The parameter μ lies in the interval [0, 4]. The graph of f_μ is given in Figure 2.12.

As with the tent map, simple numerical and graphical iterations may be carried out for varying values of the parameter μ. To avoid repetition, these tasks will be left as exercises at the end of the chapter. Instead, the analysis will be restricted to finding periodic points and plotting a bifurcation diagram.

To find points of period one, it is necessary to solve the equation given by

$$f_\mu(x) = \mu x(1 - x) = x,$$

2.3. The Logistic Map, Bifurcation Diagram, and Feigenbaum Number

Figure 2.12: A graph of the logistic map function.

which gives the points that satisfy the condition $x_{n+1} = x_n$ for all n. There are two solutions given by $x_{1,1} = 0$ and $x_{1,2} = 1 - \frac{1}{\mu}$. The stability of the critical points may be determined using the following theorem.

Theorem 1. *Suppose that the map $f_\mu(x)$ has a fixed point at x^*. Then the fixed point is stable if*

$$\left|\frac{d}{dx} f_\mu(x^*)\right| < 1,$$

and it is unstable if

$$\left|\frac{d}{dx} f_\mu(x^*)\right| > 1.$$

Using Theorem 1, $\left|\frac{df_\mu(0)}{dx}\right| = \mu$. Thus the point $x_{1,1}$ is stable for $0 < \mu < 1$ and unstable if $\mu > 1$. Since $\left|\frac{df_\mu(x_{1,2})}{dx}\right| = 2 - \mu$, this fixed point is stable for $1 < \mu < 3$ and is unstable when $\mu < 1$ or $\mu > 3$.

To find points of period two, it is necessary to solve the equation given by

(2.4) $\qquad f_\mu^2(x) = \mu(\mu x(1-x)(1-\mu x(1-x))) = x,$

which gives the points that satisfy the condition $x_{n+2} = x_n$ for all n. Two solutions for equation (2.4) are known, namely, $x_{1,1}$ and $x_{1,2}$, since points of period one repeat on every second iterate. Therefore, equation (2.4) factorizes as follows:

$$x\left(x - \left(1 - \frac{1}{\mu}\right)\right)(-\mu^3 x^2 + (\mu^2 + \mu^3)x - (\mu^2 + \mu)) = 0.$$

The equation $-\mu^3 x^2 + (\mu^2 + \mu^3)x - (\mu^2 + \mu) = 0$ has roots at

$$x_{2,1} = \frac{\mu + 1 + \sqrt{(\mu - 3)(\mu + 1)}}{2\mu} \quad \text{and} \quad x_{2,2} = \frac{\mu + 1 - \sqrt{(\mu - 3)(\mu + 1)}}{2\mu}.$$

2. Nonlinear Discrete Dynamical Systems

Thus there are two points of period two when $\mu > 3$. Let $b_1 = 3$ correspond to the first bifurcation point for the logistic map. Now

$$\frac{d}{dx} f_\mu^2(x_{2,1}) = -4\mu^3 x^3 + 6\mu^3 x^2 - 2(\mu^2 + \mu^3)x + \mu^2$$

and

$$\left| \frac{d}{dx} f_\mu^2(x_{2,1}) \right| = 1,$$

when $\mu = b_2 = 1 + \sqrt{6}$. The value b_2 corresponds to the second bifurcation point for the logistic map. Hence $x_{2,1}$ and $x_{2,2}$ lose their stability at $\mu = b_2$ (check this using MATLAB).

In summary, for $0 < \mu < 1$, the fixed point at $x = 0$ is stable and iterative paths will be attracted to that point. Physically, this would mean that the population of blowflies would die away to zero. One can think of the temperature of the tank being too low to sustain life. As μ passes through one, the trivial fixed point becomes unstable and the iterative paths are attracted to the fixed point at $x_{1,2} = 1 - \frac{1}{\mu}$. For $1 < \mu < b_1$, the fixed point of period one is stable, which means that the population stabilizes to a constant value after a sufficiently long time. As μ passes through b_1, the fixed point of period one becomes unstable and a fixed point of period two is created. For $b_1 < \mu < b_2$, the population of blowflies will alternate between two values on each iterative step after a sufficient amount of time. As μ passes through b_2, the fixed point of period two loses its stability and a fixed point of period four is created.

As with other dynamical systems, all of the information gained so far can be summarized on a bifurcation diagram. Figure 2.13 shows a bifurcation diagram for the logistic map when $0 \leq \mu \leq 3.5$. The first two bifurcation points are labeled b_1 and b_2.

Figure 2.13: The first two bifurcations for the logistic map.

2.3. The Logistic Map, Bifurcation Diagram, and Feigenbaum Number   51

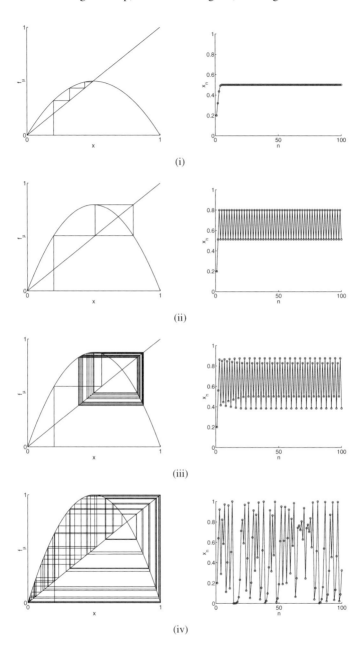

Figure 2.14: Iterative paths and time series data representing the population of blowflies at time n. The population can vary periodically or in an erratic unpredictable manner.

2. Nonlinear Discrete Dynamical Systems

For other values of μ, it is interesting to plot time series data obtained from the logistic map. Figure 2.14 shows iterative paths and time series data when $x_0 = 0.2$ (assuming the tank is initially, say, $\frac{1}{5}$ full) for the following four cases: (i) $\mu = 2$, (ii) $\mu = 3.2$, (iii) $\mu = 3.5$, and (iv) $\mu = 4$.

It is not too difficult to extend the diagram to cover the whole range of values for μ, namely $0 \leq \mu \leq 4$. The bifurcation diagram given in Figure 2.15 was produced using the MATLAB package. Thus even the simple quadratic function $f_\mu(x) = \mu x(1 - x)$ exhibits an extraordinary variety of behaviors as μ varies from 1 to 4. In the past scientists believed that to model complicated behavior, one must have complicated or many equations. One of the most exciting developments to emerge from the realm of nonlinear dynamical systems was the realization that simple equations can lead to extremely complex seemingly random behavior.

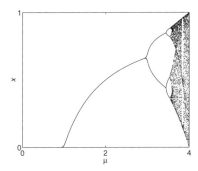

Figure 2.15: The bifurcation diagram of the logistic map produced using an iterative method.

Figure 2.15 shows *period-doubling bifurcations to chaos*. This means that as μ increases beyond b_1, points of period one become period two; at b_2 points of period two become period four, and so on. The sequence of period doublings ends at about $\mu = 3.569945\ldots$, where the system becomes chaotic. This is not the end of the story, however. Figure 2.16 clearly shows regions where the system returns to periodic behavior, even if for only a small range of μ values. These regions are called *periodic windows*.

Near to the period-three window, the logistic map can display a new type of behavior known as *intermittency*, which is almost periodic behavior interrupted by occasional chaotic bursts. A graphical iteration and time series plot are shown in Figure 2.17. The intermittent nature becomes more evident as more points are plotted.

The geometry underlying this behavior can be seen by plotting a graphical iteration for f_μ^3 when $\mu = 3.8282$, for example. This is left as an exercise for the reader. As the parameter μ is increased, the length of the intervals of chaotic bursts become larger and larger until the system becomes fully chaotic. This phenomenon

2.3. The Logistic Map, Bifurcation Diagram, and Feigenbaum Number   53

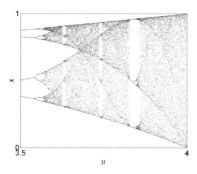

Figure 2.16: A magnification of the bifurcation diagram for the logistic map in the range $3.5 \leq \mu \leq 4$.

is known as an *intermittency route to chaos* and appears in many other physical examples.

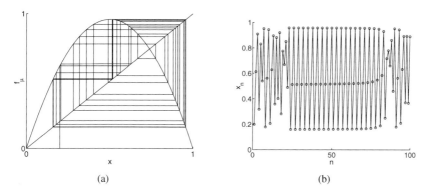

Figure 2.17: (a) Iterative paths when $\mu = 3.8282$ and $x_0 = 0.2$. (b) Time series data.

An even more remarkable discovery was made by Mitchell J. Feigenbaum in the mid-1970s and involves the concept of *universality*. The first seven bifurcation points computed numerically are given by $b_1 = 3.0$, $b_2 = 3.449490\ldots$, $b_3 = 3.544090\ldots$, $b_4 = 3.564407\ldots$, $b_5 = 3.568759\ldots$, $b_6 = 3.569692\ldots$, and $b_7 = 3.569891\ldots$. Feigenbaum discovered that if d_k is defined by $d_k = b_{k+1} - b_k$, then

$$\delta = \lim_{k \to \infty} \frac{d_k}{d_{k+1}} = 4.669202\ldots.$$

The number δ, known as the *Feigenbaum constant*, is much like the numbers π and e in that it appears throughout science. The constant δ can be found not only in iterative maps but also in certain differential equations and even in physical experiments exhibiting period-doubling cascades to chaos. Hence the Feigenbaum constant is called a universal constant.

Figure 2.15 also has fractal structure, one may see similar patterns as one zooms in to the picture. Fractals will be discussed in detail in Chapter 5.

Another method often used to determine whether or not a system is chaotic is the *Lyapunov exponent*. One of the properties of chaos is the sensitivity to initial conditions. However, it is known that an orbit on a chaotic attractor for a bounded system also returns to all accessible states with equal probability. This property is known as *ergodicity*. Thus iterates return infinitely closely infinitely often to any previous point on the chaotic attractor. The formula below may be applied to compute the Lyapunov exponent for iterates in the logistic map. It gives an indication as to whether two orbits starting close together diverge or converge.

Definition 3. The Lyapunov exponent L computed using the derivative method is defined by

$$L = \frac{1}{n}\left(\ln|f'_\mu(x_1)| + \ln|f'_\mu(x_2)| + \cdots + \ln|f'_\mu(x_n)|\right),$$

where f'_μ represents differentiation with respect to x and $x_0, x_1, x_2, \ldots, x_n$ are successive iterates. The Lyapunov exponent may be computed for a sample of points near the attractor to obtain an *average Lyapunov exponent*.

Theorem 2. *If at least one of the average Lyapunov exponents is positive, then the system is chaotic; if the average Lyapunov exponent is negative, then the orbit is periodic; and when the average Lyapunov exponent is zero, a bifurcation occurs.*

Table 2.2 lists Lyapunov exponents computed for the logistic map for several values of the parameter μ. Note that there are other methods available for determining Lyapunov exponents (see Chapter 13).

Table 2.2: The Lyapunov exponents computed to four decimal places using the first derivative method for the logistic map. A total of 50,000 iterates was used in each case.

μ	0.5	1	2.1	3	3.5	3.8282	4
Average L	-0.6932	-0.0003	-2.3025	-0.0002	-0.8720	0.2632	0.6932

The numerical results agree quite well with Theorem 2. In fact, the more chaotic a system, the higher the value of the Lyapunov exponent, as can be seen in Table 2.2. The MATLAB program is given in Section 2.6 and takes about five minutes to compile on a personal computer for each μ value. To find a better approximation of the Lyapunov exponent, a much larger number of iterates would be required. Show that for the logistic map with $\mu = 4$, the Lyapunov exponent is in fact $L = \ln(2)$.

Let us return briefly to the tent map (2.1). The Lyapunov exponent of the tent map can be found exactly since $T'(x) = \pm \mu$ for all values of x. Hence

$$L = \lim_{n \to \infty} \left(\frac{1}{n} \sum_{i=1}^{n} \ln |T'(x_i)| \right) = \ln \mu.$$

2.4 Gaussian and Hénon Maps

The Gaussian Map. Another type of nonlinear one-dimensional iterative map is the Gaussian map $G : \Re \to \Re$ defined by

$$G(x) = e^{-\alpha x^2} + \beta,$$

where α and β are constants. The graph of the Gaussian function has a general form as depicted in Figure 2.18. The parameters α and β are related to the width and height of the Gaussian curve, respectively.

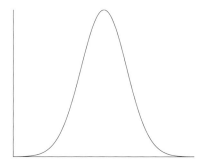

Figure 2.18: The Gaussian map function.

Define an iterative map by

$$x_{n+1} = G(x_n).$$

Since there are two parameters associated with this map, one would expect the dynamics to be more complicated than for the logistic map. All of the features that appear in the logistic map are also present for the Gaussian map. However, certain features of the latter map are not exhibited at all by the logistic map. Some of these additional phenomena may be described as *period bubblings*, *period undoublings*, and *bistability*. These features can appear in the bifurcation diagrams.

Simple numerical and graphical iterations may be carried out as for the tent and logistic maps (see the exercises at the end of the chapter). The fixed points of

56   2. Nonlinear Discrete Dynamical Systems

period one may be found by solving the iterative equation $x_{n+1} = x_n$ for all n, which is equivalent to finding the intersection points of the function $G(x)$ with the diagonal. It is not difficult to see that there can be one, two, or three intersections, as shown in Figure 2.19. For certain parameter values it is possible to have two stable fixed points of period one.

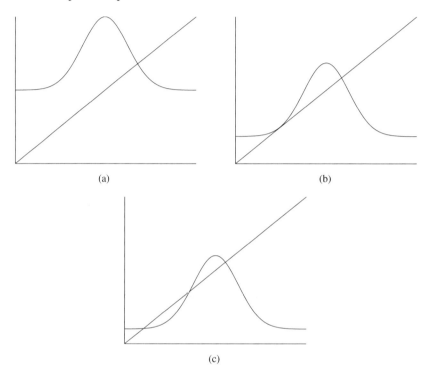

Figure 2.19: Possible intersections of the Gaussian function with the diagonal.

The Gaussian map has two points of inflection at $x = \pm\frac{1}{\sqrt{2\alpha}}$. This implies that period-one behavior can exist for two ranges of the parameters. This in turn means that a period-one point can make a transition from being stable to unstable and back to stable again, as depicted in Figure 2.20.

As the parameter β is increased from $\beta = -1$, a fixed point of period one becomes unstable and a sequence of period bubbling occurs through period-two, period-four, and back to period-two behavior. As the parameter is increased still further, the unstable fixed point of period one becomes stable again and a single branch appears once more. For higher values of the parameter α, the system can display more complex dynamics. An example is shown in Figure 2.21.

Figure 2.21 displays period-doubling and period-undoubling bifurcations and multistability. For example, when $\beta = -1$, there are two possible steady-state so-

2.4. Gaussian and Hénon Maps   57

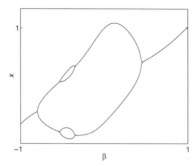

Figure 2.20: A bifurcation diagram for the Gaussian map when $\alpha = 4$ produced using an iterative method.

(a)

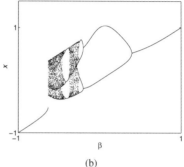
(b)

Figure 2.21: Bifurcation diagrams for the Gaussian map when $\alpha = 8$ produced using an iterative method. (a) $x_0 = 0$ and (b) $x_0 = -1$ for each value of β.

lutions. It is possible for these systems to display bistable phenomena as explained in other chapters of the book. The tent and logistic maps cannot display bistability.

The Hénon Map. Consider the two-dimensional iterated map function given by

(2.5)
$$x_{n+1} = 1 + y_n - \alpha x_n^2,$$
$$y_{n+1} = \beta x_n,$$

where $\alpha > 0$ and $|\beta| < 1$. The map was first discussed in 1976 by Hénon [11], who used it as a simple model for the Poincaré map of the Lorenz system. The Hénon map displays periodicity, mixing, and sensitivity to initial conditions. The system can also display hysteresis and bistability can be observed in the bifurcation diagrams. Each of these phenomena will now be discussed briefly in turn.

Suppose that the discrete nonlinear system

$$x_{n+1} = P(x_n, y_n), \quad y_{n+1} = Q(x_n, y_n),$$

has a fixed point at (x_1, y_1), where P and Q are at least quadratic in x_n and y_n. The fixed point can be transformed to the origin and the nonlinear terms can be discarded after taking a Taylor series expansion. The Jacobian matrix is given by

$$J(x_1, y_1) = \begin{pmatrix} \frac{\partial P}{\partial x} & \frac{\partial P}{\partial y} \\ \frac{\partial Q}{\partial x} & \frac{\partial Q}{\partial y} \end{pmatrix}\bigg|_{(x_1, y_1)}.$$

Definition 4. Suppose that the Jacobian has eigenvalues λ_1 and λ_2. A fixed point is called *hyperbolic* if both $|\lambda_1| \neq 1$ and $|\lambda_2| \neq 1$. If either $|\lambda_1| = 1$ or $|\lambda_2| = 1$, then the fixed point is called nonhyperbolic.

The type of fixed point is determined using arguments similar to those used in Chapter 8. In the discrete case, the fixed point is stable as long as $|\lambda_1| < 1$ and $|\lambda_2| < 1$, otherwise the fixed point is unstable. For example, the fixed points of period one for the Hénon map can be found by solving the equations given by $x_{n+1} = x_n$ and $y_{n+1} = y_n$ simultaneously. Therefore, period-one points satisfy the equations

$$x = 1 - \alpha x^2 + y, \quad y = \beta x.$$

The solutions are given by

$$x = \frac{(\beta - 1) \pm \sqrt{(1-\beta)^2 + 4\alpha}}{2\alpha}, \quad y = \beta \left(\frac{(\beta - 1) \pm \sqrt{(1-\beta)^2 + 4\alpha}}{2\alpha} \right).$$

Thus the Hénon map has two fixed points of period one if and only if $(1-\beta)^2 + 4\alpha > 0$. As a particular example, consider system (2.5) with $\alpha = \frac{3}{16}$ and $\beta = \frac{1}{2}$. There are two fixed points of period one given by $A = (-4, -2)$ and $B = (\frac{4}{3}, \frac{2}{3})$. The Jacobian is given by

$$J = \begin{pmatrix} -2\alpha x & 1 \\ \beta & 0 \end{pmatrix}.$$

The eigenvalues for the fixed point A are $\lambda_1 \approx -0.28$ and $\lambda_2 \approx 1.78$; therefore, A is a saddle point. The eigenvalues for the fixed point B are $\lambda_1 = -1$ and $\lambda_2 = 0.5$. Thus this critical point is nonhyperbolic.

Fix the parameter $\beta = 0.4$ in the Hénon map (2.5). There are points of periods one (when $\alpha = 0.2$), two (when $\alpha = 0.5$), and four (when $\alpha = 0.9$), for example. The reader can verify these results using the MATLAB program in Section 2.6. Some iterative plots are given in Figure 2.22.

The choice of initial conditions is important in these cases as some orbits are unbounded and move off to infinity. One must start with points that are within

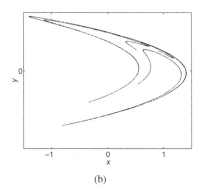

Figure 2.22: Iterative plots for system (2.5) when $\beta = 0.4$ and (a) $\alpha = 1$ and (b) $\alpha = 1.2$. In each case the initial point was $(0.1, 0)$.

the *basin of attraction* for this map. Basins of attraction will be discussed in later chapters. Of course, all of this information can be summarized on a bifurcation diagram, and this will be left as an exercise for the reader. There are the usual phenomena associated with bifurcation diagrams. However, for the Hénon map, different chaotic attractors can exist simultaneously for a range of parameter values of α. This system also displays hysteresis for certain parameter values.

To demonstrate the stretching and folding associated with this map, consider a set of initial points lying on the square of length two centered at the origin. Figure 2.23 shows how the square is stretched and folded after only two iterations. This stretching and folding is reminiscent of the Smale horseshoe discussed in Chapter 14.

The chaotic attractor formed is an invariant set and has fractal structure. Note that $\det(J)$ for the Hénon map is equal to $|\beta|$. This implies that a small area is reduced by a factor of β on each iteration since $|\beta| < 1$.

2.5 Applications

This section introduces four discrete dynamical systems taken from biology, economics, nonlinear optics, and neural networks. The reader can investigate these systems via the exercises in exercises in Section 2.7.

Biology. The average 70-liter human body contains 5 liters of blood, a small amount of which consists of *erythrocytes* or red blood cells. These cells, which are structurally the simplest in the body, can be counted to measure hematologic conditions such as *anemia*. Anemia is any condition resulting in a significant decrease in total body erythrocyte mass. The population of red blood cells oscillates in a healthy human with the average woman having 4.2–5.4 per μL, and the average man having 4.7–6.1 per μL. A simple blood cell population model was investigated

2. Nonlinear Discrete Dynamical Systems

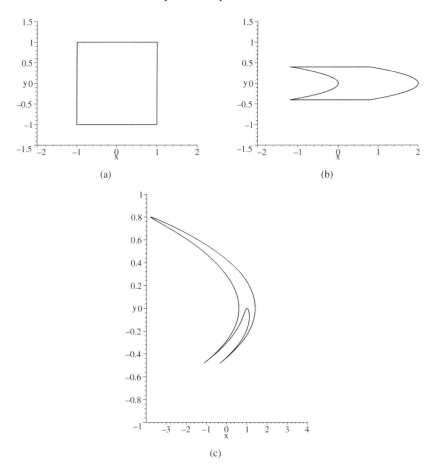

Figure 2.23: Application of the Hénon transformation to a square when $\alpha = 1.2$ and $\beta = 0.4$: (a) initial points, (b) first iterates, and (c) second iterates.

by Lasota [8] in 1977. Let c_n denote the red cell count per unit volume in the nth time interval, then

$$c_{n+1} = c_n - d_n + p_n,$$

where d_n and p_n are the number of cells destroyed and produced in one time interval, respectively. In the model considered by Lasota, $d_n = ac_n$ and $p_n = bc_n^r e^{-sc_n}$, where $0 < a \leq 1$ and $b, r, s > 0$. Hence

(2.6) $$c_{n+1} = (1-a)c_n + bc_n^r e^{-sc_n}.$$

Typical parameters used in the model are $b = 1.1 \times 10^6$, $r = 8$, and $s = 16$.

2.5. Applications

Economics. The Gross National Product (GNP) measures economic activity based on labor and production output within a country. Consider the following simple growth model investigated by Day [7] in 1982:

$$k_{t+1} = \frac{s(k_t) f(k_t)}{1 + \lambda},$$

where k_t is the capital-labor ratio, s is the savings ratio function, f is the per capita production function, and λ is the natural rate of population growth. In one case considered by Day,

$$s(k) = \sigma, \quad f(k) = \frac{Bk^\beta (m - k)^\gamma}{(1 + \lambda)},$$

where $\beta, \gamma, m > 0$. This leads to the following discrete dynamical system:

(2.7) $$k_{t+1} = \sigma \frac{Bk_t^\beta (m - k_t)^\gamma}{(1 + \lambda)},$$

which can be thought of as a highly simplified model for the GNP of a country.

Nonlinear Optics. When modeling the intracavity field of a laser in a bulk cavity ring under the assumption that saturable absorption can be ignored, Hammel, Jones, and Maloney [6] obtained the following complex one-dimensional difference equation relating the field amplitude, say, E_{n+1}, at the $(n + 1)$st cavity pass to that of a round trip earlier:

(2.8) $$E_{n+1} = A + B E_n \exp\left[i\left(\phi - \frac{C}{1 + |E_n|^2}\right)\right],$$

where ϕ is a phase angle, and A, B, C are all constant. This mapping can also be thought of as two-dimensional (one-dimensional complex). Splitting E_n into its real and imaginary parts, (2.8) becomes

(2.9) $$\begin{aligned} x_{n+1} &= A + B[x_n \cos(\theta) - y_n \sin(\theta)], \\ y_{n+1} &= B[x_n \sin(\theta) + y_n \cos(\theta)], \end{aligned}$$

where $\theta = (\phi - \frac{C}{1+|E_n|^2})$. Equations (2.8) and (2.9) are known as *Ikeda mappings*. Electromagnetic waves and optical resonators are discussed in some detail in Chapter 4.

Neural Networks. According to Pasemann and Stollenwerk [2], the activity of a recurrent two-neuron module shown in Figure 2.24 at time n is given by the vector $\mathbf{x_n} = (x_n, y_n)^T$. The discrete dynamical system used to model the neuromodule is given by

(2.10) $$\begin{aligned} x_{n+1} &= \theta_1 + w_{11}\sigma(x_n) + w_{12}\sigma(y_n), \\ y_{n+1} &= \theta_2 + w_{21}\sigma(x_n), \end{aligned}$$

2. Nonlinear Discrete Dynamical Systems

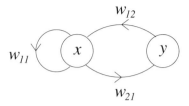

Figure 2.24: A recurrent two-neuron module with an excitory neuron with activity y_n, and a self-connected inhibitory neuron with activity x_n.

where σ defines the sigmoidal transfer function defining the output of a neuron

$$\sigma(x) = \frac{1}{1+e^{-x}};$$

θ_1 and θ_2 are the neuron biases; the w_{ij} are weights; the index i indicates the neuron destination for that weight, and the index j represents the source of the signal fed to the neuron. Neural networks are discussed in some detail in Chapter 17.

2.6 MATLAB Commands

Remember to group the for... end statements, etc.

```
% Program_2a - Graphical iteration of the tent map (Figure
% 2.7(b)).

clear
% Initial condition 0.2001, must be symbolic.
nmax=200;
t=sym(zeros(1,nmax));
t(1)=sym(2001/10000);
mu=2;
halfm=nmax/2;
axis([0 1 0 1]);
for n=2:nmax
    if (double(t(n-1)))>0 & (double(t(n-1)))<=1/2
            t(n)=sym(2*t(n-1));
        else
            if (double(t(n-1)))<1
                t(n)=sym(2*(1-t(n-1)));
            end
    end
end

for n=1:halfm
    t1(2*n-1)=t(n);
```

```
    t1(2*n)=t(n);
end

t2(1)=0;t2(2)=double(t(2));
for n=2:halfm
    t2(2*n-1)=double(t(n));
    t2(2*n)=double(t(n+1));
end
hold on
fsize=15;
plot(double(t1),double(t2),'r');
x=[0 0.5 1];y=[0 mu/2 0];
plot(x,y,'b');
x=[0 1];y=[0 1];
plot(x,y,'g');
title('Graphical iteration for the tent map')
set(gca,'xtick',[0:0.2:1],'FontSize',fsize)
set(gca,'ytick',[0:0.2:1],'FontSize',fsize)
xlabel('x','FontSize',fsize)
ylabel('T','FontSize',fsize)
hold off
% End of Program_2a.
```

```
% Program_2b - Graphical iteration of the logistic map.

clear
fsize=15;
nmax=100;halfm=nmax/2;
t=zeros(1,nmax);
t(1)=0.2;
mu=3.8282;
axis([0 1 0 1]);
for n=1:nmax
    t(n+1)=mu*t(n)*(1-t(n));
end

for n=1:halfm
    t1(2*n-1)=t(n);
    t1(2*n)=t(n);
end

t2(1)=0;t2(2)=t(2);
for n=2:halfm
    t2(2*n-1)=t(n);
    t2(2*n)=t(n+1);
end
hold on
```

2. Nonlinear Discrete Dynamical Systems

```
plot(t1,t2,'r');
fplot('3.8282*x*(1-x)',[0 1]);
x=[0 1];y=[0 1];
plot(x,y,'g');
hold off
%title('Graphical iteration for the tent map')
set(gca,'xtick',[0 1],'Fontsize',fsize)
set(gca,'ytick',[0 1],'Fontsize',fsize)
xlabel('x','Fontsize',fsize)
ylabel('f_{\mu}','Fontsize',fsize)
% End of Program_2b.
```

```
% Program_2c - Lyapunov exponent (Table 2.2).

clear;
itermax=49999;mu=4;x=0.1;xo=x;
    for n=1:itermax
        xn=mu*xo*(1-xo);
        x=[x xn];
        xo=xn;
    end

vpa(sum(log(abs(mu*(1-2*x)))))/itermax,6)
% End of Program_2c.
```

```
% Program_2d - Bifurcation diagram of the logistic map
% (Figure 2.15).

clear
itermax=100;
finalits=30;finits=itermax-(finalits-1);
for r=0:0.005:4
    x=0.4;
    xo=x;
    for n=2:itermax
        xn=r*xo*(1-xo);
        x=[x xn];
        xo=xn;
    end
    plot(r*ones(finalits),x(finits:itermax),'.','MarkerSize',1)
    hold on
end
fsize=15;
set(gca,'xtick',[0:1:4],'FontSize',fsize)
set(gca,'ytick',[0,0.5,1],'FontSize',fsize)
xlabel('{\mu}','FontSize',fsize)
```

```
ylabel('\itx','FontSize',fsize)
hold off
% End of Program_2d.
```

```
% Program_2e - Bifurcation diagram of the Gaussian map
% (Figure 2.21(a)).

clear;
itermax=160;alpha=8;min=itermax-9;
for beta=-1:0.001:1
    x=0;
    xo=x;
    for n=1:itermax
        xn=exp(-alpha*xo^2)+beta;
        x=[x xn];
        xo=xn;
    end
    plot(beta*ones(10),x(min:itermax),'.','MarkerSize',1)
    hold on
end

fsize=15;
set(gca,'xtick',[-1 1],'FontSize',fsize)
set(gca,'ytick',[-1 1],'FontSize',fsize)
xlabel('{\beta}','FontSize',fsize)
ylabel('\itx','FontSize',fsize)
hold off
% End of Program_2e.
```

```
% Program_2f - Iteration of the Henon map (Figure 2.22(b)).

a=1.2;
b=0.4;
N=6000;
x=zeros(1,N);
y=zeros(1,N);
x(1)=0.1;
y(1)=0;
for n=1:N
    x(n+1)=1+y(n)-a*(x(n))^2;
    y(n+1)=b*x(n);
end
axis([-1 2 -1 1])
plot(x(50:N),y(50:N),'.','MarkerSize',1);
fsize=15;
set(gca,'xtick',[-1:1:1],'FontSize',fsize)
```

```
set(gca,'ytick',[-1:1:2],'FontSize',fsize)
xlabel('\itx','FontSize',fsize)
ylabel('\ity','FontSize',fsize)
% End of Program_2f.
```

2.7 Exercises

1. Consider the tent map defined by

$$T(x) = \begin{cases} 2x, & 0 \leq x < \frac{1}{2}, \\ 2(1-x), & \frac{1}{2} \leq x \leq 1. \end{cases}$$

   Sketch graphical iterations for the initial conditions (i) $x_0 = \frac{1}{4}$, (ii) $x_0 = \frac{1}{6}$, (iii) $x_0 = \frac{5}{7}$, and (iv) $x_0 = \frac{1}{19}$. Find the points of periods one, two, three, and four. Give a formula for the number of points of period N.

2. (a) Let T be the function $T : [0, 1] \rightarrow [0, 1]$ defined by

$$T(x) = \begin{cases} \frac{3}{2}x, & 0 \leq x < \frac{1}{2}, \\ \frac{3}{2}(1-x), & \frac{1}{2} \leq x \leq 1. \end{cases}$$

   Sketch the graphs of $T(x)$, $T^2(x)$, and $T^3(x)$. How many points are there of periods one, two, and three, respectively?

   (b) Let T be the function $T : [0, 1] \rightarrow [0, 1]$ defined by

$$T(x) = \begin{cases} \frac{9}{5}x, & 0 \leq x < \frac{1}{2}, \\ \frac{9}{5}(1-x), & \frac{1}{2} \leq x \leq 1. \end{cases}$$

   Determine the fixed points of periods one, two, and three.

3. By editing the MATLAB program given in Section 2.6, plot a bifurcation diagram for the tent map.

4. Consider the logistic map function defined by $f_\mu(x) = \mu x(1-x)$. Determine the functions $f_\mu(x)$, $f_\mu^2(x)$, $f_\mu^3(x)$, and $f_\mu^4(x)$, and plot the graphs when $\mu = 4.0$. How many points are there of periods one, two, three, and four?

5. Consider the iterative equation

$$x_{n+1} = \mu x_n(100 - x_n),$$

   which may be used to model the population of a certain species of insect. Given that the population size periodically alternates between two distinct values, determine a value of μ that would be consistent with this behavior. Determine an equation that gives the points of period two for a general μ value.

6. Plot bifurcation diagrams for

   (a) the Gaussian map when $\alpha = 20$ for $-1 \leq \beta \leq 1$;

   (b) the Gaussian map when $\beta = -0.5$ for $0 \leq \alpha \leq 20$.

7. Find the fixed points of periods one and two for the Hénon map given by

$$x_{n+1} = \frac{3}{50} + \frac{9}{10} y_n - x_n^2, \quad y_{n+1} = x_n.$$

   Derive the inverse map.

8. (a) Show that the Hénon map given by

$$x_{n+1} = 1 - \alpha x_n^2 + y_n, \quad y_{n+1} = \beta x_n,$$

   where $\alpha > 0$ and $|\beta| < 1$ undergoes a bifurcation from period-one to period-two behavior exactly when $\alpha = \frac{3(\beta-1)^2}{4}$ for fixed β.

   (b) Investigate the bifurcation diagrams for the Hénon map by plotting the x_n values as a function of α for $\beta = 0.4$.

9. (a) Consider the blood cell iterative equation (2.6). Assuming that $b = 1.1 \times 10^6$, $r = 8$, and $s = 16$, show that there are (i) two stable and one unstable fixed points of period one when $a = 0.2$, and (ii) two unstable and one stable fixed point of period one when $a = 0.3$.

   (b) Assume that $\sigma = 0.5$, $\beta = 0.3$, $\gamma = 0.2$, $\lambda = 0.2$, $m = 1$ in the economic model (2.7). Show that there is a stable fixed point of period one at $x_{1,2} = 0.263$ when $B = 1$, and an unstable fixed point of period one at $x_{1,2} = 0.873$ when $B = 3.3$.

   (c) Show that the inverse map of equation (2.8) is given by

$$E_{n+1} = \frac{(E_n - A)}{B} \exp\left[-i\left(\phi - \frac{CB^2}{(B^2 + |E_n - A|^2)}\right)\right].$$

   (d) Consider the neuromodule model (2.10). Assume that $\theta_1 = -2$, $\theta_2 = 3$, $w_{11} = -20$, $w_{12} = 6$ and $w_{21} = -6$. Show that there is one fixed point of period one at approximately $(-1.280, 1.695)$, and that it is a saddle point.

10. According to Ahmed, El-Misiery, and Agiza [1], an inflation-unemployment model is given by

$$U_{n+1} = U_n - b(m - I_n), \quad I_{n+1} = I_n - (1-c)f(U_n) + f(U_n - b(m - I_n)),$$

where $f(U) = \beta_1 + \beta_2 e^{-U}$; U_n and I_n are measures of unemployment and inflation at time n, respectively; and b, c, β_1, and β_2 are constants. Show that the system has a unique fixed point of period one at

$$\left(\ln\left(\frac{-\beta_2}{\beta_1}\right), m\right).$$

Given that $m = 2$, $\beta_1 = -2.5$, $\beta_2 = 20$, and $c = 0.18$, show that the eigenvalues of the Jacobian matrix are given by

$$\lambda_{1,2} = 1 - \frac{5b}{4} \pm \frac{\sqrt{25b^2 - 40bc}}{4}.$$

Recommended Reading

[1] E. Ahmed, A. El-Misiery, and H. N. Agiza, On controlling chaos in an inflation-unemployment dynamical system, *Chaos Solitons Fractals*, **10**-9 (1999), 1567–1570.

[2] F. Pasemann and N. Stollenwerk, Attractor switching by neural control of chaotic neurodynamics, *Comput. Neural Systems*, **9** (1998), 549–561.

[3] H. Nagashima and Y. Baba, *Introduction to Chaos: Physics and Mathematics of Chaotic Phenomena*, Institute of Physics, Bristol, PA, 1998.

[4] R. A. Holmgrem, *A First Course in Discrete Dynamical Systems*, Springer-Verlag, New York, 1996.

[5] D. Kaplan and L. Glass, *Understanding Nonlinear Dynamics*, Springer-Verlag, New York, 1995.

[6] S. M. Hammel, C. K. R. T. Jones, and J. V. Maloney, Global dynamical behavior of the optical field in a ring cavity, *J. Optim. Soc. Amer.* B, **2** (1985), 552–564.

[7] R. H. Day, Irregular growth cycles, *Amer. Econ. Rev.*, **72** (1982), 406–414.

[8] A. Lasota, Ergodic problems in biology, *Astérisque*, **50** (1977), 239–250.

[9] T. Y. Li and J. A. Yorke, Period three implies chaos, *Amer. Math. Monthly*, **82** (1975), 985–992.

[10] R. M. May, *Stability and Complexity in Model Ecosystems*, Princeton University Press, Princeton, NJ, 1974.

[11] M. Hénon, Numerical study of quadratic area-preserving mappings, *Quart. Appl. Math.*, **27** (1969), 291–311.

3
Complex Iterative Maps

Aims and Objectives
- To introduce simple complex iterative maps.
- To introduce Julia sets and the Mandelbrot set.
- To carry out some analysis on these sets.

On completion of this chapter, the reader should be able to
- carry out simple complex iterations;
- plot Julia sets and the Mandelbrot set using simple MATLAB programs;
- determine boundaries of points with low periods;
- find basins of attraction (or domains of stability).

It is assumed that the reader is familiar with complex numbers and the Argand diagram. Julia sets are defined, and the MATLAB package is used to plot approximations of these sets.

There are an infinite number of Julia sets associated with one mapping. In one particular case, these sets are categorized by plotting a so-called Mandelbrot set. A black-and-white version of the Mandelbrot set is produced using the MATLAB package.

Applications of complex iterative maps to the real world are presented in Chapter 4.

70   3. Complex Iterative Maps

3.1 Julia Sets and the Mandelbrot Set

As a simple introduction to one-dimensional nonlinear complex iterative maps, consider the quadratic map

$$z_{n+1} = f_c(z_n) = z_n^2 + c, \tag{3.1}$$

where z_n and c are complex numbers. Although equation (3.1) is as simple as the equation of a real circle, the dynamics displayed are highly complicated. In 1919, Gaston Julia published a prize-winning lengthy article on certain types of conformal complex mappings, the images of which would not appear until the advent of computer graphics many years later. Recall that a conformal mapping preserves both the size and the sign of angles.

Definition 1. Consider a complex mapping of the form $z_{n+1} = f(z_n)$. The points that lie on the boundary between points that are bounded and those that are unbounded are collectively referred to as the *Julia set*.

The following properties of a Julia set, J say, are well known:

- The set J is a repellor.
- The set J is invariant.
- An orbit on J is either periodic or chaotic.
- All unstable periodic points are on J.
- The set J is either wholly connected or wholly disconnected.
- The set J nearly always has fractal structure (see Chapter 5).

The colorful Julia sets displayed in many textbooks and videos such as [1] to [6] are generated on powerful graphic computers. Unfortunately, the reader has to be satisfied with less detailed black-and-white figures in this text. To see the true beauty and some detail of the Julia sets and the Mandelbrot set, the author would encourage the reader to watch the video [5]. The author's students have always thoroughly enjoyed it.

To generate Julia sets, some of the properties listed above are utilized. For example, if the set J is a repellor under the forward iterative map (3.1), then the Julia set will become an attractor under an inverse mapping. For computational reasons, it is best to work with the real and imaginary parts of the complex numbers separately. However, the situation has improved with MATLAB 6.5 Release 13.[1]

---

[1] As a reminder, as this book was in production, The MathWorks, Inc. was preparing to release MATLAB 7. Please note that the program files presented in this book will work with the updated version of MATLAB.

3.1. Julia Sets and the Mandelbrot Set

For equation (3.1) it is not difficult to determine the inverse map. Now

$$z_{n+1} = z_n^2 + c,$$

and thus

$$x_{n+1} = x_n^2 - y_n^2 + a, \quad \text{and} \quad y_{n+1} = 2x_n y_n + b,$$

where $z_n = x_n + iy_n$ and $c = a + ib$. To find the inverse map, one must find expressions for x_n and y_n in terms of x_{n+1} and y_{n+1}. Now

$$x_n^2 - y_n^2 = x_{n+1} - a,$$

and note that

$$(x_n^2 + y_n^2)^2 = (x_n^2 - y_n^2)^2 + 4x_n^2 y_n^2 = (x_{n+1} - a)^2 + (y_{n+1} - b)^2.$$

Hence

$$x_n^2 + y_n^2 = +\sqrt{(x_{n+1} - a)^2 + (y_{n+1} - b)^2},$$

since $x_n^2 + y_n^2 > 0$. Suppose that

$$u = \sqrt{(x_{n+1} - a)^2 + (y_{n+1} - b)^2} \quad \text{and} \quad v = x_{n+1} - a.$$

Then

(3.2) $$x_n = \pm\sqrt{u + v} \quad \text{and} \quad y_n = \frac{y_{n+1} - b}{2x_n}.$$

In terms of the computation, there will be a problem if $x_n = 0$. To overcome this difficulty, the following simple algorithm is applied. Suppose that the two roots of equation (3.2) are given by $x_1 + iy_1$ and $x_2 + iy_2$. If $x_1 = \sqrt{u + v}$, then $y_1 = \sqrt{u - v}$ if $y > b$, or $y_1 = -\sqrt{u - v}$ if $y < b$. The other root is then given by $x_2 = -\sqrt{u + v}$ and $y_2 = -y_1$.

This transformation has a two-valued inverse, and twice as many predecessors are generated on each iteration. One of these points is chosen randomly in the computer program. Recall that all unstable periodic points are on J. It is not difficult to determine the fixed points of period one for mapping (3.1). Suppose that z is a fixed point of period one. Then $z_{n+1} = z_n = z$, and

$$z^2 - z + c = 0,$$

which gives two solutions, either

$$z_{1,1} = \frac{1 + \sqrt{1 - 4c}}{2} \quad \text{or} \quad z_{1,2} = \frac{1 - \sqrt{1 - 4c}}{2}.$$

The stability of these fixed points can be determined in the usual way. Hence the fixed point is stable if

$$\left|\frac{df_c}{dz}\right| < 1,$$

and it is unstable if

$$\left|\frac{df_c}{dz}\right| > 1.$$

By selecting an unstable fixed point of period one as an initial point, it is possible to generate a Julia set using a so-called *backward training iterative process*.

Julia sets define the border between bounded and unbounded orbits. Suppose that the Julia set associated with the point $c = a + ib$ is denoted by $J(a, b)$. As a simple example, consider the mapping

(3.3) $$z_{n+1} = z_n^2.$$

One of two fixed points of equation (3.3) lies at the origin, say, z^*. There is also a fixed point at $z = 1$. Initial points that start wholly inside the circle of radius one are attracted to z^*. An initial point starting on $|z| = 1$ will generate points that again lie on the unit circle $|z| = 1$. Initial points starting outside the unit circle will be repelled to infinity, since $|z| > 1$. Therefore, the circle $|z| = 1$ defines the Julia set $J(0, 0)$ that is a repellor (points starting near to but not on the circle are repelled), invariant (orbits that start on the circle are mapped to other points on the unit circle), and wholly connected. The interior of the unit circle defines the *basin of attraction* (or domain of stability) for the fixed point at z^*. In other words, any point starting inside the unit circle is attracted to z^*.

Suppose that $c = -0.5 + 0.3i$ in equation (3.1). Figure 3.1(a) shows a picture of the Julia set $J(-0.5, 0.3)$ containing 2^{15} points. The Julia set $J(-0.5, 0.3)$ defines the border between bounded and unbounded orbits. For example, an orbit starting inside the set $J(-0.5, 0.3)$ at $z_0 = 0 + 0i$ remains bounded, whereas an orbit starting outside the set $J(-0.5, 0.3)$ at $z = -1 - i$, for instance, is unbounded. The reader will be asked to demonstrate this in the exercises at the end of the chapter.

Four of an infinite number of Julia sets are plotted in Figure 3.1. The first three are totally connected, but $J(0, 1.1)$ is totally disconnected.

In 1979, Mandelbrot devised a way of distinguishing those Julia sets that are wholly connected from those that are wholly disconnected. He used the fact that $J(a, b)$ is connected if and only if the orbit generated by $z \to z^2 + c$ is bounded. In this way, it is not difficult to generate the now famous *Mandelbrot set*.

Assign a point on a computer screen to a coordinate position $c = (a, b)$ in the Argand plane. The point $z = 0 + 0i$ is then iterated under the mapping (3.1) to give an orbit

$$0 + 0i, c, c^2 + c, (c^2 + c)^2 + c, \ldots.$$

3.1. Julia Sets and the Mandelbrot Set   73

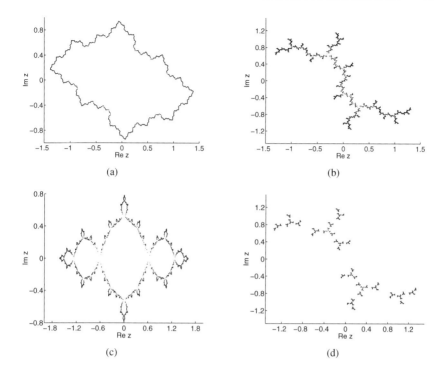

Figure 3.1: Four Julia sets for the mapping (3.1), where $J(a, b)$ denotes the Julia set associated with the point $c = a + ib$: (a) $J(-0.5, 0.3)$, (b) $J(0, 1)$, (c) $J(-1, 0)$, and (d) $J(0, 1.1)$.

If after 50 iterations, the orbit remains bounded (within a circle of radius 4 in the program used here), then the point is colored black. If the orbit leaves the circle of radius 4 after m iterations, where $1 < m < 50$, then the point is colored black if m is even and white if m is odd. In this way a black-and-white picture of the Mandelbrot set is obtained as in Figure 3.2.

Unfortunately, Figure 3.2 does no justice to the beauty and intricacy of the Mandelbrot set. This figure is a theoretical object that can be generated to an infinite amount of detail, and the set is a kind of fractal displaying self-similarity in certain parts and scaling behavior. One has to try to imagine a whole new universe that can be seen by zooming into the picture. For a video journey into the Mandelbrot set, the reader is once more directed to the video [5].

It has been found that this remarkable figure is a universal "constant" much like the Feigenbaum number introduced in Chapter 2. Some simple properties of the Mandelbrot set will be investigated in the next section.

74   3. Complex Iterative Maps

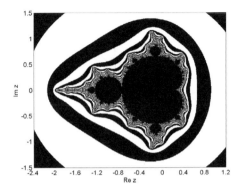

Figure 3.2: The Mandelbrot set (central black figure) produced using a personal computer.

3.2 Boundaries of Periodic Orbits

For the Mandelbrot set, the fixed points of period one may be found by solving the equation $z_{n+1} = z_n$ for all n, or equivalently,

$$f_c(z) = z^2 + c = z,$$

which is a quadratic equation of the form

(3.4) $$z^2 - z + c = 0.$$

The solutions occur at

$$z_{1,1} = \frac{1 + \sqrt{1-4c}}{2} \quad \text{and} \quad z_{1,2} = \frac{1 - \sqrt{1-4c}}{2},$$

where $z_{1,1}$ is the first fixed point of period one and $z_{1,2}$ is the second fixed point of period one using the notation introduced in Chapter 2. As with other discrete systems, the stability of each period-one point is determined from the derivative of the map at the point. Now

(3.5) $$\frac{df_c}{dz} = 2z = re^{i\theta},$$

where $r \geq 0$ and $0 \leq \theta < 2\pi$. Substituting from equation (3.5), equation (3.4) then becomes

$$\left(\frac{re^{i\theta}}{2}\right)^2 - \frac{re^{i\theta}}{2} + c = 0.$$

3.2. Boundaries of Periodic Orbits

The solution for c is

(3.6) $$c = \frac{re^{i\theta}}{2} - \frac{r^2 e^{i2\theta}}{4}.$$

One of the fixed points, say, $z_{1,1}$, is stable as long as

$$\left| \frac{df_c}{dz}(z_{1,1}) \right| < 1.$$

Therefore, using equation (3.5), the boundary of the points of period one is given by

$$\left| \frac{df_c}{dz}(z_{1,1}) \right| = |2z_{1,1}| = r = 1$$

in this particular case. Let $c = x + iy$. Then from equation (3.6), the boundary is given by the following parametric equations:

$$x = \frac{1}{2}\cos\theta - \frac{1}{4}\cos(2\theta), \qquad y = \frac{1}{2}\sin\theta - \frac{1}{4}\sin(2\theta).$$

The parametric curve is plotted in Figure 3.3 and forms a cardioid that lies at the heart of the Mandelbrot set.

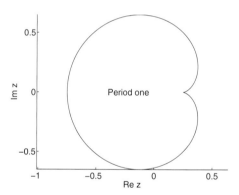

Figure 3.3: The boundary of fixed points of period one for the Mandelbrot set.

Using similar arguments to those above, it is not difficult to extend the analysis to determine the boundary for the fixed points of period two. Fixed points of period two satisfy the equation $z_{n+2} = z_n$ for all n. Therefore,

$$f_c^2(z) = (z^2 + c)^2 + c = z,$$

or, equivalently,

(3.7) $$z^4 + 2cz^2 - z + c^2 + c = 0.$$

However, since points of period one repeat on every second iterate, the points $z_{1,1}$ and $z_{1,2}$ satisfy equation (3.7). Therefore, equation (3.7) factorizes into

$$(z^2 - z + c)(z^2 + z + c + 1) = 0.$$

Hence the fixed points of period two satisfy the quadratic equation

(3.8) $$z^2 + z + c + 1 = 0,$$

which has roots at

$$z_{2,1} = \frac{-1 + \sqrt{-3 - 4c}}{2} \quad \text{and} \quad z_{2,2} = \frac{-1 - \sqrt{-3 - 4c}}{2}.$$

Once more the stability of each critical point is determined from the derivative of the map at the point, now

$$\frac{df_c^2}{dz} = 4z^3 + 4cz = 4z(z^2 + c).$$

Thus

$$\left|\frac{df_c^2}{dz}(z_{2,1})\right| = |4 + 4c|,$$

and the boundary is given by

$$|c + 1| = \frac{1}{4}.$$

The parametric curve is plotted in Figure 3.4 and forms a circle centered at $(-1, 0)$ of radius $\frac{1}{4}$ in the Argand plane. This circle forms the "head" of the Mandelbrot set, sometimes referred to as the *potato man*.

The Mandelbrot set for the nonlinear complex iterative map $z_{n+1} = z_n^2 - 2z_n + c$ is plotted in Figure 3.5.

Chapter 4 illustrates how nonlinear complex iterative maps are being applied in physical applications when modeling lasers and the propagation of light through optical fibers.

3.3 MATLAB Commands

```
% Program_3a - Julia set J(a,b) for z=z^2+c (Figure 3.1(d)).

clear
k=15;niter=2^k;
x=zeros(1,niter);y=zeros(1,niter);
```

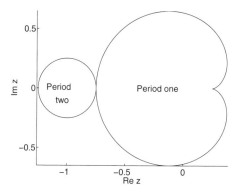

Figure 3.4: The boundary of fixed points of periods one and two for the Mandelbrot set.

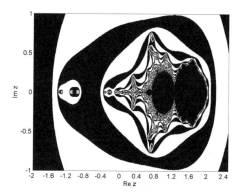

Figure 3.5: The Mandelbrot set for the mapping $z_{n+1} = z_n^2 - 2z_n + c$.

```
x1=zeros(1,niter);y1=zeros(1,niter);
a=0;b=1.1;
x(1)=real(0.5+sqrt(0.25-(a+i*b)));
y(1)=imag(0.5+sqrt(0.25-(a+i*b)));
% Check that the point is unstable.
isunstable=2*abs(x(1)+i*y(1))

hold on
for n=1:niter
    x1=x(n);y1=y(n);
    u=sqrt((x1-a)^2+(y1-b)^2)/2;v=(x1-a)/2;
    u1=sqrt(u+v);v1=sqrt(u-v);
    x(n+1)=u1;y(n+1)=v1;
```

3. Complex Iterative Maps

```
        if y(n)<b y(n+1)=-y(n+1);
        end
        if rand < .5
            x(n+1)=-u1;y(n+1)=-y(n+1);
        end
end

fsize=15;
plot(x,y,'k.','MarkerSize',1)
set(gca,'xtick',[-1.6:0.4:1.6],'FontSize',fsize)
set(gca,'ytick',[-1.2:0.4:1.2],'FontSize',fsize)
xlabel('Re z','FontSize',fsize)
ylabel('Im z','FontSize',fsize)
hold off
% End of Program_3a.
```

```
% Program_3b - The Mandelbrot set for z=z^2+c (Figure 3.2).

% Thanks to Steve Lord from The MathWorks for his help.
% Vectorized program.
Nmax = 50; scale = 0.005;
xmin = -2.4; xmax  = 1.2;
ymin = -1.5; ymax  = 1.5;

% Generate x and y coordinates and z complex values
[x,y]=meshgrid(xmin:scale:xmax,ymin:scale:ymax);
z = x+1i*y;

% Generate w accumulation matrix and k counting matrix
w = zeros(size(z));
k = zeros(size(z));

N = 0;
while N<Nmax & ~all(k(:))
    w = w.^2+z;
    N = N+1;
    k(~k & abs(w)>4) = N;
end
k(k==0) = Nmax;
figure
s = pcolor(x, y, mod(k, 2));
colormap([0 0 0;1 1 1])
set(s,'edgecolor','none')

axis([xmin xmax -ymax ymax])
fsize=15;
set(gca,'xtick',[xmin:0.4:xmax],'FontSize',fsize)
```

```
set(gca,'ytick',[-ymax:0.5:ymax],'FontSize',fsize)
xlabel('Re z','FontSize',fsize)
ylabel('Im z','FontSize',fsize)
% End of Program_3b.
```

3.4 Exercises

1. Consider the Julia set given in Figure 3.1(a). Take the mapping $z_{n+1} = z_n^2 + c$, where $c = -0.5 + 0.3i$.

    (i) Iterate the initial point $z_0 = 0 + 0i$ for 500 iterations and list the final 100. Increase the number of iterations, what can you deduce about the orbit?

    (ii) Iterate the initial point $z_0 = -1 - i$ and list z_1 to z_{10}. What can you deduce about this orbit?

2. Given that $c = -1 + i$, determine the fixed points of periods one and two for the mapping $z_{n+1} = z_n^2 + c$.

3. Consider equation (3.1); plot the Julia sets $J(0, 0)$, $J(-0.5, 0)$, $J(-0.7, 0)$, and $J(-2, 0)$.

4. Compute the fixed points of period one for the complex mapping

$$z_{n+1} = 2 + \frac{z_n e^{i|z_n|^2}}{10}.$$

5. Determine the boundaries of points of periods one and two for the mapping

$$z_{n+1} = c - z_n^2.$$

6. Plot the Mandelbrot set for the mapping

$$z_{n+1} = c - z_n^2.$$

7. Determine the fixed points of periods one and two for the mapping $z_{n+1} = z_n^2 - 2z_n + c$.

8. Modify Program 3b in Section 3.3 to plot a Mandelbrot set for the mapping $z_{n+1} = z_n^4 + c$.

9. Determine the periods of the points (i) $c = -1.3$ and (ii) $c = -0.1 + 0.8i$ for the mapping $z_{n+1} = z_n^2 + c$.

10. Plot the Mandelbrot set for the mapping $z_{n+1} = z_n^3 + c$.

Recommended Reading and Viewing

[1] R. L. Devaney, *The Mandelbrot and Julia Sets: A Tool Kit of Dynamics Activities*, Key Curriculum Press, Emeryville, CA, 2002.

[2] G. W. Flake, *The Computational Beauty of Nature: Computer Explorations of Fractals*, MIT Press, Cambridge, MA, 1998.

[3] H.-O. Peitgen (ed.), E. M. Maletsky, H. Jürgens, T. Perciante, D. Saupe, and L. Yunker, *Fractals for the Classroom: Strategic Activities*, Vol. 2, Springer-Verlag, New York, 1994.

[4] H.-O. Peitgen, H. Jürgens, and D. Saupe, *Chaos and Fractals: New Frontiers of Science*, Springer-Verlag, New York, 1992.

[5] H.-O. Peitgen, H. Jürgens, D. Saupe, and C. Zahlten, *Fractals: An Animated Discussion*, Spektrum Akademischer Verlag, Heidelberg, 1989; W. H. Freeman, New York, 1990.

[6] H.-O. Peitgen and P. H. Richter, *The Beauty of Fractals*, Springer-Verlag, New York, 1986.

4
Electromagnetic Waves and Optical Resonators

Aims and Objectives

- To introduce some theory of electromagnetic waves.
- To introduce optical bistability and show some related devices.
- To discuss possible future applications.
- To apply some of the theory of nonlinear dynamical systems to model a real physical system.

On completion of this chapter, the reader should be able to

- understand the basic theory of Maxwell's equations;
- derive the equations to model a nonlinear simple fiber ring (SFR) resonator;
- investigate some of the dynamics displayed by these devices and plot chaotic attractors;
- use a linear stability analysis to predict regions of instability and bistability;
- plot bifurcation diagrams using the first and second iterative methods;
- compare the results from four different methods of analysis.

4. Electromagnetic Waves and Optical Resonators

As an introduction to optics, electromagnetic waves are discussed via Maxwell's equations.

The reader is briefly introduced to a range of bistable optical resonators including the nonlinear Fabry–Perot interferometer, the cavity ring, the SFR, the double-coupler fiber ring, the fiber double-ring, and a nonlinear optical loop mirror (NOLM) with feedback. All of these devices can display hysteresis and all can be affected by instabilities. Possible applications are discussed in the physical world.

Linear stability analysis is applied to the nonlinear SFR resonator. The analysis gives intervals where the system is bistable and unstable but does not give any information on the dynamics involved in these regions. To use optical resonators as bistable devices, the bistable region must be isolated from any instabilities. To supplement the linear stability analysis, iterative methods are used to plot bifurcation diagrams.

For a small range of parameter values, the resonator can be used as a bistable device. Investigations are carried out to see how the bistable region is affected by the linear phase shift due to propagation of the electric field through the fiber loop.

4.1 Maxwell's Equations and Electromagnetic Waves

This section is intended to give the reader a simple general introduction to optics. Most undergraduate physics textbooks discuss *Maxwell's electromagnetic equations* in some detail. The aim of this section is to list the equations and show that Maxwell's equations can be expressed as *wave equations*. Maxwell was able to show conclusively, that just four equations could be used to interpret and explain a great deal of electromagnetic phenomena.

The four equations, collectively referred to as Maxwell's equations, did not originate entirely with him but with Ampère, Coulomb, Faraday, Gauss, and others. First, consider Faraday's law of induction, which describes how electric fields are produced from changing magnetic fields. This equation can be written as

$$\oint_C \mathbf{E} \cdot d\mathbf{r} = -\frac{\partial \phi}{\partial t},$$

where $\mathbf{E}$ is the electric field strength, $\mathbf{r}$ is a spatial vector, and ϕ is the magnetic flux. This equation may be written as

$$\oint_C \mathbf{E} \cdot d\mathbf{r} = -\frac{\partial}{\partial t} \iint_S \mathbf{B} \cdot d\mathbf{S},$$

where $\mathbf{B}$ is a magnetic field vector. Applying Stokes's theorem,

$$\iint_S \nabla \wedge \mathbf{E} \cdot d\mathbf{S} = -\frac{\partial}{\partial t} \iint_S \mathbf{B} \cdot d\mathbf{S}.$$

Therefore,

(4.1) $$\nabla \wedge \mathbf{E} = -\frac{\partial \mathbf{B}}{\partial t},$$

4.1. Maxwell's Equations and Electromagnetic Waves

which is the point form of Faraday's law of induction.

Ampère's law describes the production of magnetic fields by electric currents. Now

$$\oint_C \mathbf{H} \cdot d\mathbf{r} = \iint_S \mathbf{J} \cdot d\mathbf{S},$$

where $\mathbf{H}$ is another magnetic field vector ($\mathbf{B} = \mu \mathbf{H}$) and $\mathbf{J}$ is the current density. By Stokes's theorem

$$\oint_C \mathbf{H} \cdot d\mathbf{r} = \iint_S \nabla \wedge \mathbf{H} \cdot d\mathbf{S} = \iint_S \mathbf{J} \cdot d\mathbf{S}.$$

Therefore,

$$\nabla \wedge \mathbf{H} = \mathbf{J}.$$

Maxwell modified this equation by adding the time rate of change of the electric flux density (electric displacement) to obtain

(4.2) $$\nabla \wedge \mathbf{H} = \mathbf{J} + \frac{\partial \mathbf{D}}{\partial t},$$

where $\mathbf{D}$ is the electric displacement vector.

Gauss's law for electricity describes the electric field for electric charges, and Gauss's law for magnetism shows that magnetic field lines are continuous without end. The equations are

(4.3) $$\nabla \cdot \mathbf{E} = \frac{\rho}{\epsilon_0},$$

where ρ is the charge density and ϵ_0 is the permittivity of free space (a vacuum), and

(4.4) $$\nabla \cdot \mathbf{B} = 0.$$

In using Maxwell's equations (4.1) to (4.4) above and solving problems in electromagnetism, the three so-called constitutive relations are also used. These are

$$\mathbf{B} = \mu \mathbf{H} = \mu_r \mu_0 \mathbf{H}, \quad \mathbf{D} = \epsilon \mathbf{E} = \epsilon_r \epsilon_0 \mathbf{E}, \quad \text{and} \quad \mathbf{J} = \sigma \mathbf{E},$$

where μ_r, μ_0 are the relative permeabilities of a material and free space, respectively; ϵ_r, ϵ_0 are the relative permittivities of a material and free space, respectively; and σ is conductivity.

If $\mathbf{E}$ and $\mathbf{H}$ are sinusoidally varying functions of time, then in a region of free space, Maxwell's equations become

$$\nabla \cdot \mathbf{E} = 0, \quad \nabla \cdot \mathbf{H} = 0, \quad \nabla \wedge \mathbf{E} + i\omega\mu_0 \mathbf{H} = 0, \quad \text{and} \quad \nabla \wedge \mathbf{H} - i\omega\epsilon_0 \mathbf{E} = 0.$$

The wave equations are obtained by taking the curls of the last two equations; thus

$$\nabla^2 \mathbf{E} + \epsilon_0 \mu_0 \omega^2 \mathbf{E} = 0 \quad \text{and} \quad \nabla^2 \mathbf{H} + \epsilon_0 \mu_0 \omega^2 \mathbf{H} = 0,$$

where ω is the angular frequency of the wave. These differential equations model an unattenuated wave traveling with velocity

$$c = \frac{1}{\sqrt{\epsilon_0 \mu_0}},$$

where c is the speed of light in a vacuum. The field equation

$$\mathbf{E}(\mathbf{r}, t) = \mathbf{E_0} \exp[i(\omega t - \mathbf{k}\mathbf{r})]$$

satisfies the wave equation, where $|\mathbf{k}| = 2\pi/\lambda$ is the modulus of the wave vector and λ is the wavelength of the wave. The remarkable conclusion drawn by Maxwell is that light is an electromagnetic wave and its properties can all be deduced from his equations. The electric fields propagating through an optical fiber loop will be investigated in this chapter.

Similar equations are used to model the propagation of light waves through different media including a dielectric (a nonconducting material whose properties are isotropic); see the next section. In applications to nonlinear optics, the Maxwell–Debye or Maxwell–Bloch equations are usually used, but the theory is beyond the scope of this book. Interested readers are referred to [8], [12], and the research papers listed at the end of this chapter.

4.2 Historical Background

In recent years, there has been a great deal of interest in optical bistability because of its potential applications in high-speed all-optical signal processing and all-optical computing. Indeed, in 1984 Smith [13] published an article in *Nature* with the enthralling title "Towards the Optical Computer," and in 1999 Matthews [3] reported on work carried out by A. Wixforth and his group on the possibility of optical memories. Bistable devices can be used as logic gates, memory devices, switches, and differential amplifiers. The electronic components used nowadays can interfere with one another, need wires to guide the electronic signals, and carry information relatively slowly. Using light beams, it is possible to connect all-optical components. There is no interference; lenses and mirrors can be used to communicate thousands of channels of information in parallel; the information-carrying capacity—the bandwidth—is enormous; and there is nothing faster than the speed of light in the known universe.

In 1969, Szöke et al. [21] proposed the principle of *optical bistability* and suggested that optical devices could be superior to their electronic counterparts. As reported later in Chapter 12, the two essential ingredients for bistability are nonlinearity and feedback. For optical hysteresis, nonlinearity is provided by the medium as a refractive (or dispersive) nonlinearity or as an absorptive nonlinearity, or as both. Refractive nonlinearities alone will be considered in this chapter. The feedback is introduced through mirrors or fiber loops or by the use of an electronic circuit. The bistable optical effect was first observed in sodium vapor in 1976 at Bell Laboratories, and a theoretical explanation was provided by Felber and Marburger [20] in the same year. Nonlinearity was due to the Kerr effect (see Section 4.3), which modulated the refractive index of the medium.

Early experimental apparatus for producing optical bistability consisted of hybrid devices that contained both electronic and optical components. Materials used included indium antimonide (InSb), gallium arsenide (GaAs), and tellurium (Te). By 1979, micron-sized optical resonators had been constructed. A fundamental model of the nonlinear *Fabry–Perot interferometer* is shown in Figure 4.1.

Figure 4.1: A Fabry–Perot resonator: I, R, and T stand for incident, reflected, and transmitted intensities, respectively.

An excellent introduction to nonlinearity in optics is provided by the textbook of Agrawal [1]. Applications in nonlinear fiber optics are presented in [2].

A block diagram of the first electro-optic device is shown in Figure 4.2 and was constructed by Smith and Turner in 1977 [17]. Nonlinearity is induced by the Fabry–Perot interferometer and a He-Ne (helium-neon) laser is used at 6328Å. A bistable region is observed for a small range of parameter values. An isolated bistable region is shown in Figure 4.4(a). For input values between approximately 4 and 5 units there are two possible output values. The output is dependent upon the history of the system, that is, whether the input power is increasing or decreasing.

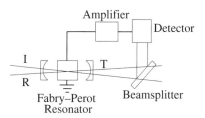

Figure 4.2: The first electro-optic device to display bistability.

In theoretical studies, Ikeda, Daido, and Akimoto [16] showed that optical circuits exhibiting bistable behavior can also contain temporal instabilities under certain conditions. The *cavity ring* (CR) resonator, first investigated by Ikeda, consists of a ring cavity comprising four mirrors that provide the feedback and containing a nonlinear dielectric material (see Figure 4.3). Light circulates around the cavity in one direction and the medium induces a nonlinear phase shift dependent on the intensity of the light. Mirrors M_1 and M_2 are partially reflective, while mirrors M_3 and M_4 are 100% reflective.

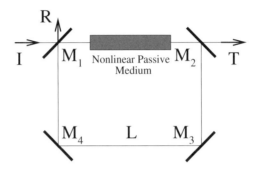

Figure 4.3: The CR resonator containing a nonlinear dielectric medium.

Possible bifurcation diagrams for this device are shown in Figure 4.4. In Figure 4.4(a), the bistable region is isolated from any instabilities, but in Figure 4.4(b) instabilities have encroached upon the bistable cycle. These figures are similar to those that would be seen if the CR were connected to an oscilloscope. However, most of the dynamics are lost; mathematically it is best to plot bifurcation diagrams using points alone (as shown later in Figure 4.14). The length L is different in the two cases and hence so is the cavity round-trip time (the time it takes light to complete one loop in the cavity).

In recent years, there has been intense research activity in the field of fiber optics. Many physics textbooks now provide an excellent introduction to the subject, and [1] provides an introduction to nonlinear optics. The interest in this chapter, however, lies solely in the application to all-optical bistability. A block diagram of the (SFR) resonator is shown in Figure 4.5. It has recently been shown that the dynamics of this device are the same as those for the CR resonator (over a limited range of initial time) apart from a scaling. The first all-optical experiment was carried out using a single-mode fiber in a simple loop arrangement, the fiber acting as the nonlinear medium [14]. In mathematical models, the input electric field is given as

$$E_{\text{in}}(t) = \xi_j(t)e^{i\omega t},$$

where ξ_j represents a complex amplitude (which may contain phase information) and ω is the circular frequency of the light.

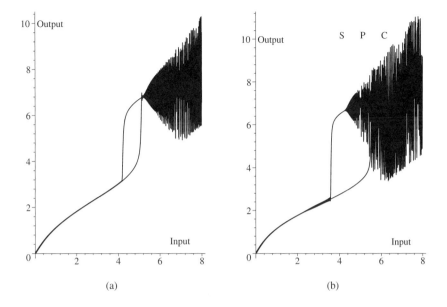

Figure 4.4: Possible bifurcation diagrams for the CR resonator: (a) an isolated bistable region and (b) instabilities within the bistable region. S represents stable behavior, P is period undoubling, and C stands for chaos.

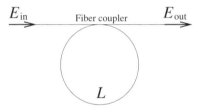

Figure 4.5: A schematic of the SFR resonator. The input electric field is E_{in} and the output electric field is E_{out}.

In experimental setups, for example, the light source could be a Q-switched YAG laser operating at 1.06 μm. The optical fiber is made of fused silica and is assumed to be lossless.

An analysis of the SFR resonator will be discussed in more detail in the next section, and the stability of the device will be investigated in Sections 4.5 and 4.6.

The *double-coupler fiber ring* resonator was investigated by Li and Ogusu [4] in 1998 (see Figure 4.6). It was found that there is a similarity between the dynamics displayed by this device and the Fabry–Perot resonator in terms of transmission and

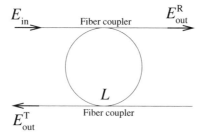

Figure 4.6: The double-coupler fiber ring resonator: E_{in} is the input field amplitude, E_{out}^R is the reflected output, and E_{out}^T is the transmitted output.

reflection bistability. It is possible to generate both clockwise and counterclockwise hysteresis loops using this device. An example of a counterclockwise bistable cycle is given in Figure 4.4(a). The reader will be asked to carry out some mathematical analysis for this device in the exercises at the end of the chapter (Section 4.8).

In 1994, Ja [9] presented a theoretical study of an *optical fiber double-ring* resonator, as shown in Figure 4.7. Ja predicted multiple bistability of the output intensity using the Kerr effect. However, instabilities were not discussed. It was proposed that this type of device could be used in new computer logic systems where more than two logic states are required. In principle, it is possible to link a number of loops of fiber, but instabilities are expected to cause some problems.

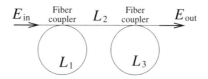

Figure 4.7: An optical fiber double-ring resonator with two couplers.

The *nonlinear optical loop mirror* (NOLM) with feedback [11] has been one of the most novel devices for demonstrating a wide range of all-optical processing functions including optical logic. The device is shown in Figure 4.8. Note that the beams of light are counterpropagating in the large loop but not in the feedback section and that there are three couplers.

All of the devices discussed thus far can display bistability and instability leading to chaos. In order to understand some of these dynamics, the SFR resonator will now be discussed in some detail.

4.3. The Nonlinear SFR Resonator

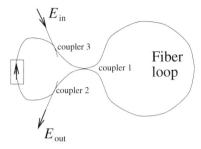

Figure 4.8: A schematic of a NOLM with feedback.

Figure 4.9: The SFR resonator. The electric field entering the fiber ring is labeled E_1 and the electric field leaving the fiber ring is labeled E_2. The coupler splits the power intensity in the ratio $\kappa : 1 - \kappa$.

4.3 The Nonlinear SFR Resonator

Consider the all-optical fiber resonator depicted in Figure 4.9 and define the slowly varying complex electric fields as indicated.

Note that the power P and intensity I are related to the electric field in the following way:

$$P \propto I \propto |E|^2.$$

If the electric field crosses the coupler, then a phase shift is induced, which is represented by a multiplication by i in the equations. Assume that there is no loss at the coupler. Then across the coupler the complex field amplitudes satisfy the following equations:

(4.5) $$E_1 = \sqrt{\kappa} E_2 + i\sqrt{1-\kappa} E_{\text{in}}$$

and

(4.6) $$E_{\text{out}} = \sqrt{\kappa} E_{\text{in}} + i\sqrt{1-\kappa} E_2,$$

4. Electromagnetic Waves and Optical Resonators

where κ is the power-splitting ratio at the coupler. Consider the propagation from E_1 to E_2. Then

(4.7) $$E_2 = E_1 e^{i\phi},$$

where the total loss in the fiber is negligible (typically about 0.2 dB/km) and

$$\phi = \phi_L + \phi_{NL}.$$

The linear phase shift is ϕ_L, and the nonlinear phase shift due to propagation is given by

$$\phi_{NL} = \frac{2\pi r_2 L}{\lambda_0 A_{\text{eff}}} |E_1|^2,$$

where λ_0 is the wavelength of propagating light in a vacuum, A_{eff} is the effective core area of the fiber, L is the length of the fiber loop, and r_2 is the *nonlinear refractive index coefficient* of the fiber. It is well known that when the optical intensity is large enough, the constant r_2 satisfies the equation

$$r = r_0 + r_2 I = r_0 + \frac{r_2 r_0}{2\eta_0} |E_1|^2 = r_0 + r_2 \frac{P}{A_{\text{eff}}},$$

where r is the refractive index of the fiber, r_0 is the linear value, I is the instantaneous optical intensity, P is the power, and A_{eff} is the effective cross-sectional area of the fiber. If the nonlinearity of the fiber is represented by this equation, then the fiber is said to be of *Kerr type*. In most applications, it is assumed that the response time of the *Kerr effect* is much less than the time taken for light to circulate once in the loop.

Substitute (4.3) into (4.1) and (4.2). Simplify to obtain

$$E_1(t) = i\sqrt{1-\kappa}\, E_{\text{in}}(t) + \sqrt{\kappa}\, E_1(t - t_R) e^{i\phi(t-t_R)},$$

where $t_R = \frac{rL}{c}$ is the time taken for the light to complete one loop, r is the refractive index, and c is the velocity of light in a vacuum. Note that this is an iterative formula for the electric field amplitude inside the ring. Take time steps of length equal to t_R. This expression can be written more conveniently as an iterative equation of the form

(4.8) $$E_{n+1} = A + B E_n \exp\left(i \left(\frac{2\pi r_2 L}{\lambda_0 A_{\text{eff}}} |E_n|^2 + \phi_L\right)\right),$$

where $A = i\sqrt{1-\kappa}\, E_{\text{in}}$, $B = \sqrt{\kappa}$, and E_j is the electric field amplitude at the jth circulation around the fiber loop. Typical fiber parameters chosen for this system are $\lambda_0 = 1.55 \times 10^{-6}$ m, $r_2 = 3.2 \times 10^{-20}$ m²W⁻¹, $A_{\text{eff}} = 30\,\mu\text{m}^2$, and $L = 80$ m.

Equation (4.8) may be scaled without loss of generality to the simplified equation

(4.9) $$E_{n+1} = A + B E_n \exp[i(|E_n|^2 + \phi_L)].$$

Some of the dynamics of equation (4.9) will be discussed in the next section.

4.4 Chaotic Attractors and Bistability

Split equation (4.5) into its real and imaginary parts by setting $E_n = x_n + iy_n$, and set $\phi_L = 0$. The equivalent real two-dimensional system is given by

(4.10)
$$x_{n+1} = A + B(x_n \cos |E_n|^2 - y_n \sin |E_n|^2),$$
$$y_{n+1} = B(x_n \sin |E_n|^2 + y_n \cos |E_n|^2),$$

where $|B| < 1$. This system is one version of the so-called *Ikeda map*. As with the Hénon map introduced in Chapter 2, the Ikeda map can have fixed points of all periods. In this particular case, system (4.10) can have many fixed points of period one depending on the parameter values A and B.

Example 1. Determine and classify the fixed points of period one for system (4.10) when $B = 0.15$ and

(i) $A = 1$;

(ii) $A = 2.2$.

Solution. The fixed points of period one satisfy the simultaneous equations

$$x = A + Bx \cos(x^2 + y^2) - By \sin(x^2 + y^2)$$

and

$$y = Bx \sin(x^2 + y^2) + By \cos(x^2 + y^2).$$

(i) When $A = 1$ and $B = 0.15$, there is one solution at $x_{1,1} \approx 1.048$, $y_{1,1} \approx 0.151$. The solution is given graphically in Figure 4.10(a). To classify the critical point $P^* = (x_{1,1}, y_{1,1})$, consider the Jacobian matrix

$$J(P^*) = \begin{pmatrix} \frac{\partial P}{\partial x} & \frac{\partial P}{\partial y} \\ \frac{\partial Q}{\partial x} & \frac{\partial Q}{\partial y} \end{pmatrix}\bigg|_{P^*}.$$

The eigenvalues of the Jacobian matrix at P^* are $\lambda_1 \approx -0.086 + 0.123i$ and $\lambda_2 \approx -0.086 - 0.123i$. Therefore, P^* is a stable fixed point of period one.

(ii) When $A = 2.2$ and $B = 0.15$, there are three points of period one, as the graphs in Figure 4.10(b) indicate. The fixed points occur approximately at the points $U = (2.562, 0.131)$, $M = (2.134, -0.317)$, and $L = (1.968, -0.185)$. Using the Jacobian matrix, the eigenvalues for U are $\lambda_{1,2} = -0.145 \pm 0.039i$; the eigenvalues for M are $\lambda_1 = 1.360$, $\lambda_2 = 0.017$; and the eigenvalues for L are $\lambda_1 = 0.555$, $\lambda_2 = 0.041$.

Therefore, U and L are stable fixed points of period one, while M is an unstable fixed point of period one. These three points are located within a bistable

92   4. Electromagnetic Waves and Optical Resonators

Figure 4.10: The fixed points of period one are determined by the intersections of the two curves, $x = A + 0.15x\cos(x^2 + y^2) - 0.15y\sin(x^2 + y^2)$ and $y = 0.15x\sin(x^2 + y^2) + 0.15y\cos(x^2 + y^2)$; (a) $A = 1$ and (b) $A = 2.2$.

region of the bifurcation diagram given later in this chapter. The point U lies on the upper branch of the hysteresis loop and the point L lies on the lower branch. Since M is unstable it does not appear in the bifurcation diagram but is located between U and L.

As the parameter A changes, the number of fixed points and the dynamics of the system change. For example, when $A = 1$, there is one fixed point of period one; when $A = 2.2$, there are two stable fixed points of period one and one unstable fixed point of period one; when $A = 2.4$, there are two stable fixed points of period two. As A increases the system displays chaotic behavior (see Example 2). All of the information can be summarized on a bifurcation diagram that will be shown later in this chapter.

Example 2. Plot iterative maps for system (4.10) when $B = 0.15$ and

(a) $A = 5$;

(b) $A = 10$.

Solution. Two chaotic attractors for system (4.10) are shown in Figure 4.11.

Theorem 1. *The circle of radius $\frac{|AB|}{1-B}$ centered at A is invariant for system (4.6).*

4.4. Chaotic Attractors and Bistability   93

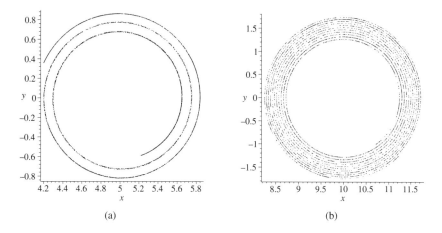

Figure 4.11: The chaotic attractors when (a) $A = 5$ (5000 iterates) and (b) $A = 10$ (5000 iterates).

Proof. Suppose that a general initial point in the Argand diagram is taken to be E_n; then the first iterate is given by

$$E_{n+1} = A + BE_n e^{i|E_n|^2}.$$

The second iterate can be written as

$$E_{n+2} = A + BE_{n+1} e^{i|E_{n+1}|^2} = A + B(A + BE_n e^{i|E_n|^2}) e^{i|E_{n+1}|^2}.$$

Thus

$$E_{n+2} = A + ABe^{i|E_{n+1}|^2} + B^2 E_n e^{i(|E_n|^2+|E_{n+1}|^2)}.$$

Using a similar argument, the third iterate is

$$E_{n+3} = A + B(A + ABe^{i|E_{n+1}|^2} + B^2 E_n e^{i(|E_n|^2+|E_{n+1}|^2)}) e^{i|E_{n+2}|^2}.$$

Therefore,

$$E_{n+3} = A + ABe^{i|E_{n+2}|^2} + AB^2 e^{i(|E_{n+1}|^2+|E_{n+2}|^2)} + B^3 E_n e^{i(|E_n|^2+|E_{n+1}|^2+|E_{n+2}|^2)}.$$

A general expression for the Nth iterate E_{n+N} is not difficult to formulate. Hence

$$E_{n+N} = A + ABe^{i|E_{n+N-1}|^2} + AB^2 e^{i(|E_{n+N-2}|^2+|E_{n+N-1}|^2)}$$
$$+ \cdots + AB^{N-1} \exp\left(i \sum_{j=1}^{N-1} |E_{n+j}|^2\right) + B^N E_n \exp\left(i \sum_{j=0}^{N-1} |E_{n+j}|^2\right).$$

94   4. Electromagnetic Waves and Optical Resonators

As $N \to \infty$, $B^N \to 0$, since $0 < B < 1$. Set $R_j = |E_{n+N-j}|^2$. Then

$$|E_{n+N} - A| = |ABe^{iR_1} + AB^2 e^{i(R_1+R_2)} + \cdots + AB^{N-1} e^{i(R_1+R_2+\cdots+R_{N-1})}|.$$

Since $|z_1 + z_2 + \cdots + z_m| \leq |z_1| + |z_2| + \cdots + |z_m|$ and $|e^{i\theta}| = 1$,

$$|E_{n+N} - A| \leq |AB| + |AB^2| + \cdots + |AB^{N-1}|.$$

This forms an infinite geometric series as $N \to \infty$. Therefore,

$$|E_{n+N} - A| \leq \frac{|AB|}{1-B}.$$

The disc given by $|E - A| = AB/(1-B)$ is positively invariant for system (4.10). The invariant disks in two cases are easily identified in Figures 4.11(a) and (b). □

4.5 Linear Stability Analysis

To investigate the stability of the nonlinear SFR resonator, a linear stability analysis (see Chapter 7) will be applied. A first-order perturbative scheme is used to predict the values of a parameter where the stationary solutions become unstable. Briefly, a small perturbation is added to a stable solution and a Taylor series expansion is carried out, the nonlinear terms are ignored, and a linear stability analysis is applied.

It was shown in Section 4.3 that the following simplified complex iterative equation can be used to model the electric field in the fiber ring:

(4.11) $$E_{n+1} = A + BE_n \exp[i(|E_n|^2 - \phi_L)],$$

where E_n is the slowly varying field amplitude; $A = i\sqrt{1-\kappa} E_{\text{in}}$, is related to the input; $B = \sqrt{\kappa}$, where κ is the power coupling ratio; and ϕ_L is the linear phase shift suffered by the electric field as it propagates through the fiber loop. To simplify the linear stability analysis, there is assumed to be no loss at the coupler and the phase shift ϕ_L is set to zero. The effect of introducing a linear phase shift will be discussed later in this chapter.

Suppose that E_S is a stable solution of the iterative equation (4.11). Then

$$E_S = A + BE_S e^{i|E_S|^2}.$$

Therefore,

$$A = E_S[1 - B(\cos(|E_S|^2) + i \sin |E_S|^2)].$$

Using the relation $|z|^2 = zz^*$, where z^* is the conjugate of z,

$$|A|^2 = (E_S[1 - B(\cos(|E_S|^2) + i \sin |E_S|^2)]) \\ \times (E_S^*[1 - B(\cos(|E_S|^2) - i \sin |E_S|^2)]).$$

4.5. Linear Stability Analysis

Hence

(4.12) $$|A|^2 = |E_S|^2(1 + B^2 - 2B\cos(|E_S|^2)).$$

The stationary solutions of system (4.11) are given as a multivalued function of A satisfying equation (4.12). This gives a bistable relationship equivalent to the *graphical method*, which is well documented in the literature; see, for example, [10], [18], and [19].

Differentiate equation (4.12) to obtain

(4.13) $$\frac{d|A|^2}{d|E_S|^2} = 1 + B^2 + 2B(|E_S|^2 \sin(|E_S|^2) - \cos(|E_S|^2)).$$

To establish where the stable solutions become unstable, consider a slight perturbation from the stable situation in the fiber ring, and let

(4.14) $$E_n(t) = E_S + \xi_n(t) \quad \text{and} \quad E_{n+1}(t) = E_S + \xi_{n+1}(t),$$

where $\xi_n(t)$ is a small time-dependent perturbation to E_S. Substitute (4.14) into (4.11) to get

$$E_S + \xi_{n+1} = A + B(E_S + \xi_n)\exp[i(E_S + \xi_n)(E_S^* + \xi_n^*)],$$

so

(4.15) $$E_S + \xi_{n+1} = A + B(E_S + \xi_n)\exp[i|E_S|^2]\exp[i(E_S\xi_n^* + \xi_n E_S^* + |\xi_n|^2)].$$

Take a Taylor series expansion of the exponential function to obtain

$$\exp[i(E_S\xi_n^* + \xi_n E_S^* + |\xi_n|^2)] = 1 + i(E_S\xi_n^* + \xi_n E_S^* + |\xi_n|^2)$$
$$+ \frac{i^2(E_S\xi_n^* + \xi_n E_S^* + |\xi_n|^2)^2}{2} + \cdots.$$

Ignore the nonlinear terms in ξ_n. Equation (4.15) then becomes

$$E_S + \xi_{n+1} = A + B(E_S + \xi_n)\exp[i|E_S|^2](1 + iE_S\xi_n^* + \xi_n E_S^*).$$

Since $A = E_S - BE_S \exp[i|E_S|^2]$, the equation simplifies to

(4.16) $$\xi_{n+1} = B(\xi_n + i|E_S|^2\xi_n + i(E_S)^2\xi_n^*)\exp(i|E_S|^2).$$

Since ξ is real, it may be split into its positive and negative frequency parts as follows:

(4.17) $$\xi_n = E_+ e^{\lambda t} + E_- e^{\lambda^* t} \quad \text{and} \quad \xi_{n+1} = E_+ e^{\lambda(t+t_R)} + E_- e^{\lambda^*(t+t_R)},$$

where $|E_+|$, $|E_-|$ are much smaller than $|E_S|$, t_R is the fiber ring round-trip time, and λ is the amplification rate of a small fluctuation added to a stable solution. Substitute equation (4.17) into (4.16). Then the validity of (4.16) at all times t requires that

$$E_+ e^{\lambda t_R} = B(E_+ + i|E_S|^2 E_+ + iE_S^2 E_-^*)\exp(i|E_S|^2),$$
$$E_-^* e^{\lambda t_R} = B(E_-^* - i|E_S|^2 E_-^* - i(E_S^*)^2 E_+)\exp(-i|E_S|^2),$$

or, equivalently,

$$\begin{pmatrix} \beta(1+i|E_S|^2) - e^{\lambda t_R} & i\beta E_S^2 \\ -i\beta^*(E_S^*)^2 & \beta^*(1-i|E_S|^2) - e^{\lambda t_R} \end{pmatrix} \begin{pmatrix} E_+ \\ E_-^* \end{pmatrix} = \begin{pmatrix} 0 \\ 0 \end{pmatrix},$$

where $\beta = B\exp(i|E_S|^2)$. To obtain a valid solution, the characteristic equation must be solved:

$$e^{2\lambda t_R} - 2e^{\lambda t_R} B(\cos|E_S|^2 - |E_S|^2 \sin|E_S|^2) + B^2 = 0.$$

Substituting from equation (4.13), the characteristic equation becomes

(4.18) $$e^{2\lambda t_R} - e^{\lambda t_R}\left(1 + B^2 - \frac{d|A|^2}{d|E_S|^2}\right) + B^2 = 0.$$

Let $D = \frac{d|A|^2}{d|E_S|^2}$. The stability edges for E_S occur where $e^{\lambda t_R} = +1$ and $e^{\lambda t_R} = -1$, since this is a discrete mapping. Using equation (4.18), this yields the conditions

$$D_{+1} = 0 \quad \text{and} \quad D_{-1} = 2(1 + B^2).$$

Thus the system is stable as long as

(4.19) $$0 < D < 2(1 + B^2).$$

The condition $D = 0$ marks the boundary between the branches of positive and negative slope on the graph of $|E_S|^2$ versus $|A|^2$ and hence defines the regions where the system is bistable. Thus the results from the graphical method match with the results from the linear stability analysis. The system becomes unstable at the boundary where $D = D_{-1}$.

It is now possible to apply four different methods of analysis to determine the stability of the electric field amplitude in the SFR resonator. Linear stability analysis may be used to determine both the unstable and bistable regions and bifurcation diagrams can be plotted. The graphical method [1] is redundant in this case.

There are two methods commonly used to plot bifurcation diagrams. The author will refer to these as the first and second iterative methods from now on.

The First Iterative Method. A parameter is fixed and one or more initial points are iterated forward. Transients are ignored and a number of the final iterates are plotted. The parameter is then increased by a suitable step length and the process is repeated. There are many points plotted for each value of the parameter. For example, the bifurcation diagrams plotted in Sections 2.3 and 2.4 were all generated using the first iterative method.

The Second Iterative Method. A parameter is varied and the solution to the previous iterate is used as the initial condition for the next iterate. In this way, a feedback mechanism is introduced. In this case, there is a history associated with the process and only one point is plotted for each value of the parameter. For example, most of the bifurcation diagrams plotted in Section 4.6 were plotted using the second iterative method.

The first and second iterative methods are used in other chapters of the book.

4.6 Instabilities and Bistability

In the previous section, the results from the linear stability analysis established that system (4.11) is stable as long as equation (4.19) is satisfied. A possible *stability diagram* for system (4.11) is given in Figure 4.12, which shows the graph of $D = \frac{d|A|^2}{d|E_S|^2}$ and the bounding lines $D_{+1} = 0$ and $D_{-1} = 2(1 + B^2)$ when $B = 0.15$.

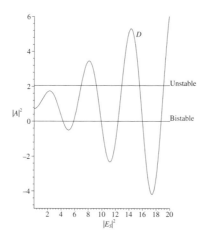

Figure 4.12: Stability diagram for the SFR resonator when $B = 0.15$ ($\kappa = 0.0225$). The system is stable as long as $0 < D < 2(1 + B^2)$.

Table 4.1 lists the first two bistable and unstable intensity regions for the SFR resonator (in Watts per meter squared in physical applications) for a range of fixed values of the parameter B.

4. Electromagnetic Waves and Optical Resonators

Table 4.1: The first two regions of bistability and instability computed for the SFR resonator to three decimal places using a linear stability analysis.

B	First bistable region A^2/Wm^{-2}	First unstable region A^2/Wm^{-2}	Second bistable region A^2/Wm^{-2}	Second unstable region A^2/Wm^{-2}
0.05	10.970 – 11.038	12.683 – 16.272	16.785 – 17.704	17.878 – 23.561
0.15	4.389 – 4.915	5.436 – 12.007	9.009 – 12.765	9.554 – 20.510
0.3	3.046 – 5.951	1.987 – 4.704	6.142 – 16.175	3.633 – 15.758
0.6	1.004 – 8.798	1.523 – 7.930	2.010 – 24.412	1.461 – 24.090
0.9	0.063 – 12.348	1.759 – 11.335	0.126 – 34.401	0.603 – 34.021

The dynamic behavior of system (4.11) may also be investigated by plotting bifurcation diagrams using either the first or second iterative methods defined in Chapter 2. In order to observe any hysteresis, one must, of course, use the second iterative method, which involves a feedback. The method developed by Bischofberger and Shen [18] in 1979 for a nonlinear Fabry–Perot interferometer is modified and used here for the SFR resonator. The input intensity is increased to a maximum and then decreased back to zero, as depicted in Figure 4.13. In this case, the simulation consists of a triangular pulse entering the ring configuration, but it is not difficult to modify the MATLAB program to investigate *Gaussian input* pulses. The input intensity is increased linearly up to 16 Wm^{-2} and then decreased back down to zero. Figure 4.13 shows the output intensity and input intensity against the number of passes around the ring, which in this particular case was 4000. To observe the bistable region, it is necessary to display the ramp-up and ramp-down parts of the diagram on the same graph, as in Figure 4.14(b).

Figure 4.14 shows a gallery of bifurcation diagrams, corresponding to some of the parameter values used in Table 4.1, produced using the second iterative method. The diagrams make interesting comparisons with the results displayed in Table 4.1.

A numerical investigation has revealed that for a small range of values close to $B = 0.15$ (see Figure 4.14(b)), the SFR resonator could be used as a bistable device. Unfortunately, for most values of B, instabilities overlap with the first bistable region. For example, when $B = 0.3$ (Figure 4.14(c)), the first unstable region between 1.987 Wm^{-2} and 4.704 Wm^{-2} intersects with the first bistable region between 3.046 Wm^{-2} and 5.951 Wm^{-2}. Clearly, the instabilities have affected the bistable operation. In fact, the hysteresis cycle has failed to materialize. Recall that $B = \sqrt{\kappa}$, where κ is the power coupling ratio. As the parameter B gets larger, more of the input power is circulated in the ring, and this causes the system to become chaotic for low input intensities.

The first iterative method can be employed to show regions of instability. Note, however, that bistable regions will not be displayed since there is no feed-

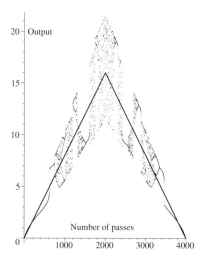

Figure 4.13: Bifurcation diagram when $B = 0.15$ using the second iterative method showing a plot of triangular input and output intensities against number of ring passes for the SFR resonator.

back in this method. It is sometimes possible for a small unstable region to be missed using the second iterative method. The steady state remains on the unstable branch until it becomes stable again. Thus in a few cases, the first iterative method gives results that may be missed using the second iterative method. As a particular example, consider system (4.1) where $B = 0.225$. Results from a linear stability analysis indicate that there should be an unstable region in the range $2.741 - 3.416$ Wm^{-2}. Figure 4.15(a) shows that this region is missed using the second iterative method, whereas the first iterative method (Figure 4.15(b)) clearly displays period-two behavior. In physical applications, one would expect relatively small unstable regions to be skipped, as in the former case.

Consider the complex iterative equation

$$(4.20) \qquad E_{n+1} = i\sqrt{1-\kappa}\, E_{\text{in}} + \sqrt{\kappa}\, E_n \exp\left[i\left(\frac{2\pi n_2 L}{\lambda_0 A_{\text{eff}}}|E_n|^2 - \phi_L\right)\right],$$

which was derived earlier. Equation (4.20) is the iterative equation that models the electric field in the SFR resonator. Typical *fiber parameters* chosen for this system are $\lambda_0 = 1.55 \times 10^{-6}$ m, $n_2 = 3.2 \times 10^{-20}$ m^{2}W^{-1}, $A_{\text{eff}} = 30\ \mu$m^2, and $L = 80$ m. Suppose that equation (4.11) was iterated 10,000 times. This would equate to hundredths of a second of elapsed time in physical applications using these values for the fiber parameters.

In the work considered so far, the linear phase shift due to propagation ϕ_L has been set to zero. Figure 4.16 shows how the bistable region is affected when

4. Electromagnetic Waves and Optical Resonators

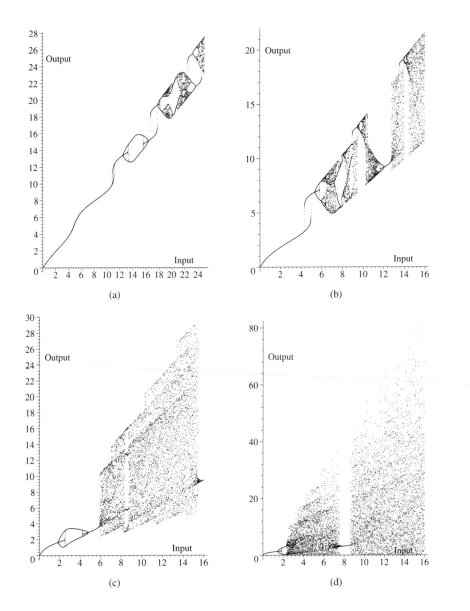

Figure 4.14: A gallery of bifurcation diagrams for the SFR resonator when (a) $B = 0.05$, (b) $B = 0.15$, (c) $B = 0.3$, and (d) $B = 0.6$. In each case, 6000 iterations were carried out.

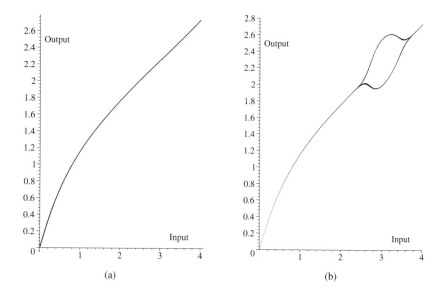

Figure 4.15: Bifurcation diagrams when $B = 0.225$ (a) using the second iterative method with feedback and (b) using the first iterative method without feedback.

ϕ_L is nonzero and $B = 0.15$. As the linear phase shift increases from zero to $\frac{\pi}{4}$, the first bistable region gets larger and shifts to the right slightly, as depicted in Figure 4.16(b). When $\phi_L = \frac{\pi}{2}$, an instability has appeared between 20 Wm^{-2} and 40 Wm^{-2} and a second unstable region has encroached on the first bistable region, as shown in Figure 4.16(c). When $\phi_L = \pi$, instabilities appear at both ends of the bistable region, as shown in Figure 4.16(d). Therefore, the linear phase shift can affect the bistable operation of the SFR resonator. Should such systems be used for bistable operation, the results indicate the need to control the feedback phase to prevent any instabilities from entering the power range in the hysteresis loop.

In conclusion, the dynamic properties of a nonlinear optical resonator have been analyzed using a graphical method, a linear stability analysis, and bifurcation diagrams. The bifurcation diagrams give a clearer insight into the dynamics than the results from the linear stability analysis and graphical method, but all four used in conjunction provide useful results.

4.7 MATLAB Commands

```
% Program_4a - Plotting chaotic attractors for Ikeda map
% (Figure 4.11(b)).
```

4. Electromagnetic Waves and Optical Resonators

Figure 4.16: Bifurcation diagrams for the SFR resonator using equation (4.10) when $B = 0.15$ and (a) $\phi_L = 0$, (b) $\phi_L = \frac{\pi}{4}$, (c) $\phi_L = \frac{\pi}{2}$, and (d) $\phi_L = \pi$.

```
clear
echo off
A=10;
B=0.15;
N=10000;
```

4.7. MATLAB Commands   103

```
E=zeros(1,N);
E(1)=A;x(1)=A;y(1)=0;
for n=1:N
    E(n+1)=A+B*E(n)*exp(i*abs(E(n))^2);
    x(n+1)=real(E(n+1));
    y(n+1)=imag(E(n+1));
end
axis([8 12 -2 2])
axis equal
plot(x(50:N),y(50:N),'.','MarkerSize',1);
fsize=20;
set(gca,'xtick',[8:1:12],'FontSize',fsize)
set(gca,'ytick',[-2:1:2],'FontSize',fsize)
xlabel('Real E','FontSize',fsize)
ylabel('Imag E','FontSize',fsize)
% End of Program_4a.
```

```
% Program_4b - Bifurcation diagrams for the SFR (Figures
% 4.13 and 4.16(a)).
% The second iterative method.

clear
format long;
halfN=1999;N=2*halfN+1;N1=1+halfN;
C=0.345913;
E1(1)=0;kappa=0.0225;Pmax=120;phi=0;

% Ramp the power up
for n=1:halfN
    E2(n+1)=E1(n)*exp(i*(abs(C*E1(n))^2-phi));
    E1(n+1)=i*sqrt(1-kappa)*sqrt(n*Pmax/N1)
            +sqrt(kappa)*E2(n+1);
    Esqr(n+1)=abs(E1(n+1))^2;
end

% Ramp the power down
for n=N1:N
    E2(n+1)=E1(n)*exp(i*(abs(C*E1(n))^2-phi));
    E1(n+1)=i*sqrt(1-kappa)*sqrt(2*Pmax-n*Pmax/N1)
            +sqrt(kappa)*E2(n+1);
    Esqr(n+1)=abs(E1(n+1))^2;
end

for n=1:halfN
    Esqr1(n)=Esqr(N+1-n);
    ptsup(n)=n*Pmax/N1;
end
```

```
% Plot the bifurcation diagrams
fsize=20;
subplot(2,1,1)
plot(Esqr(1:N),'.','MarkerSize',1)
xlabel('Number of Ring Passes','FontSize',fsize);
ylabel('Output','FontSize',fsize);

subplot(2,1,2)
hold on
plot(ptsup(1:halfN),Esqr(1:halfN),'.','MarkerSize',1);
plot(ptsup(1:halfN),Esqr1(1:halfN),'.','MarkerSize',1);
xlabel('Input Power','FontSize',fsize);
ylabel('Output Power','FontSize',fsize);
hold off
% End of Program_4b.
```

4.8 Exercises

1. Determine the number of fixed points of period one for system (4.10) when $B = 0.4$ and $A = 3.9$ by plotting the graphs of the simultaneous equations.

2. Plot iterative maps for equation (4.8), using the parameter values given in the text, when $\kappa = 0.0225$ and (i) $E_{in} = 4.5$, (ii) $E_{in} = 6.3$, and (iii) $E_{in} = 11$.

3. Given that

$$E_{n+1} = A + BE_n e^{i|E_n|^2},$$

prove that the inverse map is given by

$$E_{n+1} = \left(\frac{E_n - A}{B}\right) \exp\left(\frac{-i|E_n - A|^2}{B^2}\right).$$

4. Given the complex Ikeda mapping

$$E_{n+1} = A + BE_n \exp\left[i\left(\phi - \frac{C}{1 + |E_n|^2}\right)\right],$$

where A, B, and C are constants, show that the steady-state solution, say, $E_{n+1} = E_n = E_S$, satisfies the equation

$$\cos\left(\frac{C}{1 + |E_S|^2} - \phi\right) = \frac{1}{2B}\left(1 + B^2 - \frac{A^2}{|E_S|^2}\right).$$

5. Consider the double-coupler nonlinear fiber ring resonator as shown in Figure 4.17.

Figure 4.17: Schematic of a double-coupler nonlinear fiber ring resonator.

Suppose that

$$E_R(t) = \sqrt{\kappa} E_{\text{in}}(t) + i\sqrt{1-\kappa} E_4(t),$$
$$E_1(t) = i\sqrt{1-\kappa} E_{\text{in}}(t) + \sqrt{\kappa} E_4(t),$$
$$E_2(t) = E_1(t-t_R) e^{i\phi_1(t-t_R)},$$
$$\phi_1(t-t_R) = \frac{\pi r_2 L}{\lambda A_{\text{eff}}} |E_1(t-t_R)|^2,$$
$$E_3(t) = \sqrt{\kappa} E_2(t),$$
$$E_T(t) = i\sqrt{1-\kappa} E_2(t),$$
$$E_4(t) = E_3(t-t_R) e^{i\phi_2(t-t_R)},$$
$$\phi_2(t-t_R) = \frac{\pi r_2 L}{\lambda A_{\text{eff}}} |E_3(t-t_R)|^2,$$

where the fiber loop is of length L; both halves are of length $\frac{L}{2}$; t_R is the time taken for the electric field to complete half a fiber loop; and both couplers split the power in the ratio $\kappa : 1-\kappa$. Assuming that there are no losses in the fiber, show that

$$E_T(t) = -(1-\kappa) E_{\text{in}}(t-t_R) e^{i\phi_1(t-t_R)} + \kappa E_T(t-2t_R) e^{i(\phi_1(t-t_R)+\phi_2(t-2t_R))}.$$

6. Consider the complex iterative equation

$$E_{n+1} = A + B E_n \exp[i(|E_n|^2)],$$

used to model the SFR resonator. Use a linear stability analysis to determine the first bistable and unstable regions when (a) $B = 0.1$, (b) $B = 0.2$, and (c) $B = 0.25$ to three decimal places, respectively.

7. Plot bifurcation diagrams for Exercise 6, parts (a)–(c), when the maximum input intensity is 25 Wm^{-2} and the input pulse is triangular.

4. Electromagnetic Waves and Optical Resonators

8. Plot the bifurcation diagram for the iterative equation in Exercise 6 for $B = 0.15$ when the input pulse is Gaussian with a maximum of 25 Wm^{-2}. How is the bistable region affected by the width of the pulse?

9. Consider the complex iterative equation

$$E_{n+1} = A + BE_n \exp[i(|E_n|^2 - \phi_L)],$$

where $B = 0.15$ and ϕ_L represents a linear phase shift. Plot bifurcation diagrams for a maximum input intensity of $A = 3$ units when

   (a) $\phi_L = \frac{\pi}{4}$,
   (b) $\phi_L = \frac{\pi}{2}$,
   (c) $\phi_L = \frac{3\pi}{4}$,
   (d) $\phi_L = \pi$,
   (e) $\phi_L = \frac{5\pi}{4}$,
   (f) $\phi_L = \frac{3\pi}{2}$,
   (g) $\phi_L = \frac{7\pi}{4}$.

10. Apply the linear stability analysis to the iterative equation

$$E_{n+1} = i\sqrt{1-\kappa}\,E_{\text{in}} + \sqrt{\kappa}\,E_n \exp\left[i\left(\frac{2\pi n_2 L}{\lambda_0 A_{\text{eff}}}|E_n|^2\right)\right],$$

for the parameter values given in this chapter. Compare the results with the bifurcation diagrams.

Recommended Reading

[1] G. P. Agrawal, *Nonlinear Fiber Optics*, 3rd ed., Academic Press, New York, 2001.

[2] G. P. Agrawal, *Applications in Nonlinear Fiber Optics*, Academic Press, New York, 2001.

[3] R. Matthews, Catch the wave, *New Scientist*, **162**-2189 (1999), 27–32.

[4] H. Li and K. Ogusu, Analysis of optical instability in a double-coupler nonlinear fiber ring resonator, *Optics Comm.*, **157** (1998), 27–32.

[5] S. Lynch, A. L. Steele, and J. E. Hoad, Stability analysis of nonlinear optical resonators, *Chaos Solitons Fractals*, **9**-6 (1998), 935–946.

[6] K. Ogusu, A. L. Steele, J. E. Hoad, and S. Lynch, Corrections to and comments on "Dynamic behaviour of reflection optical bistability in a nonlinear fibre ring resonator," *IEEE J. Quantum Electron.*, **33** (1997), 2128–2129.

[7] A. L. Steele, S. Lynch, and J. E. Hoad, Analysis of optical instabilities and bistability in a nonlinear optical fiber loop mirror with feedback, *Optics Comm.*, **137** (1997), 136–142.

[8] P. Mandel, *Theoretical Problems in Cavity Nonlinear Optics*, Cambridge University Press, Cambridge, UK, 1997.

[9] Y. H. Ja, Multiple bistability in an optical-fiber double-ring resonator utilizing the Kerr effect, *IEEE J. Quantum Electron.*, **30**-2 (1994), 329–333.

[10] Chao-Xiang Shi, Nonlinear fiber loop mirror with optical feedback, *Optics Comm.*, **107** (1994), 276–280.

[11] N. J. Doran and D. Wood, Nonlinear-optical loop mirror, *Optics Lett.*, **13** (1988), 56–58.

[12] H. M. Gibbs, *Optical Bistability: Controlling Light with Light*, Academic Press, New York, 1985.

[13] S. D. Smith, Towards the optical computer, *Nature*, **307** (1984), 315–316.

[14] H. Natsuka, S. Asaka, H. Itoh, K. Ikeda, and M. Matouka, Observation of bifurcation to chaos in an all-optical bistable system, *Phys. Rev. Lett.*, **50** (1983), 109–112.

[15] W. J. Firth, Stability of nonlinear Fabry-Perot resonators, *Optics Comm.*, **39**-5 (1981), 343–346.

[16] K. Ikeda, H. Daido, and O. Akimoto, Optical turbulence: Chaotic behavior of transmitted light from a ring cavity, *Phys. Rev. Lett.*, **45**-9 (1980), 709–712.

[17] P. W. Smith and E. H. Turner, A bistable Fabry-Perot resonator, *Appl. Phys. Lett.*, **30**-6 (1977), 280–281.

[18] T. Bischofberger and Y. R. Shen, Theoretical and experimental study of the dynamic behavior of a nonlinear Fabry-Perot interferometer, *Phys. Rev. A*, **19** (1979), 1169–1176.

[19] J. H. Marburger and F. S. Felber, Theory of a lossless nonlinear Fabry-Perot interferometer, *Phys. Rev. A*, **17** (1978), 335–342.

[20] F. S. Felber and J. H. Marburger, Theory of nonresonant multistable optical devices, *Appl. Phys. Lett.*, **28** (1976), 731.

[21] A. Szöke, V. Daneu, J. Goldhar, and N. A. Kurnit, Bistable optical element and its applications, *Appl. Phys. Lett.*, **15** (1969), 376.

5
Fractals and Multifractals

Aims and Objectives
- To provide a brief introduction to fractals.
- To introduce the notion of fractal dimension.
- To provide a brief introduction to multifractals and define a multifractal formalism.
- To consider some very simple examples.

On completion of this chapter the reader should be able to

- plot early-stage generations of certain fractals using either graph paper, pencil, and rule or the MATLAB package;
- determine the fractal dimension of some mathematical fractals;
- estimate the fractal dimension using simple box-counting techniques;
- distinguish between homogeneous and heterogeneous fractals;
- appreciate how multifractal theory is being applied in the real world;
- construct multifractal Cantor sets and Koch curves and plot graphs of their respective multifractal spectra.

5. Fractals and Multifractals

Fractals are introduced by means of some simple examples, and the fractal dimension is defined. Box-counting techniques are used to approximate the fractal dimension of certain early-stage generation fractals, which can be generated using pencil, paper, and rule.

A multifractal formalism is introduced that avoids some of the more abstract pure mathematical concepts. The theory is explained in terms of box-counting dimensions, which are introduced in this chapter. This is potentially a very complicated topic, and readers new to this field are advised to look at Example 4 before attempting to understand the formalism.

Some applications of multifractal analysis to physical systems in the real world are also discussed. A few simple self-similar multifractals are constructed, and the analysis is applied to these objects.

5.1 Construction of Simple Examples

Definition 1. A *fractal* is an object that displays self-similarity under magnification and can be constructed using a simple motif (an image repeated on ever-reduced scales).

Fractals have generated a great deal of interest since the advent of the computer. Many shops now sell colorful posters and T-shirts displaying fractals, and some black-and-white fractals have been plotted in Chapter 3. Although the Julia sets and the Mandelbrot set are not true fractals, they do have fractal structure. Many objects in nature display this self-similarity at different scales; for example, cauliflower, ferns, trees, mountains, clouds, and even blood vessel networks in our own bodies have some fractal structure. These objects cannot be described using the geometry of lines, planes, and spheres. Instead, *fractal geometry* is required. Fractal analysis is being applied in many branches of science—for example, to computer graphics and image compression (take a closer look at the images on the Web) and to oil extraction from rocks using viscous fingering—and multifractal analysis has expanded rapidly over recent years (see later in this chapter).

It is important to note that all of the fractals appearing in this textbook are early-generation fractals. However, there is nothing to stop scientists from imagining an ideal mathematical fractal that is constructed to infinity. Some of these fractals will now be investigated.

The Cantor Set. The Cantor fractal was first considered by Georg Cantor in 1870. It is constructed by removing the middle third of a line segment at each stage of construction. Thus at stage 0, there is one line segment of unit length. At stage 1, the middle third is removed to leave two segments each of length $\frac{1}{3}$. At stage 3, there will be four segments each of length $\frac{1}{9}$. Continuing in this way, it is not difficult to see that at the kth stage, there will be $N = 2^k$ segments each of length $l = 3^{-k}$. An early-stage construction (up to stage 3) is shown in Figure 5.1.

5.1. Construction of Simple Examples

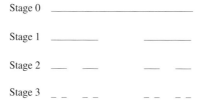

Figure 5.1: An early generation of the Cantor set.

If this process is continued to infinity, then

$$\lim_{k \to \infty} 2^k = \infty \quad \text{and} \quad \lim_{k \to \infty} 3^{-k} = 0.$$

The Cantor set will therefore consist of an infinite number of discrete points that, unfortunately, is impossible to generate on a computer screen. However, all is not lost. By using the ternary number system, it is possible to classify which points in the unit interval belong to the Cantor set and which do not. Recall that ternary proper fractions can be expanded by applying a simple algorithm: Treble the numerator of the proper fraction concerned; when this number is larger than or equal to the denominator, subtract the denominator, noting down the ternary factor above the line, and continue with the remainder. For example, $\frac{4}{7} = 0.\underline{120102}$, since

$$\begin{array}{c|cccccc}
 & 1 & 2 & 0 & 1 & 0 & 2 & 1 & \ldots \\
\hline
4 & 12 & & & & & & & \\
 & 5 & 15 & & & & & & \\
 & & 1 & 3 & 9 & & & & \\
 & & & 2 & 6 & 18 & & & \\
 & & & & 4 & 12 & & & \\
 & & & & & 5 & \ldots, & &
\end{array}$$

where the underlining after the decimal point represents a recurring decimal. It is not too difficult to show that the Cantor set can be identified by points whose ternary fractions consist of zeroes and twos only. Thus $p_1 = 0.20202$ will belong to the Cantor set, whereas $p_2 = 0.\underline{120102}$ will not.

The Koch Curve. Helge Von Koch first imagined the Koch curve in 1904. It is constructed by replacing a unit line segment with a motif consisting of four line segments each of length $\frac{1}{3}$, as depicted in Figure 5.2.

A simple MATLAB program is given in Section 5.4 to plot early generations of the Koch curve. Note that at the kth stage there are $N = 4^k$ line segments each of length $l = 3^{-k}$. Thus for the mathematical fractal constructed to infinity,

$$\lim_{k \to \infty} 4^k = \infty \quad \text{and} \quad \lim_{k \to \infty} 3^{-k} = 0,$$

so the mathematical Koch curve consists of a curve that is infinitely long.

5. Fractals and Multifractals

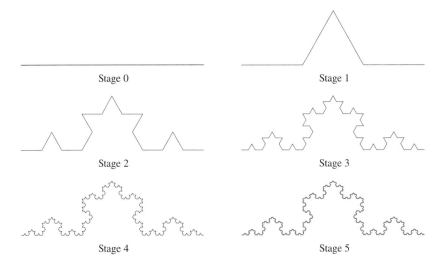

Figure 5.2: Construction of the Koch curve up to stage 5.

The Koch Square. Consider a variation of the Koch curve that is constructed by replacing one line segment with five line segments each of length $\frac{1}{3}$. Furthermore, suppose that these curves are attached to the outer edge of a unit square. The first five stages of construction are shown in Figure 5.3.

It is possible to determine the area and perimeter bounded by the Koch square in the following way. Suppose that at stage 0 the square has area $A_0 = 1$ unit^2 and the area at stage k is A_k. Then

$$A_1 = 1 + 4(3^{-2}) \text{ unit}^2.$$

At stage 2, the area is given by

$$A_2 = 1 + 4(3^{-2}) + 4 \times 5 \times (3^{-4}) \text{ unit}^2.$$

Continuing in this way, the area at the kth stage is given by

$$A_k = 1 + 4(3^{-2}) + 4 \times 5 \times (3^{-4}) + 4 \times 5^2 \times (3^{-6}) + \cdots + 4 \times 5^{k-1} \times (3^{-2k}) \text{ unit}^2.$$

Take the limit $k \to \infty$. Then

$$A_\infty = 1 + \frac{4}{9} + \sum_{i=1}^{\infty} 4 \times 5^i \times (9^{-(i+1)}) \text{ unit}^2.$$

This is the sum of an infinite geometric series, and hence

$$A_\infty = 1 + \frac{4}{9} + \frac{\frac{4 \times 5}{9^2}}{1 - \frac{5}{3^2}} = 2 \text{ unit}^2.$$

5.1. Construction of Simple Examples 113

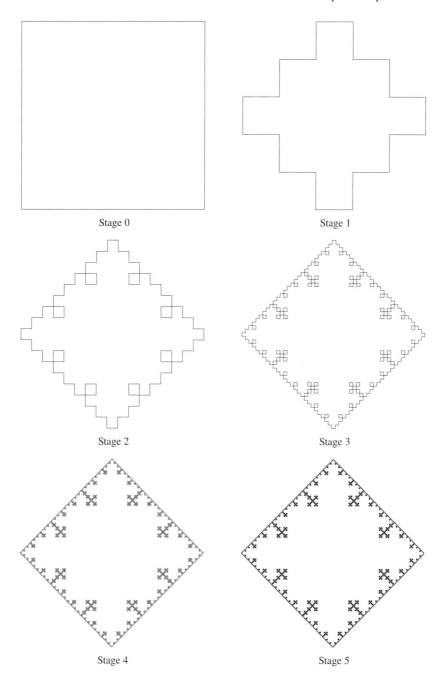

Figure 5.3: The Koch square fractal constructed to stage 5.

5. Fractals and Multifractals

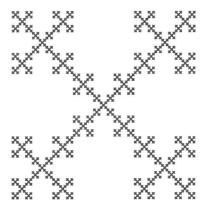

Figure 5.4: The inverted Koch square at stage 5.

It is not difficult to show that the perimeter P_k at the kth stage is given by

$$P_k = 4 \times \left(\frac{5}{3}\right)^k,$$

and $P_\infty = \infty$. Therefore, the Koch square has infinite perimeter and finite area.

It is possible to construct an inverted Koch square fractal by attaching the Koch curves to the inner edge of the unit square. The result up to stage 5 is shown in Figure 5.4.

The Sierpiński Triangle. This fractal may be constructed in a number of ways; see the exercises at the end of the chapter (Section 5.5). One way is to play a so-called chaos game with a die. Consider an equilateral triangle with vertices A, B, and C, as depicted in Figure 5.5.

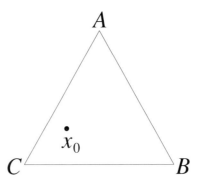

Figure 5.5: A triangle used in the chaos game with an initial point x_0.

5.1. Construction of Simple Examples   115

The rules of the chaos game are very simple. Start with an initial point x_0 somewhere inside the triangle.

Step 1. Cast an ordinary cubic die with six faces.

Step 2. If the number is either 1 or 2, move halfway to the point A and plot a point.

Step 2. Else, if the number is either 3 or 4, move halfway to the point B and plot a point.

Step 2. Else, if the number is either 5 or 6, move halfway to the point C and plot a point.

Step 3. Starting with the new point generated in Step 2, return to Step 1.

The die is cast again and again to generate a sequence of points $\{x_0, x_1, x_2, x_3, \ldots\}$. As with the other fractals considered here, the mathematical fractal would consist of an infinite number of points. In this way, a chaotic attractor is formed, as depicted in Figure 5.6. A MATLAB program is given in Section 5.5.

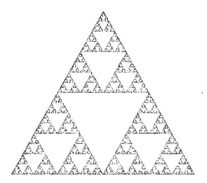

Figure 5.6: An early-stage-generation Sierpiński triangle plotted using the chaos game. There are 50000 points plotted.

The first few initial points are omitted to reveal the chaotic attractor. This object is known as the Sierpiński triangle.

Stochastic processes can be introduced to obtain fractals that look more like objects in nature. We restrict ourselves to two-dimensional figures only in this chapter.

Definition 2. An *iterated function system* (IFS) is a finite set $T_1, T_2, T_3, \ldots, T_n$ of affine linear transformations of $\Re^2$, where

$$T_j(x, y) = (a_j x + b_j y + c_j, d_j x + e_j y + f_j).$$

116   5. Fractals and Multifractals

Furthermore, a *hyperbolic iterated function system* is a collection of affine linear transformations that are also contractions.

The IFSs follow basic rules, as in the case of the chaos game used to generate the Sierpiński triangle. The rules of the chaos game can be generalized to allow greater freedom as follows:

Step 1. Create two or more affine linear transformations.

Step 2. Assign probabilities to each of the transformations.

Step 3. Start with an initial point.

Step 4. Select a random transformation to get a second point.

Step 5. Repeat the process.

An iterated function system consisting of four transformations was used to generate Figure 5.7. This figure resembles a fern in nature and is known as *Barnsley's fern*. A MATLAB program is listed in Section 5.5.

Figure 5.7: A fractal attractor of an IFS, Barnsley's fern, generated using 60,000 points.

The affine linear transformations may be found by taking reflections, rotations, scalings, and translations of triangles that represent the fronds of the fern.

5.2 Calculating Fractal Dimensions

Definition 3. A self-similar fractal has fractal dimension (or *Hausdorff index*) D_f given by
$$D_f = \frac{\ln N(l)}{-\ln l},$$

5.2. Calculating Fractal Dimensions

where l represents a scaling and $N(l)$ denotes the number of segments of length l. Thus the relationship

(5.1) $$N(l) \propto (l)^{-D_f}$$

is also valid. The number D_f, which need not be an integer, gives a measure of how the density of the fractal object varies with respect to length scale.

Definition 4. A fractal is an object that has noninteger fractal dimension. (This is an alternative to Definition 1.)

Example 1. Determine the fractal dimension of

(i) the Cantor set,

(ii) the Koch curve,

(iii) the Koch square, and

(iv) the Sierpiński triangle.

Solution. (i) A construction of the Cantor set up to stage 3 is depicted in Figure 5.1. At each stage, one segment is replaced with two segments that are $\frac{1}{3}$ the length of the previous segment. Thus in this case, $N(l) = 2$ and $l = \frac{1}{3}$. The mathematical self-similar Cantor set fractal constructed to infinity will therefore have dimension given by

$$D_f = \frac{\ln 2}{\ln 3} \approx 0.6309.$$

Note that a point is defined to have dimension zero and a line dimension one. Hence the Cantor set is more dense than a point but less dense than a line.

(ii) The Koch curve is constructed up to stage 5 in Figure 5.2. In this case, one segment is replaced with four segments which are scaled by $\frac{1}{3}$; therefore, $N(l) = 4$ and $l = \frac{1}{3}$. The mathematical self-similar Koch fractal generated to infinity will have dimension

$$D_f = \frac{\ln 4}{\ln 3} \approx 1.2619.$$

Thus the Koch curve is more dense than a line but less dense than a plane, which is defined to have dimension two.

(iii) The Koch square generated to stage 5 is shown in Figure 5.3. Note that this object is not strictly self-similar; magnification will not reveal smaller Koch squares. However, it is possible to define a fractal dimension, since there is a scaling behavior. For the Koch square,

$$D_f = \frac{\ln 5}{\ln 3} \approx 1.4650.$$

5. Fractals and Multifractals

Hence the Koch square is more dense than the Koch curve but is still less dense than the plane. Note that the inverted Koch square will have exactly the same fractal dimension.

(iv) The mathematical Sierpiński triangle fractal (see Figure 5.6) may be constructed by removing the central triangle from equilateral triangles to infinity. A motif is shown in Figure 5.8.

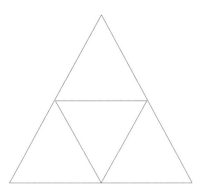

Figure 5.8: The motif used to generate the Sierpiński triangle.

It is important to note that the scaling l referred to in Definition 2 is linear. Thus the linear scale is $\frac{1}{2}$ since the sides of the smaller triangles are half as long as the sides of the original triangle in the motif. At each stage, one triangle is replaced with three triangles, so $l = \frac{1}{2}$ and $N(l) = 3$. The fractal dimension of the mathematical Sierpiński triangle generated to infinity is

$$D_f = \frac{\ln 3}{\ln 2} \approx 1.5850.$$

The Sierpiński triangle has the highest dimension in examples (i)–(iv) and is therefore the most dense.

Box-Counting Dimensions. The fractal dimensions calculated so far have been for hypothetical fractal objects that cannot exist in the real world. Mandelbrot [18] shows how fractals appear throughout science and nature. Trees, clouds, rocks, and the fractals generated in earlier chapters can display a certain type of scaling and self-similarity. Mandelbrot showed that these objects obey a power law as described in equation (5.1) over a certain range of scales. By covering the object with boxes of varying sizes and counting the number of boxes that contain the object, it is possible to estimate a so-called *box-counting* dimension, which is equivalent to the fractal dimension. Mandelbrot defined the fractal dimension to be

$$D_f = \lim_{l \to 0} \frac{\ln N(l)}{-\ln l},$$

where $N(l)$ boxes of length l cover the fractal object. These boxes need not be square.

Consider the following two examples.

Example 2. The Koch curve is covered with boxes of varying scales, as shown in Figure 5.9. Use a box-counting technique to show that the object obeys the power law given in equation (5.1) and hence estimate the box-counting dimension.

Table 5.1 gives the box count $N(l)$ for the different scalings l, and the natural logs are calculated.

Table 5.1: Box-count data for the Koch curve generated to stage 6.

l	12^{-1}	15^{-1}	18^{-1}	24^{-1}	38^{-1}	44^{-1}
$N(l)$	14	24	28	34	60	83
$-\ln l$	2.4849	2.7081	2.8904	3.1781	3.6376	3.7842
$\ln N(l)$	2.6391	3.1781	3.3322	3.5264	4.0943	4.4188

Using the least-squares method of regression, the line of best fit on a log–log plot is given by $y \approx 1.2246x - 0.2817$, and the correlation coefficient is approximately 0.9857. The line of best fit is shown in Figure 5.10.

Therefore, the box-counting dimension of the Koch curve generated to stage 6 is approximately 1.2246. There is obviously a scaling restriction with this object since the smallest segment is of length $3^{-6} \approx 0.0014$ units and the box-counting algorithm will break down as boxes approach this dimension. There is always some kind of scaling restriction with physical images as there are a limited number of pixels on a computer screen. It is interesting to note that the mathematical Koch curve has a higher dimension of approximately 1.2619. This is to be expected as true mathematical fractal is a lot more dense.

Example 3. A chaotic attractor comprising 5000 points for the Hénon map

$$x_{n+1} = 1.2 + 0.4y_n - x_n^2, \quad y_{n+1} = x_n$$

is covered with boxes of varying scales, as shown in Figure 5.11. Use a box-counting technique to show that the object obeys the power law given in equation (5.1) and hence estimate the box-counting dimension.

Table 5.2 gives the box count $N(l)$ for the different scalings l, and the natural logs are calculated.

Using the least-squares method of regression, the line of best fit on a log–log plot is given by $y \approx 1.0562x + 1.1810$, and the correlation coefficient is approximately 0.9961. The line of best fit is shown in Figure 5.12.

Therefore, the box-counting dimension of the Hénon attractor with 5000 points is approximately 1.0562. There is a scaling restriction in this case as there

120   5. Fractals and Multifractals

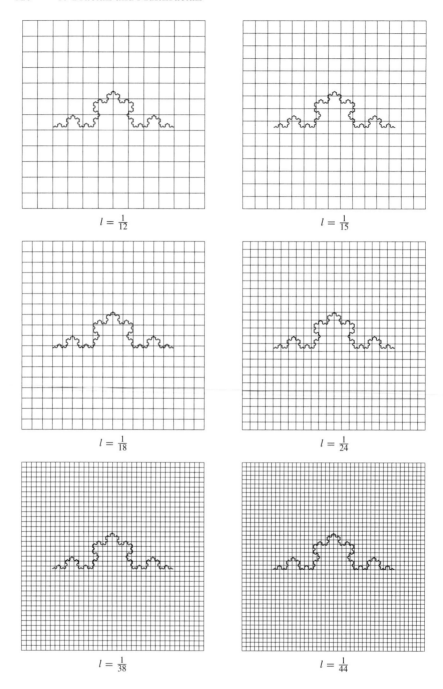

Figure 5.9: Different coarse coverings of the Koch curve generated to stage 6.

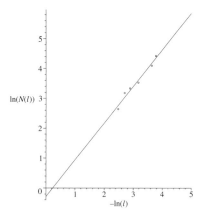

Figure 5.10: The line of best fit on a log–log plot for the early-generation Koch curve. The correlation coefficient is 0.9857.

Table 5.2: Box-count data for the Hénon map with 5000 points.

l	12^{-1}	16^{-1}	20^{-1}	24^{-1}	28^{-1}	32^{-1}
$N(l)$	47	58	76	93	109	131
$-\ln l$	2.4849	2.7726	2.9957	3.1781	3.3322	3.4657
$\ln N(l)$	3.8501	4.0604	4.3307	4.5326	4.6914	4.8752

are only 5000 data points. Once more, the dimension of the mathematical fractal with an infinite number of data points will be larger.

The Hénon map is not self-similar and is, in fact, a multifractal. See the next section.

5.3 A Multifractal Formalism

In the previous section, it was shown that a fractal object can be characterized by its fractal dimension D_f, which gives a measure of how the density varies with respect to length scale. Most of the fractals appearing earlier in this chapter can be constructed to the infinite stage in the minds of mathematicians. They are *homogeneous* since the fractals consist of a geometrical figure repeated on an ever-reduced scale. For these objects, the fractal dimension is the same on all scales. Unfortunately, in the real world, fractals are not homogeneous; there is rarely an identical motif repeated on all scales. Two objects might have the same fractal dimension and yet look completely different. It has been found that real-world fractals are *heterogeneous*; that is, there is a nonuniformity possessing rich scaling and self-similarity properties that can change from point to point. Put plainly, the

122   5. Fractals and Multifractals

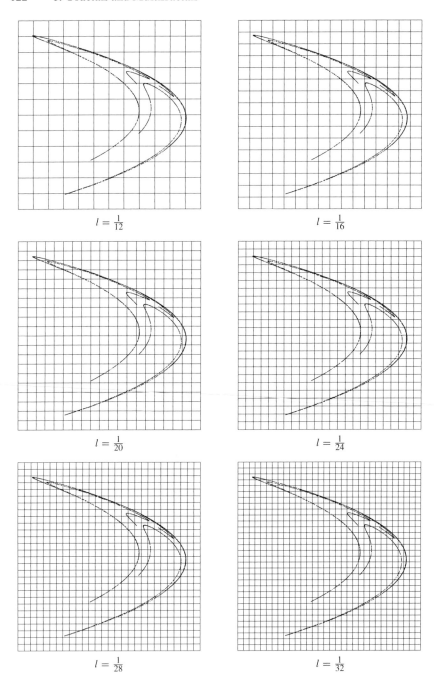

Figure 5.11: Different coarse coverings of the Hénon chaotic attractor.

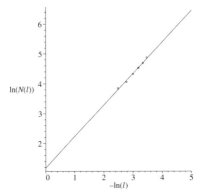

Figure 5.12: The line of best fit on a log–log plot for the early-generation Hénon attractor. The correlation coefficient is 0.9961.

object can have different dimensions at different scales. It should also be pointed out that there is always some kind of scaling restriction with physical fractals. These more complicated objects are known as *multifractals*, and it is necessary to define continuous spectra of dimensions to classify them.

There are many different ways in which a mathematician can define *dimension*, and the subject can become very complicated and abstract. For example, there is Hausdorff dimension, topological dimension, Euclidean dimension, and box-counting dimension, to name but a few. More details on the pure mathematical approach to multifractals are presented in [5] and [13]. The most widely used method of determining multifractal spectra is that of Falconer [5], which is described briefly below.

Let μ be a self-similar probability measure defined on an object $S \subset \Re^d$, where $\mu(B)$ is a probability measure determined from the probability of hitting the object in the box $B_i(l)$ and $N \propto \frac{1}{l^2}$ is the number of boxes in the grid. The *generalized fractal dimensions* D_q or, alternatively, the $f(\alpha)$ *spectrum of singularities* may be computed using box-counting techniques. First, consider the generalized fractal dimensions. Cover the object S with a grid of boxes $(B_i(l))_{i=1}^N$ of size l as in Section 5.1. The qth moment (or *partition function*) Z_q is defined by

$$(5.2) \qquad Z_q(l) = \sum_{\mu(B) \neq 0} [\mu(B)]^q = \sum_{i=1}^{N} p_i^q(l).$$

For self-similar multifractals, given a real number q, $\tau(q)$ may be defined as the positive number satisfying

$$(5.3) \qquad \sum_{i=1}^{N} p_i^q r_i^{\tau(q)} = 1,$$

5. Fractals and Multifractals

where p_i represent probabilities ($\sum_{i=1}^{N} p_i = 1$) with r_i fragmentation ratios. The function $\tau : \Re \to \Re$ is a decreasing real analytic function with

$$\lim_{q \to -\infty} \tau(q) = \infty \quad \text{and} \quad \lim_{q \to \infty} \tau(q) = -\infty.$$

The generalized dimensions D_q and the scaling function $\tau(q)$ are defined by

(5.4) $$\tau(q) = D_q(1-q) = \lim_{l \to 0} \frac{\ln Z_q(l)}{-\ln l}.$$

The generalized dimensions are obtained from an assumed power-law behavior of the partition function in the limit as $l \to 0$ and $N \to \infty$,

$$Z_q \propto l^{D_q(q-1)}.$$

Definition 5. The generalized (box-counting) fractal dimensions D_q, where $q \in \Re$, are defined by

(5.5) $$D_q = \lim_{l \to 0} \frac{1}{1-q} \frac{\ln \sum_{i=1}^{N} p_i^q(l)}{-\ln l},$$

where the index i labels the individual boxes of size l and $p_i(l)$ denotes the relative weight of the ith box or the probability of the object lying in the box. Hence

$$p_i(l) = \frac{N_i(l)}{N},$$

where $N_i(l)$ is the weight of the ith box and N is the total weight of the object. When $q = 0$,

$$D_0 = D_f = \lim_{l \to 0} \frac{\ln N(l)}{-\ln(l)},$$

where $N(l)$ is the number of boxes contained in the minimal cover. When $q = 1$, L'Hôpital's rule can be applied (see the exercises in Section 5.6) to give

$$D_1 = \lim_{l \to 0} \frac{\sum_{i=1}^{N} p_i \ln(p_i)}{-\ln(l)},$$

which is known as the *information dimension*. This gives an indication of how the morphology increases as $l \to 0$. The quantity D_2 is known as the *correlation dimension* and indicates the correlation between pairs of points in each box. The generalized dimensions $D_3, D_4, \ldots$ are associated with correlations between triples, quadruples, etc., of points in each box.

Now consider the so-called $f(\alpha)$ *spectrum of dimensions*. The weight p_s of segments of type s scales with the size l of a box as follows:

$$p_s(l) \propto (l)^{\alpha_s},$$

5.3. A Multifractal Formalism

where α_s is the so-called *coarse Hölder exponent* defined by

$$\alpha_s = \frac{\ln p_s(l)}{\ln l}.$$

The number of segments N_s of type s scales with the size l of a box according to

$$N_s(l) \propto (l)^{-f_s}.$$

The exponents α_s and f_s can then be used to determine $f(\alpha)$, as demonstrated in the examples in the next section.

In many cases, $f(\alpha) = \dim_H S_\alpha$ is related to the Hausdorff–Besicovich dimension of the set $\mathbf{x} \in S$; see [5] for more information. In most cases, a multifractal spectrum $f(\alpha)$ may be obtained from $\tau(q)$ by a so-called *Legendre transformation*, which is described here briefly for completeness. Hence

$$f(\alpha) = \inf_{-\infty < q < \infty} (\tau(q) + \alpha q).$$

The $f(\alpha)$ can be derived from $\tau(q)$, and vice versa, by the identities

(5.6) $$f(\alpha(q)) = q\alpha(q) + \tau(q) \quad \text{and} \quad \alpha = -\frac{\partial \tau}{\partial q}.$$

It is known that the function $f(\alpha)$ is strictly cap convex (see Figure 5.13(c)) and that $\alpha(q)$ is a decreasing function of q.

In practice, to compute $\tau(q)$ using the partition function, the following three steps are required:

- Cover the object with boxes $(B_i(l))_{i=1}^N$ of size l and compute the corresponding box measures $\mu_i = \mu(B_i(l)) = p_i(l)$.

- Compute the partition function Z_q for various values of l.

- Check that the log–log plots for Z_q against l are straight lines. If so, then $\tau(q)$ is the slope of the line corresponding to the exponent q.

In summary, $\tau(q)$ and D_q can be obtained from equations (5.2) and (5.4), and the $f(\alpha)$ values can be determined as above or computed (see [16]) using the expressions

(5.7) $$f(q) = \lim_{l \to 0} \frac{\sum_{i=1}^{N} \mu_i(q, l) \ln \mu_i(q, l)}{\ln l}$$

and

(5.8) $$\alpha(q) = \lim_{l \to 0} \frac{\sum_{i=1}^{N} \mu_i(q, l) \ln p_i(l)}{\ln l},$$

where $\mu_i(q, l)$ are the normalized probabilities

$$\mu_i(q, l) = \frac{p_i^q(l)}{\sum_{j=1}^{N} p_j^q(l)}.$$

In physical applications, an image on a computer screen of 512 × 512 pixels is typically used. A problem arises with negative values of q; boxes with very low measure may contribute disproportionately. Several papers have been published addressing this *clipping problem*; see [2], for example. This is not a problem with some of the physical applications discussed here since most of the useful results are obtained for $0 \leq q \leq 5$.

The multifractal functions $\tau(q)$, D_q, and $f(\alpha)$ have typical forms for self-similar measures. For example, consider $f : [\alpha_{\min}, \alpha_{\max}] \to \Re$, then $-\alpha_{\min}$ and $-\alpha_{\max}$ are the slopes of the asymptotes of the strictly convex function τ. The geometry of the Legendre transform determines that f is continuous on $[\alpha_{\min}, \alpha_{\max}]$ and $f(\alpha_{\min}) = f(\alpha_{\max}) = 0$. It is not difficult to show that $\tau(0) = D_0$ and $q = 0$ corresponds to the maximum of $f(\alpha)$. When $q = 1$, $\tau(q) = 0$, and so $f(\alpha) = \alpha$. Moreover, $\frac{d}{d\alpha}(f(\alpha) - \alpha)) = q - 1 = 0$. Thus $f(\alpha)$ is tangent to $f(\alpha) = \alpha$ at $q = 1$.

Typical $\tau(q)$, D_q and $f(\alpha)$ curves and some of their properties are shown in Figure 5.13. Note that in Figure 5.13(a) the line asymptotic to the curve as $q \to \infty$ has slope $-\alpha_{\min}$, and the line asymptotic to the curve as $q \to -\infty$ has slope $-\alpha_{\max}$.

There are major limitations associated with this so-called *fixed-size box-counting algorithm*, and in many applications, results are only reliable for a narrow range of q values, typically $0 \leq q \leq 5$. In [9], Mach, Mas, and Sagués also consider a *fixed-weight box-counting algorithm*, where the measure quantities p_i are fixed and the size factors r_i vary; see equation (5.3). They show that the fixed-size box-counting algorithm gives good results for small positive q, and the fixed-weight box-counting algorithm can be used to give good results for small negative q. Alber and Peinke [2] recently developed an improved multifractal box-counting algorithm using so-called *fuzzy discs* and a symmetric scaling-error compensation. They apply their method to the Hénon map with great success.

Some simple multifractals are constructed in the next section, and a multifractal analysis is applied to determine multifractal spectra.

5.4 Multifractals in the Real World and Some Simple Examples

Since the publication of the seminal paper by Halsey et al. [17] on multifractals, there has been intense research activity, and numerous papers have been published in many diverse fields of science. A small selection of this research material will be discussed to demonstrate how the analysis is being applied in the physical world.

5.4. Multifractals in the Real World and Some Simple Examples

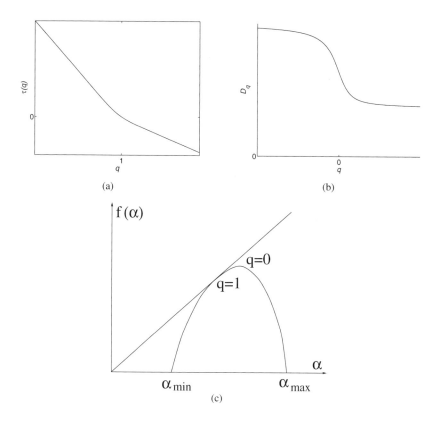

Figure 5.13: Typical curves of (a) the $\tau(q)$ function, (b) the D_q spectrum, and (c) the $f(\alpha)$ spectrum. In case (c), points on the curve near $\alpha_{\min}$ correspond to values of $q \to \infty$, and points on the curve near $\alpha_{\max}$ correspond to values of $q \to -\infty$.

In 1989, Chhabra et al. [16] used equations (5.7) and (5.8) to determine the $f(\alpha)$ spectrum for fully developed turbulence in laboratory and atmospheric flows directly from experimental data. The same methods were employed by Blacher et al. [12] in 1993 and Mills et al. [1] in 2002 when characterizing the morphology of multicomponent polymer systems. They found that there was a correlation between the mechanical properties of the samples and their respective $f(\alpha)$ curves. There have been many other studies into the mechanical properties of plastics and rubbers using image analysis techniques. A very useful tool is the multifractal analysis of density distributions. The analysis is usually applied to elemental dot maps produced by scanning electron microscopy coupled with energy dispersive X-ray spectroscopy. The analysis is used to produce generalized dimensions, and

128   5. Fractals and Multifractals

it has been found that $w = D_0 - D_5$ is related to factors such as tensile strength, elongation at break, and energy to break. The quantity w is a measure of the nonuniformity of the structure. The smaller the value of w, the more homogeneous the structure and the stronger the material. Multifractals are being applied in image compression techniques and signal processing. In [11], Sarkar and Chaudhuri estimate fractal and multifractal dimensions of grey-tone digital images. Generalized dimensions are being applied extensively in the geosciences to classify sedimentary rocks. In [10], Muller, Huseby, and Saucier relate porosity and permeability to the multifractal spectra of the relevant samples. The analysis is also often applied to *diffusion-limited aggregates* (DLA) clusters. For example, Mach, Mas, and Sagués [9] consider the electrodeposition of zinc sulphate on an electrode and apply the fixed-size and fixed-weight box algorithms to obtain the generalized dimensions. Multifractal characteristics are displayed by propagating cracks in brittle materials, as reported by Silberschmidt in [7]. In physics, the box-counting method was applied to show the multifractality of secondary-electron emission sites in silicon in [6].

Other examples of multifractal phenomena can be found in, for example, stock market analysis, rainfall, and even the distribution of stars and galaxies in the universe. The examples listed above are by no means exhaustive, but the author hopes that the reader will be encouraged to look for more examples in his or her own particular field of specialization.

In the following examples, simple multifractals are constructed using nonuniform generalizations of the Cantor set and the Koch curve. Multifractal spectra curves are plotted in both cases.

Example 4. A Cantor multifractal set is constructed by removing the middle third segment at each stage and distributing a weight so that each of the remaining two segments receive a fraction p_1 and p_2 units, respectively, and such that $p_1 + p_2 = 1$. Illustrate how the weight is distributed after the first two stages of construction. Plot $\tau(q)$ curves, D_q spectra, and $f(\alpha)$ spectra when

(i) $p_1 = \frac{1}{3}$ and $p_2 = \frac{2}{3}$,

(ii) $p_1 = \frac{1}{9}$ and $p_2 = \frac{8}{9}$.

Which of the multifractals is more heterogeneous?

Solution. Figure 5.14 illustrates how the weight is distributed up to the second stage of construction.

At stage k, each segment is of length $(\frac{1}{3})^k$ and there are $N = 2^k$ segments. Assign a unit weight to the original line. Then for $k = 1$, one line segment has weight p_1 and the other has weight p_2. For $k = 2$, there are four segments: one with weight p_1^2, two with weight $p_1 p_2$, and one with weight p_2^2. At stage 3, there are eight segments: one with weight p_1^3, three with weight $p_1^2 p_2$, three with weight

5.4. Multifractals in the Real World and Some Simple Examples

Stage 0 $P_0 = 1$

Stage 1 p_1 p_2

Stage 2 p_1^2 $p_1 p_2$ $p_2 p_1$ p_2^2

Figure 5.14: The weight distribution on a Cantor multifractal set up to stage 2.

$p_1 p_2^2$, and one with weight p_2^3. It is not difficult to see that at stage k, there will be

$$N_s(l) = \binom{k}{s}$$

segments of weight $p_1^s p_2^{k-s}$. From equation (5.2), the partition function $Z_q(l)$ is given by

$$Z_q(3^{-k}) = \sum_{s=0}^{k} \binom{k}{s} p_1^{qs} p_2^{q(k-s)} = (p_1^q + p_2^q)^k,$$

from the binomial theorem. Therefore, from equation (5.4),

$$\tau(q) = D_q(1-q) = \lim_{l \to 0} \frac{\ln(p_1^q + p_2^q)^k}{-\ln 3^{-k}},$$

so

$$\tau(q) = \frac{\ln(p_1^q + p_2^q)}{\ln 3}.$$

The D_q spectrum can be plotted using continuity at $q = 1$.

To construct an $f(\alpha)$ spectrum, consider how the weight p_s and the number of segments N_s, each of type s, scale with segment size l. Now

$$p_s(l) \propto (l)^{\alpha_s} \quad \text{and} \quad N_s(l) \propto (l)^{-f_s},$$

where $s = 0, 1, \ldots, k$. Now $p_s = p_1^s p_2^{k-s}$ and $l = 3^{-k}$. Hence

$$\alpha_s = \frac{s \ln p_1 + (k-s) \ln p_2}{\ln 3^{-k}}.$$

The number of segments of weight p_s at the kth stage is

$$N_s = \binom{k}{s}.$$

Hence

$$-f_s = \frac{\ln\left(\frac{k}{s}\right)}{\ln 3^{-k}}.$$

These parametric curves may be plotted to produce $f(\alpha)$ using the MATLAB package. The programs are listed in the next section.

(i) Suppose that $p_1 = \frac{1}{3}$ and $p_2 = \frac{2}{3}$. The multifractal curves are given in Figure 5.15.

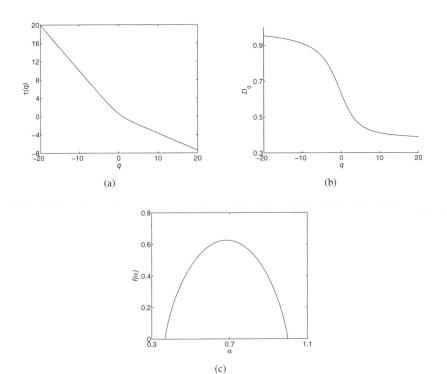

Figure 5.15: Multifractal spectra for Example 1(i) when $p_1 = \frac{1}{3}$ and $p_2 = \frac{2}{3}$. (a) $\tau(q)$ curve; (b) D_q spectrum; (c) $f(\alpha)$ spectrum when $k = 500$.

(ii) Suppose that $p_1 = \frac{1}{9}$ and $p_2 = \frac{8}{9}$. The multifractal curves are given in Figure 5.16.

Notice that in all cases, $D_0 = D_f = \frac{\ln 2}{\ln 3} \approx 0.63$. The multifractal in case (ii) is more heterogeneous. The $f(\alpha)$ curve is broader and the generalized dimensions D_q cover a wider range of values.

5.4. Multifractals in the Real World and Some Simple Examples

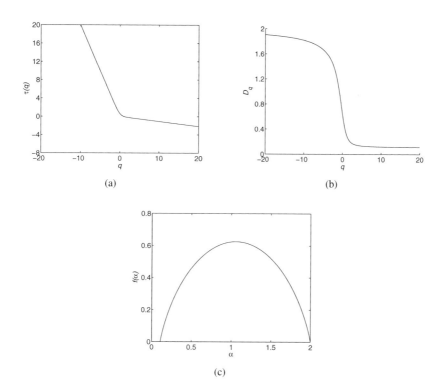

Figure 5.16: Multifractal spectra for part (ii) of Example 1 when $p_1 = \frac{1}{9}$ and $p_2 = \frac{8}{9}$: (a) $\tau(q)$ curve, (b) D_q spectrum, and (c) $f(\alpha)$ spectrum when $k = 500$.

The following images and plots have been supplied by one of the author's Ph.D. students, Dr. Steve Mills from BICC cables in Wrexham, for which the author is very grateful. This work is beyond the scope of the book, but the results are included as a matter of interest. Note that these are early-generation fractals and that the results have been obtained using powerful computers and image-analysis techniques.

Example 5. Consider the image in Figure 5.17(a), produced by applying the weight distribution as indicated. Using the computer algorithms described in various papers, it is possible to compute the D_q and $f(\alpha)$ spectra. The theoretical multifractal spectra may be derived analytically using methods similar to those used in [3]. However, it is beyond the scope of this book.

Solution. The computed multifractal spectra are plotted in Figure 5.18.

5. Fractals and Multifractals

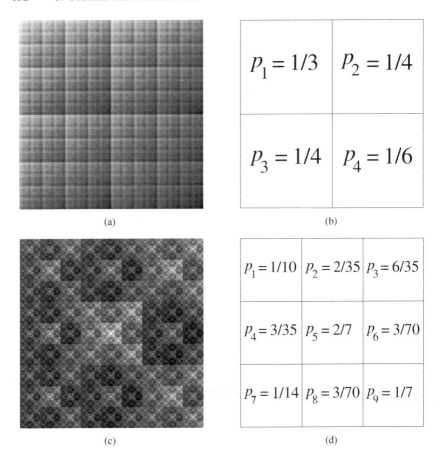

Figure 5.17: Multifractal images and the weight distribution motifs. The weights are related to the grey scale; for example, $p_1 = 1$ would be white and $p_1 = 0$ would be black on this scale.

Example 6. Consider the image in Figure 5.17(c), produced by applying the weight distribution as indicated. Using the computer algorithms described in various papers, it is possible to compute the D_q and $f(\alpha)$ spectra.

Solution. The computed multifractal spectra are plotted in Figure 5.19.

The plots in Figures 5.18 and 5.19 are typical of those displayed in the research literature. Note that the latter image is more nonuniformly distributed and is a more heterogeneous fractal.

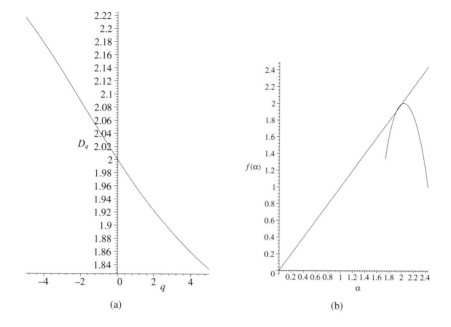

Figure 5.18: Plots for Example 5: (a) the D_q spectrum when $-5 \leq q \leq 5$ and (b) the $f(\alpha)$ spectrum.

The quantity $w = D_0 - D_5$ may be used to measure dispersion. Thus for Example 5, $w \approx 0.13$ and for Example 6, $w \approx 0.6$. Therefore, the dispersion is better in Example 5 since the fractal is more homogeneous.

Mulitifractal generalized Sierpiński triangles are considered in [3], and a multifractal spectrum of the Hénon map is discussed in [2].

5.5 MATLAB Commands

```
% Program_5a - The Koch Curve (Figure 5.2).
clear
k=6;
mmax=4^k;
x=zeros(1,mmax);
y=zeros(1,mmax);
h=3^(-k);
x(1)=0;y(1)=0;
angle(1)=0;angle(2)=pi/3;angle(3)=-pi/3;angle(4)=0;
for a=1:mmax
```

134   5. Fractals and Multifractals

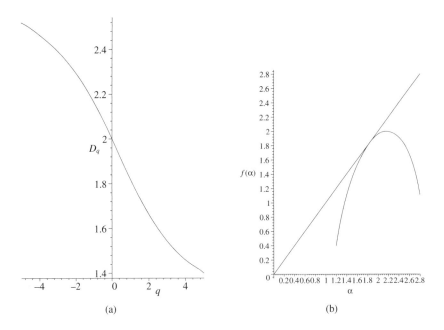

Figure 5.19: Plots for Example 6: (a) the D_q spectrum and (b) the $f(\alpha)$ spectrum.

```
    m=a-1;ang=0;
    for b=1:k
        segment(b)=mod(m,4);
        m=floor(m/4);
        r=segment(b)+1;
        ang=ang+angle(r);
    end
    x(a+1)=x(a)+h*cos(ang);
    y(a+1)=y(a)+h*sin(ang);
end

plot(x,y,'b');
axis equal
% End of Program_5a.
```

```
% Program_5b - The Sierpinski Triangle (Figure 5.6).

% The vertices of the triangle.
A=[0,0];B=[4,0];C=[2,2*sqrt(3)];
% The number of points to be plotted.
Nmax=50000;
P=zeros(Nmax,2);
```

```
scale=1/2;
for n=1:Nmax
    r=rand;
    if r<1/3;
        P(n+1,:)=P(n,:)+(A-P(n,:)).*scale;
    elseif r<2/3;
        P(n+1,:)=P(n,:)+(B-P(n,:)).*scale;
    else
        P(n+1,:)=P(n,:)+(C-P(n,:)).*scale;
    end
end

plot(P(:,1),P(:,2),'.','MarkerSize',1);
axis([0 4 0 2*sqrt(3)])
set(gca,'Position',[0 0 1 1])
% End of Program_5b.
```

```
% Program_5c - Barnsley's fern (Figure 5.7).

function Program_5c(N)
% This function plots Barnsley's fern with N points.
% The transformations are in the form
% T(x,y) = (a*x+b*y+c, d*x+e*y+f).
echo on
N=50000;
close all
P=zeros(N,2);
P(1,:)=[0.5,0.5];

% The main loop where the iterations are performed.
for k=1:N-1
r=rand;
if r<.05;
P(k+1,:)=T(P(k,:),0,0,0,0,.2,0);
elseif r<.86;
P(k+1,:)=T(P(k,:),.85,.05,0,-.04,.85,1.6);
elseif r<.93;
P(k+1,:)=T(P(k,:),.2,-.26,0,.23,.22,1.6);
else
P(k+1,:)=T(P(k,:),-.15,.28,0,.26,.24,.44);
end
end

plot(P(:,1),P(:,2),'.','MarkerSize',1);
axis([-2.5 3.5 0 11]);
set(gca,'Position',[0 0 1 1])
```

136 5. Fractals and Multifractals

```
% The transformation T
function F=T(P,a,b,c,d,e,f)
F=zeros(1,2);
F(1)=a*P(1)+b*P(2)+c;
F(2)=d*P(1)+e*P(2)+f;
% End of Program_5c.
```

```
% Program_5d - Tau curve (Figure 5.16(a)).

ezplot('(log((1/9)^x+(8/9)^x))/(log(3))',[-20,20])
axis([-20 20 -8 20])
fsize=15;
set(gca,'xtick',[-20:10:20],'FontSize',fsize)
set(gca,'ytick',[-8:4:20],'FontSize',fsize)
xlabel('\itq','FontSize',fsize)
ylabel('\it{\tau(q)}','FontSize',fsize)
title(' ')
% End of Program_5d.
```

```
% Program_5e - Dq curve (Figure 5.16(b)).

hold on
ezplot('(log((1/9)^x+(8/9)^x))/(log(3))*(1/(1-x))',[-20,0.99])
ezplot('(log((1/9)^x+(8/9)^x))/(log(3))*(1/(1-x))',[0.99,20])
axis([-20 20 0 2])
fsize=15;
set(gca,'xtick',[-20:10:20],'FontSize',fsize)
set(gca,'ytick',[0:.4:2],'FontSize',fsize)
xlabel('\itq','FontSize',fsize)
ylabel('\it{D_q}','FontSize',fsize)
title(' ')
hold off
% End of Program_5e.
```

```
% Program_5f - falpha curve (Figure 5.16(c)).

k=sym('500');p1=sym('1/9');p2=sym('8/9');
t=0:500;
x=(t*log(p1)+(k-t)*log(p2))/(k*log(1/3));
Y(length(t))=sym('0');
for i=1:length(t)
    Y(i)=-(log(maple('binomial',k,t(i)))/(k*log(1/3)));
end
plot(double(x),double(Y))

axis([0 2 0 .8])
```

```
fsize=15;
set(gca,'xtick',[0:.5:2],'FontSize',fsize)
set(gca,'ytick',[0:.2:.8],'FontSize',fsize)
xlabel('\it{\alpha}','FontSize',fsize)
ylabel('\it{f(\alpha)}','FontSize',fsize)
title(' ')
hold off
% End of Program_5f.
```

5.6 Exercises

1. (a) Consider the unit interval. A variation of the Cantor set is constructed by removing two line segments each of length $\frac{1}{5}$. Thus at stage 1, remove the segments between $\{\frac{1}{5} \ldots \frac{2}{5}\}$ and $\{\frac{3}{5} \ldots \frac{4}{5}\}$ from the unit interval, leaving three line segments remaining. Continuing in this way, construct the fractal up to stage 3 either on graph paper or on a computer screen. Find the length of segment remaining at stage k. Determine the fractal dimension of the mathematical fractal constructed to infinity.

    (b) A Lévy fractal is constructed by replacing a line segment with a try square. Thus at each stage one line segment of length, 1, say, is replaced by two of length $\frac{1}{\sqrt{2}}$. Construct the fractal up to stage 7 either on graph paper or on a computer screen. When using graph paper it is best to draw a skeleton (dotted line) of the previous stage. What is the true fractal dimension of the object generated to infinity?

    (c) A *Koch snowflake* is constructed by adjoining the Koch curve to the outer edges of a unit length equilateral triangle. Construct this fractal up to stage 4 either on graph paper or on a computer screen and show that the area bounded by the true fractal A_∞ is equal to

    $$A_\infty = \frac{2\sqrt{3}}{5} \text{ units}^2.$$

    (d) The inverted Koch snowflake is constructed in the same way as in Exercise 1(c) but the Koch curve is adjoined to the inner edges of an equilateral triangle. Construct the fractal up to stage 4 on graph paper or stage 6 on the computer.

2. The Sierpiński triangle can be constructed by removing the central inverted equilateral triangle from an upright triangle; a motif is given in this chapter. Construct the Sierpiński triangle up to stage 4 on graph paper using this method.

3. Consider Pascal's triangle given below. Cover the odd numbers with small black discs (or shade the numbers). What do you notice about the pattern obtained?

5. Fractals and Multifractals

```
                              1
                           1     1
                        1     2     1
                     1     3     3     1
                  1     4     6     4     1
               1     5    10    10    5     1
            1     6    15    20    15    6     1
         1     7    21    35    35    21    7     1
      1     8    28    56    70    56    28    8     1
   1     9    36    84   126   126   84    36    9     1
1    10    45   120   210   252   210  120    45   10   1
1  11   55   165  330  462  462  330  165  55   11   1
1  12   66  220  495  792  924  792  495  220  66   12   1
1  13   78  286  715   x0   x1   x1   x0  715  286   78   13   1
1  14   91  364   x2   x3   x4   x5   x4   x3   x2  364   91   14   1
1  15  105  455   x6   x7   x8   x9   x9   x8   x7   x6  455  105  15   1
```

where $x0 = 1287$, $x1 = 1716$, $x2 = 1001$, $x3 = 2002$, $x4 = 3003$, $x5 = 3432$, $x6 = 1365$, $x7 = 3003$, $x8 = 5005$, and $x9 = 6435$.

4. A Sierpiński square is constructed by removing a central square at each stage. Construct this fractal up to stage 3 and determine the fractal dimension of the theoretical object generated to infinity.

5. Use the box-counting algorithm to approximate the fractal dimension of Barnsley's fern. The MATLAB program for plotting the fern is given in Section 5.5.

6. Prove that

$$D_1 = \lim_{l \to 0} \frac{\sum_{i=1}^{N} p_i \ln(p_i)}{-\ln(l)},$$

by applying L'Hôpital's rule to equation (5.5).

5.6. Exercises   139

7. Plot $\tau(q)$ curves and D_q and $f(\alpha)$ spectra for the multifractal Cantor set described in Example 1 when (i) $p_1 = \frac{1}{2}$ and $p_2 = \frac{1}{2}$, (ii) $p_1 = \frac{1}{4}$ and $p_2 = \frac{3}{4}$, and (iii) $p_1 = \frac{2}{5}$ and $p_2 = \frac{3}{5}$.

8. A multifractal Koch curve is constructed and the weight is distributed as depicted in Figure 5.20. Plot the $f(\alpha)$ spectrum when $p_1 = \frac{1}{3}$ and $p_2 = \frac{1}{6}$.

Figure 5.20: The motif used to construct the Koch curve multifractal, where $2p_1 + 2p_2 = 1$.

9. A multifractal square Koch curve is constructed and a weight is distributed as depicted in Figure 5.21. Plot the $\tau(q)$ curve and the D_q and $f(\alpha)$ spectra when $p_1 = \frac{1}{9}$ and $p_2 = \frac{1}{3}$.

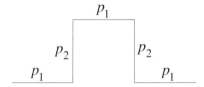

Figure 5.21: The motif used to construct the Koch curve multifractal, where $3p_1 + 2p_2 = 1$.

10. A multifractal Koch curve is constructed and a weight is distributed as depicted in Figure 5.22, where $p_1 + p_2 + p_3 + p_4 = 1$. Determine α_s and f_s.

Figure 5.22: The motif used to construct the Koch curve multifractal, where $p_1 + p_2 + p_3 + p_4 = 1$.

Recommended Reading

[1] S. L. Mills, G. Lees, C. Liauw, and S. Lynch, Dispersion assessment of flame retardant filler/polymer systems using a combination of X-ray mapping and multifractal analysis, *Polymer Testing*, **21**-8 (2002), 941–948.

[2] M. Alber and J. Peinke, Improved multifractal box-counting algorithm, virtual phase transitions, and negative dimensions, *Phys. Rev.* E, **57**-5 (1998), 5489–5493.

[3] K. J. Falconer and B. Lammering, Fractal properties of generalized Sierpiński triangles, *Fractals*, **6**-1 (1998), 31–41.

[4] P. S. Addison, *Fractals and Chaos: An Illustrated Course*, Institute of Physics, Bristol, PA, 1997.

[5] K. J. Falconer, *Techniques in Fractal Geometry*, Wiley, New York, 1997.

[6] Li Hua, D. Ze-jun, and Wu Ziqin, Multifractal analysis of the spatial distribution of secondary-electron emission sites, *Phys. Rev.* B, **53**-24 (1996), 16631–16636.

[7] V. Silberschmidt, Fractal and multifractal characteristics of propagating cracks, *J. Phys.* IV, **6** (1996), 287–294.

[8] R. M. Crownover, *Introduction to Fractals and Chaos*, Jones and Bartlett, Sudbury, MA, 1995.

[9] J. Mach, F. Mas, and F. Sagués, Two representations in multifractal analysis, *J. Phys.* A *Math. Gen.*, **28** (1995), 5607–5622.

[10] J. Muller, O. K. Huseby, and A. Saucier, Influence of multifractal scaling of pore geometry on permeabilities of sedimentary rocks, *Chaos Solitons Fractals*, **5**-8 (1995), 1485–1492.

[11] N. Sarkar and B. B. Chaudhuri, Multifractal and generalized dimensions of gray-tone digital images, *Signal Processing*, **42** (1995), 181–190.

[12] S. Blacher, F. Brouers, R. Fayt, and P. Teyssié, Multifractal analysis: A new method for the characterization of the morphology of multicomponent polymer systems, *J. Polymer Sci.* B *Polymer Phys.*, **31** (1993), 655–662.

[13] H.-O. Peitgen, H. Jürgens, and D. Saupe, *Chaos and Fractals: New Frontiers of Science*, Springer-Verlag, New York, 1992.

[14] H. A. Lauwerier, *Fractals: Images of Chaos*, Penguin, New York, 1991.

[15] H.-O. Peitgen (ed.), E. M. Maletsky, H. Jürgens, T. Perciante, D. Saupe, and L. Yunker, *Fractals for the Classroom: Strategic Activities*, Vol. 1, Springer-Verlag, New York, 1991.

[16] A. B. Chhabra, C. Meneveau, R. V. Jensen, and K. R. Sreenivasan, Direct determination of the $f(\alpha)$ singularity spectrum and its application to fully developed turbulence, *Phys. Rev.* A, **40**-9 (1989), 5284–5294.

[17] T. C. Halsey, M. H. Jensen, L. P. Kadanoff, I. Procaccia, and B. I. Shraiman, Fractal measures and their singularities, *Phys. Rev.* A, **33** (1986), 1141.

[18] B. B. Mandelbrot, *The Fractal Geometry of Nature*, W. H. Freeman, New York, 1983.

6
Controlling Chaos

Aims and Objectives

- To provide a brief historical introduction to chaos control.
- To introduce two methods of chaos control for one- and two-dimensional discrete maps.

On completion of this chapter the reader should be able to

- control chaos in the logistic and Hénon maps;
- plot time series data to illustrate the control;
- appreciate how chaos control is being applied in the real world.

This chapter is intended to give the reader a brief introduction into the new and exciting field of chaos control and to show how some of the theory is being applied to physical systems. There has been considerable research effort into chaos control over the past decade, and practical methods have been applied in, for example, biochemistry, cardiology, communications, physics laboratories, and turbulence. Chaos control has been achieved using many different methods, but this chapter will concentrate on two procedures only.

It is possible to control chaos in both continuous and discrete nonlinear dynamical systems. However, our analysis will be restricted to discrete systems in this chapter.

6.1 Historical Background

Even simple, well-defined discrete and continuous nonlinear dynamical systems without random terms can display highly complex, seemingly random behavior. Some of these systems have been investigated in this book, and mathematicians have labeled this phenomenon *determinisitc chaos*. *Nondeterministic chaos*, where the underlying equations are not known, such as that observed in a lottery or on a roulette wheel, will not be discussed in this text. Throughout history, dynamical systems have been used to model both the natural and technological sciences. In the early years of investigations, deterministic chaos was nearly always attributed to random external influences and was designed out if possible. The French mathematician and philosopher Henri Poincaré laid down the foundations of the qualitative theory of dynamical systems at the turn of the century and is regarded by many as being the first *chaologist*. Poincaré devoted much of his life to attempting to determine whether or not the solar system is stable. Despite knowing the exact form of the equations defining the motions of just three celestial bodies, he could not always predict the long-term future of the system. In fact, it was Poincaré who first introduced the notion of sensitivity to initial conditions and long-term unpredictability.

In recent years, deterministic chaos has been observed when applying simple models to cardiology, chemical reactions, electronic circuits, laser technology, population dynamics, turbulence, and weather forecasting. In the past, scientists have attempted to remove the chaos when applying the theory to physical models, and it is only in the last decade that they have come to realize the potential uses for systems displaying chaotic phenomena. For some systems, scientists are beginning to replace the maxim "stability good, chaos bad" with "stability good, chaos better." It has been found that the existence of chaotic behavior may even be desirable for certain systems.

Since the publication of the seminal paper of Ott, Grebogi, and Yorke [13] in 1990, there has been a great deal of progress in the development of techniques for the control of chaotic phenomena. Basic methods of controlling chaos along with several reprints of fundamental contributions to this topic may be found in the excellent textbook of Kapitaniak [5]. Some of these methods will now be discussed very briefly, and then a selection of early applications of chaos control in the real world will be listed.

I. *Changing the system parameters*. The simplest way to suppress chaos is to change the system parameters in such a way as to produce the desired result. In this respect, bifurcation diagrams can be used to determine the parameter values. For example, in Chapter 4 bifurcation diagrams were used to determine regions of bistability for nonlinear bistable optical resonators. It was found that isolated bistable regions existed for only a narrow range of parameter values. However, the major drawback with this procedure is that large parameter variations may be required, which could mean redesigning the apparatus and changing the dimensions of the physical system. In many practical situations, such changes are highly undesirable.

II. *Applying a damper.* A common method for suppressing chaotic oscillations is to apply some kind of damper to the system. In mechanical systems, this would be a shock absorber, and for electronic systems, one might use a shunt capacitor. Once more, this method would mean a large change to the physical system and might not be practical.

III. *Pyragas's method.* This method can be divided into two feedback-controlling mechanisms: linear feedback control and time-delay feedback control. In the first case, a periodic external force is applied whose period is equal to the period of one of the unstable periodic orbits contained in the chaotic attractor. In the second case, self-controlling delayed feedback is used in a similar manner. This method has been very successful in controlling chaos in electronic circuits such as the Duffing system and Chua's circuit. A simple linear feedback method has been applied to the logistic map in Section 6.2.

IV. *Stabilizing unstable periodic orbits (the Ott, Grebogi, and Yorke (OGY) method).* The method relies on the fact that the chaotic attractors contain an infinite number of unstable periodic orbits. By making small time-dependent perturbations to a control parameter of the system, it is possible to stabilize one or more of the unstable periodic orbits. The method has been very successful in applications but there are some drawbacks. This method will be discussed in some detail at the end of this section.

V. *Occasional proportional feedback (OPF).* Developed by Hunt [8] in 1991, this is one of the most promising control techniques for real applications. It is a one-dimensional version of the OGY method and has been successful in suppressing chaos for many physical systems. The feedback consists of a series of kicks, whose amplitude is determined from the difference of the chaotic output signal from a relaxation oscillation embedded in the signal, applied to the input signal at periodic intervals.

VI. *Synchronization.* The possibility of synchronization of two chaotic systems was first proposed by Pecora and Carroll [12] in 1990 with applications in communications. By feeding the output from one chaotic oscillator (the transmitter) into another chaotic oscillator (the receiver), they were able to synchronize certain chaotic systems for certain parameter choices. The method opens up the possibilities for secure information transmission. An example of chaos synchronization between two Lorenz systems is given in Section 18.5.

Before summarizing the OGY method, it is worthwhile to highlight some of the other major results not mentioned above. The first experimental suppression of chaos was performed by Ditto, Rausseo, and Spano [10] using the OGY algorithm. By making small adjustments to the amplitude of an external magnetic field, they were able to stabilize a gravitationally buckled magnetostrictive ribbon that oscillated chaotically in a magnetic field. They produced period-one and period-two behavior, and the procedure proved to be remarkably robust. Using both experimental and theoretical results, Singer, Wang, and Bau [9] applied a simple on–off strategy to laminarize (suppress) chaotic flow of a fluid in a thermal convection loop. The on–off controller was applied to the Lorenz equations, and the numerical results were in good agreement with the experimental results. Shortly afterwards, Hunt [8] applied a modified version of the OGY algorithm called OPF to the chaotic dynamics of a nonlinear diode resonator. Small perturbations were used to stabilize orbits of low period, but larger perturbations were required to stabilize orbits of high periods. By changing the level, width, and gain of the feedback signal, Hunt was able to stabilize orbits with periods as high as 23. Using the OPF algorithm developed by Hunt, Roy, et al. [7] were able to stabilize a weakly chaotic green laser. In recent years, the implementation of the control algorithm has been carried out electronically using either digital signals or analog hardware. The hope for the future is that all-optical processors and feedback can be used in order to increase speed. The first experimental control in a biological system was performed by Garfinkel et al. [6] in 1992. They were able to stabilize arrhythmic behavior in eight of 11 rabbit hearts using a feedback-control mechanism. It has been reported in [3] that a company has been set up to manufacture small defibrillators that can monitor the heart and deliver tiny electrical pulses to move the heart away from fibrillation and back to normality. It was also conjectured in the same article that the chaotic heart is more healthy than a regularly beating periodic heart. The OGY algorithm was implemented theoretically by the author and Steele [2] to control the chaos within a hysteresis cycle of a nonlinear bistable optical resonator using the real and imaginary parts of the electrical field amplitude. The same authors have recently managed to control the chaos using feedback of the electric field. This quantity is easy to monitor and measure continuously and should lead to physical applications in the future.

Methods I–VI and results given above are by no means exhaustive. This section is intended to provide a brief introduction to the subject and to encourage further reading.

The OGY Method. Consider the n-dimensional map

(6.1) $$\mathbf{Z}_{n+1} = \mathbf{f}(\mathbf{Z}_n, p),$$

where p is some accessible system parameter that can be changed in a small neighborhood of its nominal value, say, p_0. In the case of continuous-time systems, such a map can be constructed by introducing a transversal surface of section and setting up a Poincaré map.

6.1. Historical Background

It is well known that a chaotic attractor is densely filled with unstable periodic orbits and that ergodicity guarantees that any small region on the chaotic attractor will be visited by a chaotic orbit. The OGY method hinges on the existence of stable manifolds around unstable periodic points. The basic idea is to make small time-dependent linear perturbations to the control parameter p in order to nudge the state towards the stable manifold of the desired fixed point. Note that this can only be achieved if the orbit is in a small neighborhood, or *control region*, of the fixed point.

Suppose that $\mathbf{Z}_S(p)$ is an unstable fixed point of equation (6.1). The position of this fixed point moves smoothly as the parameter p is varied. For values of p close to p_0 in a small neighborhood of $\mathbf{Z}_S(p_0)$, the map can be approximated by a linear map given by

$$\mathbf{Z}_{n+1} - \mathbf{Z}_S(p_0) = \mathbf{J}(\mathbf{Z}_n - \mathbf{Z}_S(p_0)) + \mathbf{C}(p - p_0), \tag{6.2}$$

where $\mathbf{J}$ is the Jacobian and $\mathbf{C} = \frac{\partial \mathbf{f}}{\partial p}$. All partial derivatives are evaluated at $\mathbf{Z}_S(p_0)$ and p_0.

Assume that in a small neighborhood around the fixed point

$$p - p_0 = -\mathbf{K}(\mathbf{Z}_n - \mathbf{Z}_S(p_0)), \tag{6.3}$$

where $\mathbf{K}$ is a constant vector of dimension n to be determined. Substitute (6.3) into (6.2) to obtain

$$\mathbf{Z}_{n+1} - \mathbf{Z}_S(p_0) = (\mathbf{J} - \mathbf{CK})(\mathbf{Z}_n - \mathbf{Z}_S(p_0)). \tag{6.4}$$

The fixed point is then stable as long as the eigenvalues, or *regulator poles*, have modulus less than unity. The pole-placement technique from control theory can be applied to find the vector $\mathbf{K}$. A specific example is given in Section 6.3.

A simple schematic diagram is given in Figure 6.1 to demonstrate the action of the OGY algorithm. Physically, one can think of a marble placed on a saddle. If the marble is rolled towards the center (where the fixed point lies) then it will roll off as depicted in Figure 6.1(a). If, however, the saddle is moved slightly from side to side, by applying small perturbations, then the marble can be made to balance at the center of the saddle, as depicted in Figure 6.1(b).

Some useful points to note:

- The OGY technique is a feedback-control method.

- If the equations are unknown, sometimes delay-coordinate embedding techniques using a single variable time series can be used (the map can be constructed from experimental data).

- There may be more than one control parameter available.

- Noise may affect the control algorithm. If the noise is relatively small the control algorithm will still work in general.

148   6. Controlling Chaos

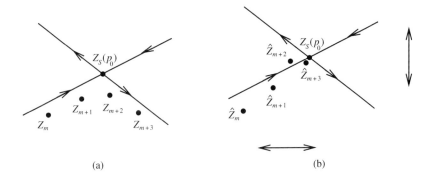

Figure 6.1: Possible iterations near the fixed point (a) without control and (b) with control. The double-ended arrows to represent small perturbations to the system dynamics. The iterates $\hat{Z}_j$ represent perturbed orbits.

It should also be pointed out that the OGY algorithm can only be applied once the orbit has entered a small control region around the fixed point. For certain nonlinear systems the number of iterations required—and hence the time—for the orbit to enter this control region may be too many to be practical. Shinbrot et al. [11] solved this problem by targeting trajectories to the desired control regions in only a small number of iterations. The method has also been successfully applied in physical systems.

6.2 Controlling Chaos in the Logistic Map

Consider the logistic map given by

(6.5) $$x_{n+1} = f_\mu(x_n) = \mu x_n(1 - x_n)$$

as introduced in Chapter 2. There are many methods available to control the chaos in this one-dimensional system, but the analysis is restricted to periodic proportional pulses in this section. For more details on the method and its application to the Hénon map the reader is directed to [4]. To control the chaos in this system, instantaneous pulses will be applied to the system variables x_n once every p iterations such that

$$x_i \to kx_i,$$

where k is a constant to be determined and p denotes the period.

Recall that a fixed point of period one, say, x_S, of equation (6.5) satisfies the equation

$$x_S = f_\mu(x_S),$$

6.2. Controlling Chaos in the Logistic Map

and this fixed point is stable if and only if

$$\left|\frac{df_\mu(x_S)}{dx}\right| < 1.$$

Define the composite function $F_\mu(x)$ by

$$F_\mu(x) = kf_\mu^p(x).$$

A fixed point of the function F_μ satisfies the equation

(6.6) $$kf_\mu^p(x_S) = x_S,$$

where the fixed point x_S is stable if

(6.7) $$\left|k\frac{df_\mu^p(x_S)}{dx}\right| < 1.$$

Define the function $C^p(x)$ by

$$C^p(x) = \frac{x}{f_\mu^p(x)}\frac{df_\mu^p(x_S)}{dx}.$$

Substituting from (6.6), equation (6.7) becomes

(6.8) $$|C^p(x_S)| < 1.$$

A fixed point of this composite map is a stable point of period p for the original logistic map when the control is switched on, providing condition (6.8) holds. In practice, chaos control always deals with periodic orbits of low periods, say, $p = 1$ to 4, and this method can be easily applied.

To illustrate the method, consider the logistic map when $\mu = 4$ and the system is chaotic. The functions $C^1(x)$, $C^2(x)$, $C^3(x)$, and $C^4(x)$ are shown in Figure 6.2.

Figure 6.2(a) shows that fixed points of period one can be stabilized for every x_S in the range between zero and approximately 0.67. When $p = 2$, Figure 6.2(b) shows that fixed points of period two can only be stabilized in three ranges of x_S values. Figures 6.2(c) and (d) indicate that there are seven and 14 acceptable ranges for fixed points of periods three and four, respectively. Notice that the control ranges are getting smaller and smaller as the periodicity increases.

Figure 6.3 shows time series data for specific examples when the chaos is controlled to period-one, period-two, period-three, and period-four behavior, respectively.

The values of x_S chosen in Figure 6.3 were derived from Figure 6.2. The values of k were calculated using equation (6.6). Note that the system can be

6. Controlling Chaos

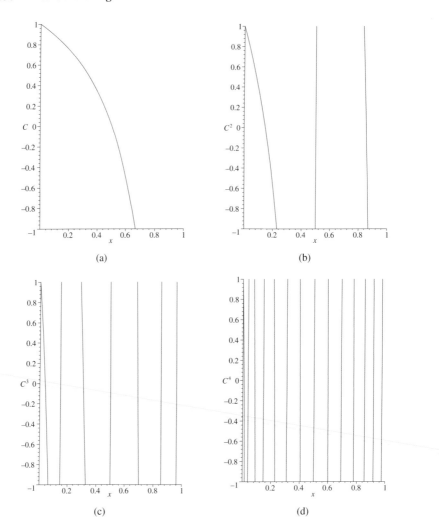

Figure 6.2: Control curves C^i, $i = 1, 2, 3, 4$, for the logistic map when $\mu = 4$. The range is restricted to $-1 < C^p(x_S) < 1$ in each case.

stabilized to many different points on and even off the chaotic attractor (see the work of Chau [4]). A MATLAB program is listed in Section 6.4.

This method of chaos control by periodic proportional pulses can also be applied to the two-dimensional discrete Hénon map. The interested reader is again directed to [4]. The OGY algorithm will be applied to the Hénon map in the next section.

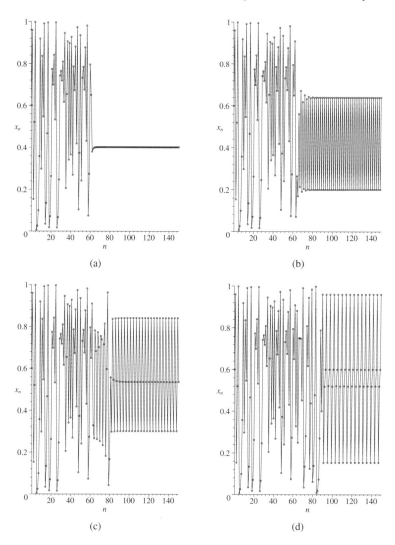

Figure 6.3: Stabilization of points of periods one, two, three, and four for the logistic map when $\mu = 4$; (a) $x_S = 0.4$, $k = 0.417$; (b) $x_S = 0.2$, $k = 0.217$; (c) $x_S = 0.3$, $k = 0.302$; and (d) $x_S = 0.6$, $k = 0.601$. In each case k is computed to three decimal places.

6.3 Controlling Chaos in the Hénon Map

Ott, Grebogi, and Yorke [13] used the Hénon map to illustrate the control method. A simple example will be given here. Consider the Hénon map as introduced in

6. Controlling Chaos

Chapter 2. The two-dimensional iterated map function is given by

(6.9) $$X_{n+1} = 1 + Y_n - \alpha X_n^2, \quad Y_{n+1} = \beta X_n,$$

where $\alpha > 0$ and $|\beta| < 1$. Take a transformation $X_n = \frac{1}{\alpha} x_n$ and $Y_n = \frac{\beta}{\alpha} y_n$, then system (6.9) becomes

(6.10) $$x_{n+1} = \alpha + \beta y_n - x_n^2, \quad y_{n+1} = x_n.$$

The proof that system (6.9) can be transformed into system (6.10) will be left to the reader in the exercises at the end of this chapter. The Hénon map is now in the form considered in [13], and the control algorithm given in Section 6.1 will now be applied to this map. Set $\beta = 0.4$ and allow the control parameter, in this case α, to vary around a nominal value, say, $\alpha_0 = 1.2$, for which the map has a chaotic attractor.

The fixed points of period one are determined by solving the simultaneous equations

$$\alpha_0 + \beta y - x^2 - x = 0 \quad \text{and} \quad x - y = 0.$$

In Chapter 2, it was shown that the Hénon map has two fixed points of period one if and only if $(1 - \beta)^2 + 4\alpha_0 > 0$. In this particular case, the fixed points of period one are located approximately at $A = (x_{1,1}, y_{1,1}) = (0.8358, 0.8358)$ and $B = (x_{1,2}, y_{1,2}) = (-1.4358, -1.4358)$. The chaotic attractor and points of period one are shown in Figure 6.4.

The Jacobian matrix of partial derivatives of the map is given by

$$J = \begin{pmatrix} \frac{\partial P}{\partial x} & \frac{\partial P}{\partial y} \\ \frac{\partial Q}{\partial x} & \frac{\partial Q}{\partial y} \end{pmatrix},$$

where $P(x, y) = \alpha_0 + \beta y - x^2$ and $Q(x, y) = x$. Thus

$$J = \begin{pmatrix} -2x & \beta \\ 1 & 0 \end{pmatrix}.$$

Consider the fixed point at A; the fixed point is a saddle point. Using the notation introduced in Section 6.1, for values of α close to α_0 in a small neighborhood of A, the map can be approximated by a linear map

(6.11) $$\mathbf{Z}_{n+1} - \mathbf{Z}_S(\alpha_0) = \mathbf{J}(\mathbf{Z}_n - \mathbf{Z}_S(\alpha_0)) + \mathbf{C}(\alpha - \alpha_0),$$

where $\mathbf{Z}_n = (x_n, y_n)^T$, $A = \mathbf{Z}_S(\alpha_0)$, $\mathbf{J}$ is the Jacobian, and

$$\mathbf{C} = \begin{pmatrix} \frac{\partial P}{\partial \alpha} \\ \frac{\partial Q}{\partial \alpha} \end{pmatrix},$$

6.3. Controlling Chaos in the Hénon Map

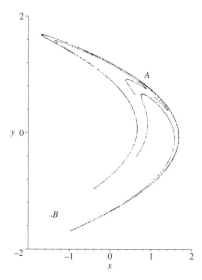

Figure 6.4: Iterative plot for the Hénon map (3000 iterations) when $\alpha_0 = 1.2$ and $\beta = 0.4$. The two fixed points of period one are labeled A and B.

and all partial derivatives are evaluated at α_0 and $\mathbf{Z}_S(\alpha_0)$. Assume in a small neighborhood of A,

$$(6.12) \qquad \alpha - \alpha_0 = -\mathbf{K}(\mathbf{Z}_n - \mathbf{Z}_S(\alpha_0)),$$

where

$$\mathbf{K} = \begin{pmatrix} k_1 \\ k_2 \end{pmatrix}.$$

Substitute (6.12) into (6.11) to obtain

$$\mathbf{Z}_{n+1} - \mathbf{Z}_S(\alpha_0) = (\mathbf{J} - \mathbf{CK})(\mathbf{Z}_n - \mathbf{Z}_S(\alpha_0)).$$

Therefore, the fixed point at $A = \mathbf{Z}_S(\alpha_0)$ is stable if the matrix $\mathbf{J} - \mathbf{CK}$ has eigenvalues (or regulator poles) with modulus less than unity. In this particular case,

$$\mathbf{J} - \mathbf{CK} \approx \begin{pmatrix} -1.671563338 - k_1 & 0.4 - k_2 \\ 1 & 0 \end{pmatrix},$$

and the characteristic polynomial is given by

$$\lambda^2 + \lambda(1.671563338 + k_1) + (k_2 - 0.4) = 0.$$

Suppose that the eigenvalues (regulator poles) are given by λ_1 and λ_2; then

$$\lambda_1 \lambda_2 = k_2 - 0.4 \quad \text{and} \quad -(\lambda_1 + \lambda_2) = 1.671563338 + k_1.$$

6. Controlling Chaos

The lines of marginal stability are determined by solving the equations $\lambda_1 = \pm 1$ and $\lambda_1 \lambda_2 = 1$. These conditions guarantee that the eigenvalues λ_1 and λ_2 have modulus less than unity. Suppose that $\lambda_1 \lambda_2 = 1$. Then

$$k_2 = 1.4.$$

Suppose that $\lambda_1 = +1$. Then

$$\lambda_2 = k_2 - 0.4 \quad \text{and} \quad \lambda_2 = -2.671563338 - k_1.$$

Therefore,

$$k_2 = -k_1 - 2.271563338.$$

If $\lambda_1 = -1$, then

$$\lambda_2 = -(k_2 - 0.4) \quad \text{and} \quad \lambda_2 = -0.671563338 - k_1.$$

Therefore,

$$k_2 = k_1 + 1.071563338.$$

The stable eigenvalues (regulator poles) lie within a triangular region as depicted in Figure 6.5.

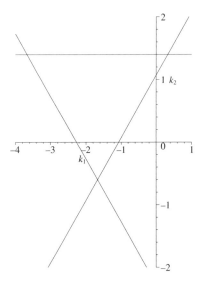

Figure 6.5: The bounded region where the regulator poles are stable.

Select $k_1 = -1.5$ and $k_2 = 0.5$. This point lies well inside the triangular region as depicted in Figure 6.5. The perturbed Hénon map becomes

(6.13)
$$x_{n+1} = (-k_1(x_n - x_{1,1}) - k_2(y_n - y_{1,1}) + \alpha_0) + \beta y_n - x_n^2, \qquad y_{n+1} = x_n.$$

Applying equations (6.10) and (6.13) without and with control, respectively, it is possible to plot time series data for these maps. Figure 6.6(a) shows a time series plot when the control is switched on after the 200th iterate; the control is left switched on until the 500th iterate. In Figure 6.6(b), the control is switched on after the 200th iterate and then switched off after the 300th iterate. Remember to check that the point is in the control region before switching on the control.

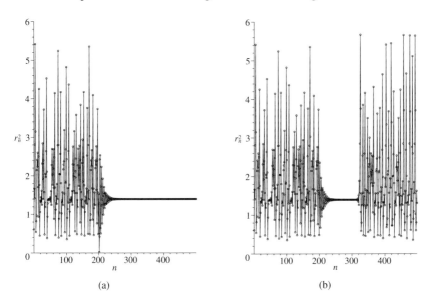

Figure 6.6: Time series data for the Hénon map with and without control, $r^2 = x^2 + y^2$. In case (a), the control is activated after the 200th iterate, and in case (b), the control is switched off after the 300th iterate.

Once again, the MATLAB program is listed in Section 6.4.

6.4 MATLAB Commands

```
% Program_6a - Chaos Control in the Logistic Map (Figure
% 6.3(b)).
```

6. Controlling Chaos

```
clear
itermax=200;
mu=4;k=0.217;
x(1)=0.6;
for n=1:2:itermax
    x(n+1)=mu*x(n)*(1-x(n));
    x(n+2)=mu*x(n+1)*(1-x(n+1));
    if n>60                          % Switch on the control
                                     % when n>60.
        x(n+1)=k*mu*x(n)*(1-x(n));   % Nudge the system every
                                     % second iterate.
        x(n+2)=mu*x(n+1)*(1-x(n+1));
    end
end
hold on
    plot(1:itermax,x(1:itermax))
    plot(1:itermax,x(1:itermax),'o')
fsize=15;
set(gca,'xtick',[0:50:itermax],'FontSize',fsize)
set(gca,'ytick',[0,1],'FontSize',fsize)
xlabel('n','FontSize',fsize)
ylabel('\itx_n','FontSize',fsize)
hold off
% End of Program_6a.
```

```
% Program_6b - Chaos Control in the Henon Map (Figure
% 6.6(a)).

clear
alpha=1.2;beta=0.4;
k1=-1.8;k2=1.2;                 % The regulator poles.
xstar=0.8358;ystar=xstar;       % The point to be stabilized.
N=400;
x(1)=0.5;y(1)=0.6;              % Initial values.
rsqr(1)=(x(1))^2+(y(1))^2;
for n=1:N
    if n>198                    % Check point is in control
                                % region.
        x(n+1)=(-k1*(x(n)-xstar)-k2*(y(n)-ystar)+alpha)
               +beta*y(n)-(x(n))^2;
        y(n+1)=x(n);
    else
        x(n+1)=alpha+beta*y(n)-(x(n))^2;
        y(n+1)=x(n);
    end
    rsqr(n+1)=(x(n+1))^2+(y(n+1))^2;
end
```

```
hold on
axis([0 N 0 6])
plot(1:N,rsqr(1:N))
plot(1:N,rsqr(1:N),'o')
fsize=15;
set(gca,'xtick',[0:50:N],'FontSize',fsize)
set(gca,'ytick',[0,6],'FontSize',fsize)
xlabel('n','FontSize',fsize)
ylabel('\it{r^2}','FontSize',fsize)
hold off
% End of Program_6b.
```

6.5 Exercises

1. Show that the map defined by

$$x_{n+1} = 1 + y_n - ax_n^2, \quad y_{n+1} = bx_n$$

can be written as

$$u_{n+1} = a + bv_n - u_n^2, \quad v_{n+1} = u_n$$

using a suitable transformation.

2. Apply the method of chaos control by periodic proportional pulses (see Section 6.2) to the logistic map

$$x_{n+1} = \mu x_n(1 - x_n)$$

when $\mu = 3.9$. Sketch the graphs $C^i(x), i = 1$ to 4. Plot time series data to illustrate control of fixed points of periods one, two, three, and four.

3. Find the points of periods one and two for the Hénon map given by

$$x_{n+1} = a + by_n - x_n^2, \quad y_{n+1} = x_n$$

when $a = 1.4$ and $b = 0.4$, and determine their type.

4. Apply the method of chaos control by periodic proportional pulses (see Section 6.2) to the two-dimensional Hénon map

$$x_{n+1} = a + by_n - x_n^2, \quad y_{n+1} = x_n$$

when $a = 1.4$ and $b = 0.4$. (In this case, you must multiply x_m by k_1 and y_m by k_2 once every p iterations.) Plot time series data to illustrate the control of points of periods one, two, and three.

6. Controlling Chaos

5. Use the OGY algorithm given in Section 6.3 to stabilize a point of period one in the Hénon map

$$x_{n+1} = a + by_n - x_n^2, \quad y_{n+1} = x_n$$

when $a = 1.4$ and $b = 0.4$. Display the control using a time series graph.

6. Consider the Ikeda map, introduced in Chapter 4, given by

$$E_{n+1} = A + BE_n e^{i|E_n|^2}.$$

Suppose that $E_n = x_n + iy_n$ and rewrite the Ikeda map as a two-dimensional map in x_n and y_n.

7. Plot the chaotic attractor for the Ikeda map

$$E_{n+1} = A + BE_n e^{i|E_n|^2}$$

when $A = 2.7$ and $B = 0.15$. How many points are there of period one? Indicate where these points are with respect to the attractor.

8. Plot the chaotic attractor for the Ikeda map

$$E_{n+1} = A + BE_n e^{i|E_n|^2}$$

when

(i) $A = 4$ and $B = 0.15$;

(ii) $A = 7$ and $B = 0.15$.

How many points are there of period one in each case? Indicate where these points are for each of the attractors on the figures.

9. Use the OGY method (see Section 6.3) with the parameter A to control the chaos to a point of period one in the Ikeda map

$$E_{n+1} = A + BE_n e^{i|E_n|^2}$$

when $A_0 = 2.7$ and $B = 0.15$. Display the control on a time series plot. (N.B.: Use a two-dimensional map.)

10. Try the same procedure of control to period one for the Ikeda map as in Exercise 9. but with the parameters $A_0 = 7$ and $B = 0.15$. Investigate the size of the control region around one of the fixed points in this case and state how it compares to the control region in Exercise 9. What can you say about flexibility and controllability?

Recommended Reading

[1] T. Kapitaniak, *Chaos for Engineers: Theory, Applications and Control*, 2nd ed., Springer-Verlag, New York, 2000.

[2] S. Lynch and A. L. Steele, Controlling chaos in nonlinear bistable optical resonators, *Chaos Solitons Fractals*, **11**-5 (2000), 721–728.

[3] M. Buchanan, Fascinating rhythm, *New Scientist*, 3 January (1998), 20–25.

[4] N. P. Chau, Controlling chaos by periodic proportional pulses, *Phys. Lett. A*, **234** (1997), 193–197.

[5] T. Kapitaniak, *Controlling Chaos: Theoretical and Practical Methods in Non-linear Dynamics*, Academic Press, New York, 1996.

[6] A. Garfinkel, M. L. Spano, W. L. Ditto, and J. N. Weiss, Controlling cardiac chaos, *Science*, **257** (1992), 1230–1235.

[7] R. Roy, T. W. Murphy, T. D. Maier, Z. Gills, and E. R. Hunt, Dynamical control of a chaotic laser: Experimental stabilization of a globally coupled system, *Phys. Rev. Lett.*, **68** (1992), 1259–1262.

[8] E. R. Hunt, Stabilizing high-period orbits in a chaotic system: The diode resonator, *Phys. Rev. Lett.*, **67** (1991), 1953–1955.

[9] J. Singer, Y.-Z. Wang, and H. H. Bau, Controlling a chaotic system, *Phys. Rev. Lett.*, **66** (1991), 1123–1125.

[10] W. L. Ditto, S. N. Rausseo, and M. L. Spano, Experimental control of chaos. *Phys. Rev. Lett.*, **65** (1990), 3211–3214.

[11] T. Shinbrot, C. Grebogi, E. Ott, and J. A. Yorke, Using chaos to direct trajectories to targets, *Phys. Rev. Lett.*, **65** (1990), 3215–3218.

[12] L. M. Pecora and T. L. Carroll, Synchronization in chaotic systems, *Phys. Rev. Lett.*, **64** (1990), 821–824.

[13] E. Ott, C. Grebogi, and J. A. Yorke, Controlling chaos, *Phys. Rev. Lett.*, **64** (1990), 1196–1199.

7
Differential Equations

Aims and Objectives

- To review basic methods for solving some differential equations.
- To apply the theory to simple mathematical models.
- To introduce an existence and uniqueness theorem.

On completion of this chapter the reader should be able to

- solve certain first- and second-order differential equations;
- apply the theory to chemical kinetics and electric circuits;
- interpret the solutions in physical terms;
- understand the existence and uniqueness theorem and its implications.

Basic theory of ODEs and analytical methods for solving some types of ODEs are reviewed. This chapter is not intended to be a comprehensive study on differential equations, but more an introduction to the theory that will be used in later chapters. Most of the material will be covered in first- and second-year undergraduate mathematics courses. The differential equations are applied to all kinds of models, but this chapter concentrates on chemical kinetics and electric circuits in particular.

The chapter ends with the existence and uniqueness theorem and some analysis.

7.1 Simple Differential Equations and Applications

Definition 1. A differential equation that involves only one independent variable is called an *ordinary differential equation* (ODE). Those involving two or more independent variables are called *partial differential equations*. This chapter will be concerned with ODEs only.

The subject of ODEs encompasses analytical, computational, and applicable fields of interest. There are many textbooks written from the elementary to the most advanced, with some focusing on applications and others concentrating on existence theorems and rigorous methods of solution. This chapter is intended to introduce the reader to both branches of the subject.

Separable Differential Equations. Consider the differential equation

$$(7.1) \qquad \frac{dx}{dt} = f(t, x)$$

and suppose that the function $f(t, x)$ can be factored into a product $f(t, x) = g(t)h(x)$, where $g(t)$ is a function of t and $h(x)$ is a function of x. If f can be factored in this way, then equation (7.1) can be solved by the method of *separation of variables*.

To solve the equation, divide both sides by $h(x)$ to obtain

$$\frac{1}{h(x)} \frac{dx}{dt} = g(t);$$

and integration with respect to t gives

$$\int \frac{1}{h(x)} \frac{dx}{dt} dt = \int g(t) dt.$$

Changing the variables in the integral gives

$$\int \frac{dx}{h(x)} = \int g(t) dt.$$

An analytic solution to (7.1) can be found only if both integrals can be evaluated. The method can be illustrated with some simple examples.

Example 1. Solve the differential equation $\frac{dx}{dt} = -\frac{t}{x}$.

Solution. The differential equation is separable. Separate the variables and integrate both sides with respect to t. Therefore,

$$\int x \frac{dx}{dt} dt = -\int t \, dt,$$

and so

$$\int x\,dx = -\int t\,dt.$$

Integration of both sides yields

$$t^2 + x^2 = r^2,$$

where r^2 is a constant. There are an infinite number of solutions. The *solution curves* are concentric circles of radius r centered at the origin. There are an infinite number of solution curves that would fill the plane if they were all plotted. Three such solution curves are plotted in Figure 7.1.

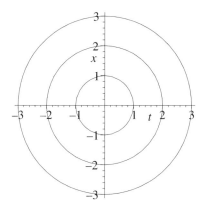

Figure 7.1: Three solution curves for Example 1.

Example 2. Solve the differential equation $\frac{dx}{dt} = -\frac{t}{x^2}$.

Solution. The differential equation is separable. Separate the variables and integrate both sides with respect to t to give

$$\int x^2\,dx = \int t\,dt.$$

Integration of both sides yields

$$\frac{x^3}{3} = \frac{t^2}{2} + C,$$

where C is a constant. Six of an infinite number of solution curves are plotted in Figure 7.2.

7. Differential Equations

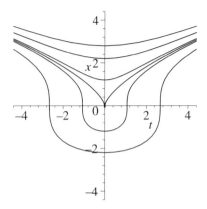

Figure 7.2: Six solution curves for Example 2.

Example 3. The population of a certain species of fish living in a large lake at time t can be modeled using *Verhulst's equation*, otherwise known as the *logistic equation*,

$$\frac{dP}{dt} = P(\beta - \delta P),$$

where $P(t)$ is the population of fish measured in tens of thousands, and β and δ are constants representing the birth and death rates of the fish living in the lake, respectively. Suppose that $\beta = 0.1, \delta = 10^{-3}$, and the initial population is 50×10^4. Solve this *initial value problem* and interpret the results in physical terms.

Solution. Using the method of separation of variables gives

$$\int \frac{dP}{P(\beta - \delta P)} = \int dt.$$

The solution to the integral on the left may be determined using partial fractions. The general solution is

$$\ln \left| \frac{P}{\beta - \delta P} \right| = \beta t + C,$$

or

$$P(t) = \frac{\beta}{\delta + k\beta e^{-\beta t}},$$

computed using MATLAB, where C and k are constants. Substituting the initial conditions, the solution is

$$P(t) = \frac{100}{1 + e^{-0.1t}}.$$

7.1. Simple Differential Equations and Applications

Thus as time increases, the population of fish tends to a value of 100×10^4. The solution curve is plotted in Figure 7.3.

Figure 7.3: The solution curve for the initial value problem in Example 3.

Note the following:

- The quantity $\frac{\beta}{\delta}$ is the ratio of births to deaths and is called the *carrying capacity* of the environment.

- Take care when interpreting the solutions. This and similar continuous models only work for large species populations. The solutions give approximate numbers. Even though time is continuous, the population size is not. For example, you cannot have a fractional living fish, so population sizes have to be rounded out to whole numbers in applications.

- Discrete models can also be applied to population dynamics (see Chapter 1).

Exact Differential Equations. A differential equation of the form

$$(7.2) \qquad M(t, x) + N(t, x)\frac{dx}{dt} = 0$$

is said to be *exact* if there exists a function, say, $F(t, x)$, with continuous second partial derivatives such that

$$\frac{\partial F}{\partial t} = M(t, x) \quad \text{and} \quad \frac{\partial F}{\partial x} = N(t, x).$$

7. Differential Equations

Such a function exists as long as

$$\frac{\partial M}{\partial x} = \frac{\partial N}{\partial t},$$

and then the solution to (7.2) satisfies the equation

$$F(t, x) = C,$$

where C is a constant. Differentiate this equation with respect to t to obtain (7.2).

Example 4. Solve the differential equation

$$\frac{dx}{dt} = \frac{9 - 12t - 5x}{5t + 2x - 4}.$$

Solution. In this case, $M(t, x) = -9 + 12t + 5x$ and $N(t, x) = 5t + 2x - 4$. Now

$$\frac{\partial M}{\partial x} = \frac{\partial N}{\partial t} = 5$$

and integration gives the solution $F(t, x) = x^2 + 6t^2 + 5tx - 9t - 4x = C$. There are an infinite number of solution curves, some of which are shown in Figure 7.4.

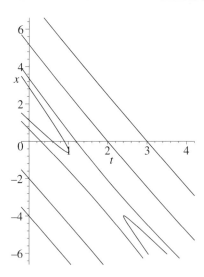

Figure 7.4: Some solution curves for Example 4.

7.1. Simple Differential Equations and Applications

Homogeneous Differential Equations. Consider differential equations of the form

(7.3) $$\frac{dx}{dt} = f\left(\frac{x}{t}\right).$$

Substitute $v = \frac{x}{t}$ into (7.3) to obtain

$$\frac{d}{dt}(vt) = f(v).$$

Therefore,

$$v + t\frac{dv}{dt} = f(v),$$

and so

$$\frac{dv}{dt} = \frac{f(v) - v}{t},$$

which is separable. A complete solution can be found as long as the equations are integrable, and then v may be replaced with $\frac{x}{t}$.

Example 5. Solve the differential equation

$$\frac{dx}{dt} = \frac{t-x}{t+x}.$$

Solution. The equation may be rewritten as

(7.4) $$\frac{dx}{dt} = \frac{1-\frac{x}{t}}{1+\frac{x}{t}}.$$

Let $v = \frac{x}{t}$. Then (7.4) becomes

$$\frac{dv}{dt} = \frac{1 - 2v - v^2}{t(1+v)}.$$

This is a separable differential equation. The general solution is given by

$$x^2 + 2tx - t^2 = C,$$

where C is a constant. Some solution curves are plotted in Figure 7.5.

7. Differential Equations

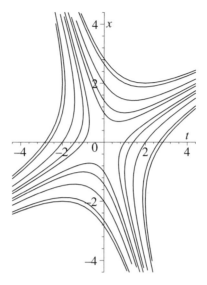

Figure 7.5: Some solution curves for Example 5.

Linear Differential Equations. Consider differential equations of the form

(7.5) $$\frac{dx}{dt} + P(t)x = Q(t).$$

Multiplying through by an integrating factor, say, $J(t)$, equation (7.5) becomes

(7.6) $$J\frac{dx}{dt} + JPx = JQ.$$

Find J such that (7.6) can be written as

$$\frac{d}{dt}(Jx) = J\frac{dx}{dt} + x\frac{dJ}{dt} = JQ.$$

In order to achieve this, set

$$\frac{dJ}{dt} = JP$$

and integrate to get

$$J(t) = \exp\left(\int P(t)dt\right).$$

Thus the solution to system (7.5) may be found by solving the differential equation

$$\frac{d}{dt}(Jx) = JQ,$$

as long as the right-hand side is integrable.

Example 6. A chemical company pumps v liters of solution containing mass m grams of solute into a large lake of volume V per day. The inflow and outflow of the water is constant. The concentration of solute in the lake, say, σ, satisfies the differential equation

(7.7) $$\frac{d\sigma}{dt} + \frac{v}{V}\sigma = \frac{m}{V}.$$

Determine the concentration of solute in the lake at time t assuming that $\sigma = 0$ when $t = 0$. What happens to the concentration in the long term?

Solution. This is a linear differential equation, and the integrating factor is given by

$$J = \exp\left(\int \frac{v}{V} dt\right) = e^{\frac{vt}{V}}.$$

Multiply (7.7) by the integrating factor to obtain

$$\frac{d}{dt}(e^{\frac{vt}{V}}\sigma) = e^{\frac{vt}{V}}\frac{m}{V}.$$

Integration gives

$$\sigma(t) = \frac{m}{v} - ke^{-\frac{vt}{V}},$$

where k is a constant. Substituting the initial conditions, the final solution is

$$\sigma(t) = \frac{m}{v}(1 - e^{-\frac{vt}{V}}).$$

As $t \to \infty$, the concentration settles to $\frac{m}{v}\text{gl}^{-1}$.

7.2 Applications to Chemical Kinetics

Even the simplest chemical reactions can be highly complex and difficult to model. Physical parameters such as temperature, pressure, and mixing, for example, are ignored in this text, and differential equations are constructed that are dependent only on the concentrations of the chemicals involved in the reaction. This is potentially a very difficult subject and some assumptions have to be made to make progress.

The Chemical Law of Mass Action. The rates at which the concentrations of the various chemical species change with time are proportional to their concentrations.

Consider the following example, where one molecule of hydrogen reacts with one molecule of oxygen to produce two molecules of hydroxyl (OH):

$$H_2 + O_2 \to 2OH.$$

7. Differential Equations

Suppose that the concentration of hydrogen is [H$_2$] and the concentration of oxygen is [O$_2$]. Then from the chemical law of mass action, the rate equation is given by

$$\text{Rate} = k[\text{H}_2][\text{O}_2],$$

where k is called the *rate constant*, and the reaction rate equation is

$$\frac{d[\text{OH}]}{dt} = 2k[\text{H}_2][\text{O}_2].$$

Unfortunately, it is not possible to write down the reaction rate equations based on the stoichiometric (balanced) chemical equations alone. There may be many mechanisms involved in producing OH from hydrogen and oxygen in the above example. Even simple chemical equations can involve a large number of steps and different rate constants. Suppose in this text that the chemical equations give the *rate-determining steps*.

Suppose that species A, B, C, and D have concentrations $a(t)$, $b(t)$, $c(t)$, and $d(t)$ at time t and initial concentrations a_0, b_0, c_0, and d_0, respectively. Table 7.1 lists some reversible chemical reactions and one of the corresponding reaction rate equations, where k_f and k_r are the forward and reverse rate constants, respectively.

Table 7.1: One of the possible reaction rate equations for each chemical reaction.

Chemical reaction	The reaction rate equation for one species may be expressed as follows:
A+B → C	$\dot{c} = k_f ab = k_f(a_0 - c)(b_0 - c)$
2A ⇌ B	$\dot{b} = k_f(a_0 - 2b)^2 - k_r b$
A ⇌ 2B	$\dot{b} = k_f\left(a_0 - \frac{b}{2}\right) - k_r b^2$
A ⇌ B+C	$\dot{c} = k_f(a_0 - c) - k_r(b_0 + c)(c_0 + c)$
A+B ⇌ C	$\dot{c} = k_f(a_0 - c)(b_0 - c) - k_r c$
A+B ⇌ C+D	$\dot{c} = k_f(a_0 - c)(b_0 - c) - k_r(c_0 + c)(d_0 + c)$

Example 7. A reaction equation for sulphate and hydrogen ions to form bisulphite ions is given by

$$\text{SO}_3^{2-} + \text{H}^+ \rightleftharpoons \text{HSO}_3^-,$$

where k_f and k_r are the forward and reverse rate constants, respectively. Denote the concentrations by $a = [\text{SO}_3^{2-}]$, $b = [\text{H}^+]$, and $c = [\text{HSO}_3^-]$, and let the initial concentrations be a_0, b_0, and c_0. Assume that there is much more of species H$^+$ than the other two species, so that its concentration b can be regarded as constant. The reaction rate equation for $c(t)$ is given by

$$\frac{dc}{dt} = k_f(a_0 - c)b - k_r(c_0 + c).$$

Find a general solution for $c(t)$.

Solution. The differential equation is separable and

$$\int \frac{dc}{k_f(a_0 - c)b - k_r(c_0 + c)} = \int dt.$$

Integration yields

$$c(t) = \frac{k_f a_0 b - k_r c_0}{k_f b + k_r} - \frac{k_r c_0}{k_f b + k_r} + A e^{(-k_f a_0 - k_r)t},$$

where A is a constant.

Example 8. The chemical equation for the reaction between nitrous oxide and oxygen to form nitrogen dioxide at 25°C,

$$2\mathrm{NO} + \mathrm{O}_2 \to 2\mathrm{NO}_2,$$

obeys the law of mass action. The rate equation is given by

$$\frac{dc}{dt} = k(a_0 - c)^2 \left(b_0 - \frac{c}{2}\right),$$

where $c = [\mathrm{NO}_2]$ is the concentration of nitrogen dioxide, k is the rate constant, a_0 is the initial concentration of NO, and b_0 is the initial concentration of O_2. Find the concentration of nitrogen dioxide after time t given that $k = 0.00713\mathrm{l}^2\mathrm{M}^{-2}\mathrm{s}^{-1}$, $a_0 = 4\mathrm{Ml}^{-1}$, $b_0 = 1\mathrm{Ml}^{-1}$, and $c(0) = 0\mathrm{Ml}^{-1}$.

Solution. The differential equation is separable and

$$\int \frac{dc}{(4 - c)^2(1 - c/2)} = \int k\, dt.$$

Integrating using partial fractions gives

$$kt = \frac{1}{c - 4} + \frac{1}{2}\ln|c - 4| - \frac{1}{2}\ln|c - 2| + \frac{1}{4} - \frac{1}{2}\ln 2.$$

It is not possible to obtain $c(t)$ explicitly, so numerical methods are employed using MATLAB. The concentration of nitrogen dioxide levels off at two moles per liter as time increases, as depicted in Figure 7.6.

Chemical reactions displaying periodic behavior will be discussed in Chapter 13. There may be a wide range of time scales involved in chemical reactions and this can lead to *stiff* systems. Loosely speaking, a stiff system of differential equations is one in which the velocity or magnitude of the vector field changes rapidly in phase space. An example is presented in Section 13.6.

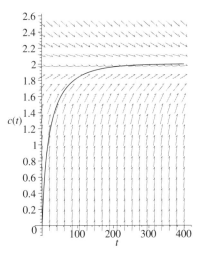

Figure 7.6: The concentration of NO_2 in moles per liter against time in seconds.

7.3 Applications to Electric Circuits

For many years, differential equations have been applied to model simple electrical and electronic circuits. If an oscilloscope is connected to the circuit, then the results from the analysis can be seen to match very well with what happens physically. As a simple introduction to electric circuits, linear systems will be considered in this chapter and the basic definitions and theory will be introduced.

Current and Voltage. The *current* I flowing through a conductor is proportional to the number of positive charge carriers that pass a given point per second. The unit of current is the *ampere* A. A *coulomb* is defined to be the amount of charge that flows through a cross-section of wire in 1 second when a current of $1A$ is flowing, so 1 amp is 1 coulomb per second. As the current passes through a circuit element, the charge carriers exchange energy with the circuit elements, and there is a *voltage drop* or *potential difference* measured in joules per coulomb, or *volts* V.

Consider simple electric circuits consisting of voltage sources, resistors, inductors, and capacitors, or *RLC* circuits. A series *RLC* circuit is shown schematically in Figure 7.7. The voltage drop across a resistor and the current flowing through it are related by Ohm's law.

Ohm's Law. The voltage drop V across a resistor is proportional to the current I flowing through it:

$$V = IR,$$

where R is the *resistance* of the resistor measured in ohms (Ω).

7.3. Applications to Electric Circuits

Figure 7.7: Schematic of a simple *RLC* series circuit.

A changing electric current can create a changing magnetic field that induces a voltage drop across a circuit element, such as a coil.

Faraday's Law. The voltage drop across an inductor is proportional to the rate of change of the current:

$$V = L\frac{dI}{dt},$$

where L is the *inductance* of the inductor measured in henries (H).

A capacitor consists of two plates insulated by some medium. When connected to a voltage source, charges of opposite sign build up on the two plates, and the total charge on the capacitor is given by

$$q(t) = q_0 + \int_{t_0}^{t} I(s)ds,$$

where q_0 is the initial charge.

Coulomb's Law. The voltage drop across a capacitor is proportional to the charge on the capacitor:

$$V(t) = \frac{1}{C}q(t) = \frac{1}{C}\left(q_0 + \int_{t_0}^{t} I(s)ds\right),$$

where C is the *capacitance* of the capacitor measured in farads (F).

The physical laws governing electric circuits were derived by G. R. Kirchhoff in 1859.

Kirchhoff's Current Law. The algebraic sum of the currents flowing into any junction of an electric circuit must be zero.

7. Differential Equations

Kirchhoff's Voltage Law. The algebraic sum of the voltage drops around any closed loop in an electric circuit must be zero.

Applying Kirchhoff's voltage law to the RLC circuit gives

$$V_L + V_R + V_C = E(t),$$

where V_R, V_L, and V_C are the voltage drops across R, L, and C, respectively, and $E(t)$ is the voltage source, or applied electromotive force (EMF). Substituting for the voltages across the circuit components gives

$$L\frac{dI}{dt} + RI + \frac{1}{C}q = E(t).$$

Since the current is the instantaneous rate of change in charge, $I = \frac{dq}{dt}$, this equation becomes

(7.8) $$L\frac{d^2q}{dt^2} + R\frac{dq}{dt} + \frac{1}{C}q = E(t).$$

This differential equation is called a *linear second-order differential equation*. This equation can be solved by the method of *Laplace transforms* [5]; there are other methods available, and readers should use whichever method they feel most comfortable with. The method of Laplace transforms can be broken down into four distinct steps when finding the solution of a differential equation:

- rewrite equation (7.8) in terms of Laplace transforms;
- insert any given initial conditions;
- rearrange the equation to give the transform of the solution;
- find the inverse transform.

The method is illustrated in the following examples.

Example 9. Consider a series resistor–inductor electrical circuit. Kirchhoff's voltage law gives

$$L\frac{dI}{dt} + RI = E.$$

Given that $L = 10H$, $R = 2\Omega$, and $E = 50\sin(t)V$, find an expression for the current in the circuit if $I(0) = 0$.

Solution. Take Laplace transforms of both sides. Then

$$10(s\bar{I} - I(0)) + 2\bar{I} = \frac{50}{s^2 + 1}.$$

7.3. Applications to Electric Circuits

Inserting the initial condition and rearranging,

$$\bar{I}(5s+1) = \frac{25}{s^2+1},$$

and splitting into partial fractions,

$$\bar{I} = \frac{25}{26}\frac{1}{s^2+1} - \frac{125}{26}\frac{s}{s^2+1} - \frac{125}{126}\frac{1}{(s-1/5)}.$$

Take inverse Laplace transforms to give

$$I(t) = \frac{25}{26}\sin(t) - \frac{125}{26}\cos(t) - \frac{125}{126}e^{-\frac{t}{5}}.$$

The periodic expression $\frac{25}{26}\sin(t) - \frac{125}{26}\cos(t)$ is called the *steady state*, and the term $\frac{125}{126}e^{-\frac{t}{5}}$ is called the *transient*. Note that the transient decays to zero as $t \to \infty$.

Example 10. Differentiate equation (7.8) with respect to time and substitute for $\frac{dq}{dt}$ to obtain

$$L\frac{d^2I}{dt^2} + R\frac{dI}{dt} + \frac{1}{C}I = \frac{dE}{dt}.$$

The second-order differential equation for a certain RLC circuit is given by

$$\frac{d^2I}{dt^2} + 5\frac{dI}{dt} + 6I = 10\sin(t).$$

Solve this differential equation given that $I(0) = \dot{I}(0) = 0$ (a *passive circuit*).

Solution. Take Laplace transforms of both sides:

$$(s^2\bar{I} - sI(0) - \dot{I}(0)) + 5(s\bar{I} - I(0)) + 6\bar{I} = \frac{10}{s^2+1}.$$

Substitute the initial conditions to obtain

$$\bar{I}(s^2 + 5s + 6) = \frac{10}{s^2+1}.$$

Splitting into partial fractions gives

$$\bar{I} = \frac{2}{s+2} - \frac{1}{s+3} + \frac{1}{s^2+1} - \frac{s}{s^2+1}.$$

Take inverse transforms to get

$$I(t) = 2e^{-2t} - e^{-3t} + \sin(t) - \cos(t).$$

176   7. Differential Equations

7.4 Existence and Uniqueness Theorem

Definition 2. A function **f(x)** with $\mathbf{f} : \Re^n \to \Re^n$ is said to satisfy a *Lipschitz condition* in a domain $D \subset \Re^n$ if there exists a constant, say, L, such that

$$\| \mathbf{f}(\mathbf{x}_1) - \mathbf{f}(\mathbf{x}_2) \| \leq L \| \mathbf{x}_1 - \mathbf{x}_2 \|,$$

where $\mathbf{x}_1, \mathbf{x}_2 \in D$.

If the function **f** satisfies the Lipschitz condition, then it is said to be *Lipschitz continuous*. Note that Lipschitz continuity in **x** implies continuity in **x**, but the converse is not always true.

Existence and Uniqueness Theorem. *Suppose that* **f** *is continuously Lipschitz; then for an initial point* $\mathbf{x}_0 \in D$, *the autonomous differential equation*

(7.9) $$\frac{d\mathbf{x}}{dt} = \dot{\mathbf{x}} = \mathbf{f}(\mathbf{x})$$

has a unique solution, say, $\phi_t(\mathbf{x}_0)$, *that is defined on the maximal interval of existence.*

Note that (7.9) is called autonomous as long as **f** is independent of t. The proof of this theorem can be found in most textbooks that specialize in the theory of ODEs. As far as the reader is concerned, this theorem implies that as long as **f** is continuously differentiable, i.e., $\mathbf{f} \in C^1(D)$, then two distinct solutions cannot intersect in finite time.

The following simple examples involving first-order ODEs illustrate the theorem quite well.

Example 11. Solve the following linear differential equations and state the maximal interval of existence for each solution:

(a) $\dot{x} = x$, $x(0) = 1$;

(b) $\dot{x} = x^2$, $x(0) = 1$;

(c) $\dot{x} = 3x^{\frac{2}{3}}$, $x(0) = 0$.

Solutions.

(a) The solution to this elementary differential equation is, $x(t) = e^t$, which is unique and defined for all t. The maximal interval of existence in this case is $-\infty < t < \infty$. Note that $f(x) = x$ is continuously differentiable.

(b) The solution is given by

$$x(t) = \frac{1}{1-t},$$

which is not defined for $t = 1$. Therefore, there is a unique solution on the maximal interval of existence given by $-\infty < t < 1$.

7.4. Existence and Uniqueness Theorem

(c) The function $f(x) = 3x^{\frac{2}{3}}$ is not continuously differentiable and does not satisfy the Lipschitz condition at $x = 0$; $\frac{\partial f}{\partial x} = 2x^{-\frac{1}{3}}$ is not continuous at $x = 0$. Integration gives

$$\int \frac{1}{3} x^{-\frac{2}{3}} dx = \int dt,$$

with general solution $x(t) = t^3 + C$. The solution to the initial value problem is therefore $x(t) = t^3$. The point $x = 0$ is zero when $\dot{x} = 0$. This means that a solution starting at this point should stay there for all t. Thus there are two solutions starting at $x_0 = 0$, namely $\phi_1(t) = t^3$ and $\phi_2(t) = 0$. In fact, there are infinitely many solutions starting at $x_0 = 0$. In this case, there exist solutions but they are not unique.

Note that the solution would be unique on the maximal interval of existence $0 < t < \infty$ if the initial condition were $x(1) = 1$.

Consider autonomous differential equations of the form

(7.10) $$\dot{\mathbf{x}} = \mathbf{f}(\mathbf{x}),$$

where $\mathbf{x} \in \Re^n$.

Definition 3. A *critical point (equilibrium point, fixed point, stationary point)* is a point that satisfies the equation $\dot{\mathbf{x}} = \mathbf{f}(\mathbf{x}) = 0$. If a solution starts at this point, it remains there forever.

Definition 4. A critical point, say, $\mathbf{x_0}$, of the differential equation (7.10) is called *stable* if given $\epsilon > 0$, there is a $\delta > 0$, such that for all $t \geq t_0$, $\| \mathbf{x}(t) - \mathbf{x_0}(t) \| < \epsilon$, whenever $\| \mathbf{x}(t_0) - \mathbf{x_0}(t_0) \| < \delta$, where $\mathbf{x}(t)$ is a solution of (7.10).

A critical point that is not stable is called an *unstable* critical point.

Example 12. Find and classify the critical points for the following one-dimensional differential equations.

(a) $\dot{x} = x$;

(b) $\dot{x} = -x$;

(c) $\dot{x} = x^2 - 1$.

Solutions.

(a) There is one critical point at $x_0 = 0$. If $x < 0$, then $\dot{x} < 0$, and if $x > 0$, then $\dot{x} > 0$. Therefore, x_0 is an unstable critical point. Solutions starting either side of x_0 are repelled away from it.

(b) There is one critical point at $x_0 = 0$. If $x < 0$, then $\dot{x} > 0$, and if $x > 0$, then $\dot{x} < 0$. Solutions starting either side of x_0 are attracted towards it. The critical point is stable.

(c) There are two critical points, one at $x_1 = -1$ and the other at $x_2 = 1$. If $x > 1$, then $\dot{x} > 0$; if $-1 < x < 1$, then $\dot{x} < 0$; and if $x < -1$, then $\dot{x} > 0$. Therefore, solutions starting near to x_1 but not on it are attracted towards this point, and x_1 is a stable critical point. Solutions starting near x_2 but not on it move away from this point, and x_2 is an unstable critical point.

By linearizing near a critical point, one can obtain a quantitative measure of stability as demonstrated below. Consider one-dimensional systems here; higher-dimensional systems will be investigated in other chapters.

Linear Stability Analysis. Let x^* be a critical point of $\dot{x} = f(x), x \in \Re$. Consider a *small perturbation*, say, $\xi(t)$, away from the critical point at x^* to give $x(t) = x^* + \xi(t)$. A simple analysis is now applied to determine whether the perturbation grows or decays as time evolves. Now

$$\dot{\xi} = \dot{x} = f(x) = f(x^* + \xi),$$

and after a Taylor series expansion,

$$\dot{\xi} = f(x^*) + \xi f'(x^*) + \frac{\xi^2}{2} f''(x^*) + \cdots.$$

In order to apply a linear stability analysis, the nonlinear terms are ignored. Hence

$$\dot{\xi} = \xi f'(x^*),$$

since $f(x^*) = 0$. Therefore, the perturbation $\xi(t)$ grows exponentially if $f'(x^*) > 0$ and decays exponentially if $f'(x^*) < 0$. If $f'(x^*) = 0$, then higher-order derivatives must be considered to determine the stability of the critical point.

A linear stability analysis is used extensively throughout the realms of non-linear dynamics and will appear in other chapters of this book.

Example 13. Use a linear stability analysis to determine the stability of the critical points for the following differential equations:

(a) $\dot{x} = \sin(x)$;

(b) $\dot{x} = x^2$;

(c) $\dot{x} = e^{-x} - 1$.

Solutions.

(a) There are critical points at $x_n = n\pi$, where n is an integer. When n is even, $f'(x_n) = 1 > 0$, and these critical points are unstable. When n is odd, $f'(x_n) = -1 < 0$, and these critical points are stable.

(b) There is one critical point at $x_0 = 0$ and $f'(x) = 2x$ in this case. Now $f'(0) = 0$ and $f''(0) = 2 > 0$. Therefore, x_0 is attracting when $x < 0$ and repelling when $x > 0$. The critical point is called *semistable*.

(c) There is one critical point at $x_0 = 0$. Now $f'(0) = -1 < 0$, and therefore the critical point at the origin is stable.

The theory of autonomous systems of ODEs in two dimensions will be discussed in the next chapter.

7.5 MATLAB Commands

```
% Program_7a - Solving simple differential equations
% symbolically.
% A separable ODE (Example 1).
soln1=dsolve('Dx=-t/x')

% Chemical kinetics (Example 8).
a=4;b=1;k=.00713;
soln2=dsolve('Dc=k*(a-c)^2*(b-c/2)')

% A 3D system (Exercise 7).
[x,y,z]=dsolve('Dx=-alpha*x','Dy=alpha*x-beta*y','Dz=beta*y')
% End of Program_7a.

% Program_7b - Solution curves, initial value problem
% (Example 3).

clear
deqn1=inline('P(1)*(100-P(1))/1000','t','P');
[t,P1]=ode45(deqn1,[0 100],50);
[t1,P2]=ode45(deqn1,[0 100],150);
hold on
plot(t,P1(:,1))
plot(t1,P2(:,1))
axis([0 100 0 200])
fsize=15;
set(gca,'xtick',[0:20:100],'FontSize',fsize)
set(gca,'ytick',[0:50:200],'FontSize',fsize)
```

```
xlabel('t','FontSize',fsize)
ylabel('P','FontSize',fsize)
hold off
% End of Program_7b.
```

```
% Program_7c  -  Solving simple differential equations
% (Example 8).

deqn=inline('.00713*(4-c(1))^2*(1-c(1)/2)','t','c');
[t,ca]=ode45(deqn,[0 400],0);
plot(t,ca(:,1))
axis([0 400 0 3])
fsize=15;
set(gca,'xtick',[0:100:400],'FontSize',fsize)
set(gca,'ytick',[0:1:3],'FontSize',fsize)
xlabel('t','FontSize',fsize)
ylabel('c(t)','FontSize',fsize)
hold off
% End of Program_7c.
```

7.6 Exercises

1. Sketch some solution curves for the following differential equations:

    (a) $\frac{dy}{dx} = -\frac{y}{x}$;

    (b) $\frac{dy}{dx} = \frac{2y}{x}$;

    (c) $\frac{dy}{dx} = \frac{y}{2x}$;

    (d) $\frac{dy}{dx} = \frac{y^2}{x}$;

    (e) $\frac{dy}{dx} = -\frac{xy}{x^2+y^2}$;

    (f) $\frac{dy}{dx} = \frac{y}{x^2}$.

2. Fossils are often dated using the differential equation

$$\frac{dA}{dt} = -\alpha A,$$

    where A is the amount of radioactive substance remaining, α is a constant, and t is measured in years. Assuming that $\alpha = 1.5 \times 10^{-7}$, determine the age of a fossil containing radioactive substance A if only 30% of the substance remains.

3. Write down the chemical reaction rate equations for the reversible reaction equations

(a) $A + B + C \rightleftharpoons D$,

(b) $A + A + A \rightleftharpoons A_3$,

given that the forward rate constant is k_f and the reverse rate constant is k_r, in each case. Assume that the chemical equations are the rate-determining steps.

4. (a) Consider a series resistor–inductor circuit with $L = 2H$, $R = 10\Omega$ and an applied EMF of $E = 100\sin(t)$. Use an integrating factor to solve the differential equation, and find the current in the circuit after 0.2 seconds given that $I(0) = 0$.

   (b) The differential equation use to model a series resistor–capacitor circuit is given by

   $$R\frac{dQ}{dt} + \frac{Q}{C} = E,$$

   where Q is the charge across the capacitor. If a variable resistance $R = 1/(5+t)\Omega$ and a capacitance $C = 0.5F$ are connected in series with an applied EMF, $E = 100V$, find the charge on the capacitor given that $Q(0) = 0$.

5. A forensic scientist is called to the scene of a murder. The temperature of the corpse is found to be 75°F and one hour later the temperature has dropped to 70°F. If the temperature of the room in which the body was discovered is a constant 68°F, how long before the first temperature reading was taken did the murder occur? Assume that the body obeys Newton's law of cooling,

$$\frac{dT}{dt} = \beta(T - T_R),$$

where T is the temperature of the corpse, β is a constant, and T_R is room temperature.

6. The differential equation used to model the concentration of glucose in the blood, say, $g(t)$, when it is being fed intravenously into the body, is given by

$$\frac{dg}{dt} + kg = \frac{G}{100V},$$

where k is a constant, G is the rate at which glucose is admitted, and V is the volume of blood in the body. Solve the differential equation and discuss the results.

7. A chemical substance A changes into substance B at a rate α times the amount of A present. Substance B changes into C at a rate β times the amount of B

present. If initially only substance A is present and its amount is M, show that the amount of C present at time t is

$$M + M\left(\frac{\beta e^{-\alpha t} - \alpha e^{-\beta t}}{\alpha - \beta}\right).$$

8. Two tanks A and B, each of volume V, are filled with water at time $t = 0$. For $t > 0$, volume v of solution containing mass m of solute flows into tank A per second; mixture flows from tank A to tank B at the same rate; and mixture flows away from tank B at the same rate. The differential equations used to model this system are given by

$$\frac{d\sigma_A}{dt} + \frac{v}{V}\sigma_A = \frac{m}{V}, \qquad \frac{d\sigma_B}{dt} + \frac{v}{V}\sigma_B = \frac{v}{V}\sigma_A,$$

where $\sigma_{A,B}$ are the concentrations of solute in tanks A and B, respectively. Show that the mass of solute in tank B is given by

$$\frac{mV}{v}(1 - e^{-vt/V}) - mte^{-vt/V}.$$

9. In an epidemic the rate at which healthy people become infected is a times their number. The rates of recovery and death are, respectively, b and c times the number of infected people. If initially there are N healthy people and no sick people, find the number of deaths up to time t. Is this a realistic model? What other factors should be taken into account?

10. (a) Determine the maximal interval of existence for each of the following initial value problems:
   (i) $\dot{x} = x^4$, $x(0) = 1$;
   (ii) $\dot{x} = \frac{x^2-1}{2}$, $x(0) = 2$;
   (iii) $\dot{x} = x(x - 2)$, $x(0) = 3$.

   (b) For what values of t_0 and x_0 does the initial value problem

   $$\dot{x} = 2\sqrt{x}, \qquad x(t_0) = x_0,$$

   have a unique solution?

Recommended Textbooks

[1] G. Turrell, *Mathematics for Chemistry and Physics*, Academic Press, New York, 2001.

[2] P. W. Davis, *Differential Equations: Modeling with MATLAB*, Prentice–Hall, Upper Saddle River, NJ, 1999.

[3] W. R. Derrick and S. I. Grossman, *Elementary Differential Equations*, 4th ed., Addison–Wesley, Reading, MA, 1997.

[4] R. Williams, *Introduction to Differential Equations and Dynamical Systems*, McGraw–Hill, New York, 1997.

[5] M. R. Spiegel, *Schaum's Outline of Laplace Transforms*, McGraw–Hill, New York, 1965.

8
Planar Systems

Aims and Objectives

- To introduce the theory of planar autonomous linear differential equations.
- To extend the theory of linear systems to that of nonlinear systems.

On completion of this chapter the reader should be able to

- find and classify critical points in the plane;
- carry out simple linear transformations;
- construct phase plane diagrams using isoclines, direction fields, and eigenvectors;
- apply the theory to simple modeling problems.

Basic analytical methods for solving two-dimensional linear autonomous differential equations are reviewed and simple phase portraits are constructed in the plane.

The method of linearization is introduced and both hyperbolic and nonhyperbolic critical points are defined. Phase portraits are constructed using *Hartman's theorem*. The linearization technique used here is based on a linear stability analysis.

8.1 Canonical Forms

Consider linear two-dimensional autonomous systems of the form

(8.1) $\quad \dfrac{dx}{dt} = \dot{x} = a_{11}x + a_{12}y, \qquad \dfrac{dy}{dt} = \dot{y} = a_{21}x + a_{22}y,$

where the a_{ij} are constants. The system is linear as the terms in x, y, $\dot{x}$, and $\dot{y}$ are all linear. System (8.1) can be written in the equivalent matrix form as

(8.2) $\qquad\qquad\qquad\qquad \dot{\mathbf{x}} = A\mathbf{x},$

where $\mathbf{x} \in \Re^2$ and

$$A = \begin{pmatrix} a_{11} & a_{12} \\ a_{21} & a_{22} \end{pmatrix}.$$

Definition 1. Every solution of (8.1) and (8.2), say, $\phi(t) = (x(t), y(t))$, can be represented as a curve in the plane. The solution curves are called *trajectories* or *orbits*.

The existence and uniqueness theorem guarantees that trajectories do not cross. Note that there are an infinite number of trajectories that would fill the plane if they were all plotted. However, the *qualitative behavior* can be determined by plotting just a few of the trajectories given the appropriate number of initial conditions.

Definition 2. The *phase portrait* is a two-dimensional figure showing how the qualitative behavior of system (8.1) is determined as x and y vary with t.

With the appropriate number of trajectories plotted, it should be possible to determine where any trajectory will end up from any given initial condition.

Definition 3. The *direction field* or *vector field* gives the gradients $\dfrac{dy}{dx}$ and direction vectors of the trajectories in the phase plane.

The slope of the trajectories can be determined using the chain rule,

$$\dfrac{dy}{dx} = \dfrac{\dot{y}}{\dot{x}},$$

and the direction of the vector field is given by $\dot{x}$ and $\dot{y}$ at each point in the xy plane.

Definition 4. The contour lines for which $\dfrac{dy}{dx}$ is a constant are called *isoclines*.

Isoclines may be used to help with the construction of the phase portrait. For example, the isoclines for which $\dot{x} = 0$ and $\dot{y} = 0$ are used to determine where the trajectories have vertical and horizontal tangent lines, respectively. If $\dot{x} = 0$, then there is no motion horizontally, and trajectories are either stationary or move vertically. A similar argument is used when $\dot{y} = 0$.

Using linear algebra, the phase portrait of any linear system of the form (8.2) can be transformed to a so-called *canonical form* $\dot{\mathbf{y}} = J\mathbf{y}$ by applying a transformation $\mathbf{x} = P\mathbf{y}$, where P is to be determined and $J = P^{-1}AP$ is of one of the following forms:

$$J_1 = \begin{pmatrix} \lambda_1 & 0 \\ 0 & \lambda_2 \end{pmatrix}, \qquad J_2 = \begin{pmatrix} \alpha & \beta \\ -\beta & \alpha \end{pmatrix},$$

$$J_3 = \begin{pmatrix} \lambda_1 & 0 \\ 0 & \lambda_1 \end{pmatrix}, \qquad J_4 = \begin{pmatrix} \lambda_1 & \mu \\ 0 & \lambda_1 \end{pmatrix},$$

where $\lambda_{1,2}$, α, β, and μ are real constants. Matrix J_1 has two real distinct eigenvalues, matrix J_2 has complex eigenvalues, and matrices J_3 and J_4 have repeated eigenvalues. The qualitative type of phase portrait is determined from each of these canonical forms.

Nonsimple Canonical Systems. The linear system (8.2) is *nonsimple* if the matrix A is singular (i.e., $\det(A) = 0$, and at least one of the eigenvalues is zero). The system then has critical points other than the origin.

Example 1. Sketch a phase portrait of the system $\dot{x} = x$, $\dot{y} = 0$.

Solution. The critical points are found by solving the equations $\dot{x} = \dot{y} = 0$, which has the solution $x = 0$. Thus there are an infinite number of critical points lying along the y-axis. The direction field has gradient given by

$$\frac{dy}{dx} = \frac{\dot{y}}{\dot{x}} = \frac{0}{x} = 0$$

for $x \neq 0$. This implies that the direction field is horizontal for points not on the y-axis. The direction vectors may be determined from the equation $\dot{x} = x$ since if $x > 0$, then $\dot{x} > 0$, and the trajectories move from left to right; and if $x < 0$, then $\dot{x} < 0$, and trajectories move from right to left. A phase portrait is plotted in Figure 8.1.

Simple Canonical Systems. System (8.2) is *simple* if $\det(A) \neq 0$, and the origin is then the only critical point. The critical points may be classified depending upon the type of eigenvalues.

8.1.I Real Distinct Eigenvalues. Suppose that system (8.2) can be diagonalized to obtain

$$\dot{x} = \lambda_1 x, \qquad \dot{y} = \lambda_2 y.$$

The solutions to this system are $x(t) = C_1 e^{\lambda_1 t}$ and $y(t) = C_2 e^{\lambda_2 t}$, where C_1 and C_2 are constant. The solution curves may be found by solving the differential

8. Planar Systems

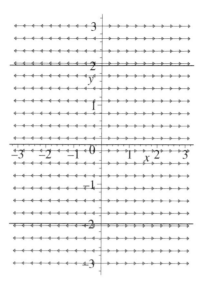

Figure 8.1: Six trajectories and a vector field plot for Example 1. Note that there are an infinite number of critical points lying on the y-axis.

equation given by

$$\frac{dy}{dx} = \frac{\dot{y}}{\dot{x}} = \frac{\lambda_2 y}{\lambda_1 x},$$

which is integrable. The solution curves are given by $|y|^{\lambda_1} = K|x|^{\lambda_2}$. The type of phase portrait depends on the values of λ_1 and λ_2, as summarized below:

- If the eigenvalues are distinct, real, and positive, then the critical point is called an *unstable node*.

- If the eigenvalues are distinct, real, and negative, then the critical point is called an *stable node*.

- If one eigenvalue is positive and the other negative, then the critical point is called a *saddle point* or *col*.

Possible phase portraits for these canonical systems along with vector fields superimposed are shown in Figure 8.2.

8.1.II Complex Eigenvalues ($\lambda = \alpha \pm i\beta$). Consider a canonical system of the form

(8.3) $\qquad \dot{x} = \alpha x + \beta y, \qquad \dot{y} = -\beta x + \alpha y.$

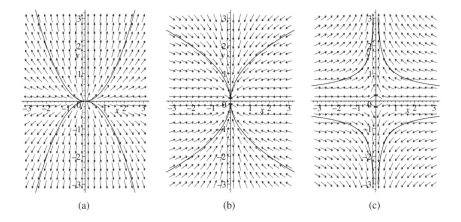

Figure 8.2: Possible phase portraits for canonical systems with two real distinct eigenvalues: (a) unstable node; (b) stable node; (c) saddle point or col.

Convert to *polar coordinates* by making the transformations $x = r\cos\theta$ and $y = r\sin\theta$. Then elementary calculus gives

$$r\dot{r} = x\dot{x} + y\dot{y}, \qquad r^2\dot{\theta} = x\dot{y} - y\dot{x}.$$

System (8.3) becomes

$$\dot{r} = \alpha r, \qquad \dot{\theta} = -\beta.$$

The type of phase portrait depends on the values of α and β:

- If $\alpha > 0$, then the critical point is called an unstable focus.
- If $\alpha = 0$, then the critical point is called a center.
- If $\alpha < 0$, then the critical point is called a stable focus.
- If $\dot{\theta} > 0$, then the trajectories spiral counterclockwise around the origin.
- If $\dot{\theta} < 0$, then the trajectories spiral clockwise around the origin.

Phase portraits of the canonical systems with the vector fields superimposed are shown in Figure 8.3.

8.1.III Repeated Real Eigenvalues. Suppose that the canonical matrices are of the form J_3 or J_4. The type of phase portrait is determined by the following:

- If there are two linearly independent eigenvectors, then the critical point is called a *singular node*.

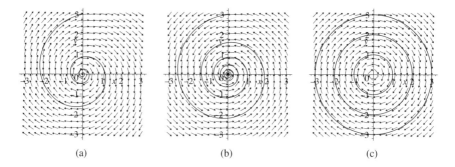

Figure 8.3: Possible phase portraits for canonical systems with complex eigenvalues: (a) unstable focus; (b) stable focus; (c) center.

- If there is one linearly independent eigenvector, then the critical point is called a *degenerate node*.

Possible phase portraits with vector fields superimposed are shown in Figure 8.4.

The classifications given in this section may be summarized using the trace and determinant of the matrix A as defined in system (8.2). If the eigenvalues are $\lambda_{1,2}$, then the characteristic equation is given by $(\lambda - \lambda_1)(\lambda - \lambda_2) = \lambda^2 - (\lambda_1 + \lambda_2)\lambda + \lambda_1\lambda_2 = \lambda^2 - \text{trace}(A)\lambda + \det(A) = 0$. Therefore,

$$\lambda_{1,2} = \frac{\text{trace}(A) \pm \sqrt{(\text{trace}(A))^2 - 4\det(A)}}{2}.$$

The summary is depicted in Figure 8.5.

8.2 Eigenvectors Defining Stable and Unstable Manifolds

Consider Figure 8.5. Apart from the region $T^2 - 4D > 0$, where the trajectories spiral, the phase portraits of the canonical forms of (8.2) all contain straight line trajectories that remain on the coordinate axes forever and exhibit exponential growth or decay along it. These special trajectories are determined by the eigenvectors of the matrix A and are called the *manifolds*. If the trajectories move towards the critical point at the origin as $t \to \infty$ along the axis, then there is exponential decay and the axis is called a *stable manifold*. If trajectories move away from the critical point as $t \to \infty$, then the axis is called an *unstable manifold*.

In the general case, the manifolds do not lie along the axes. Suppose that a trajectory is of the form

$$\mathbf{x}(t) = \exp(\lambda t)\mathbf{e},$$

8.2. Eigenvectors Defining Stable and Unstable Manifolds

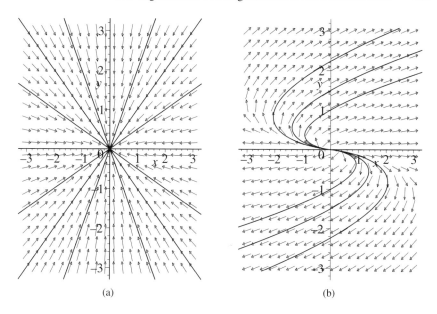

Figure 8.4: Possible phase portraits for canonical systems with repeated eigenvalues: (a) a stable singular node; (b) an unstable degenerate node.

where $\mathbf{e} \neq 0$ is a vector and λ is a constant. This trajectory satisfies equation (8.2) since it is a solution curve. Therefore, substituting into (8.2),

$$\lambda \exp(\lambda t)\mathbf{e} = \exp(\lambda t) A\mathbf{e}$$

or

$$\lambda \mathbf{e} = A\mathbf{e}.$$

From elementary linear algebra, if there exists a nonzero column vector $\mathbf{e}$ satisfying this equation, then λ is called an eigenvalue of A and $\mathbf{e}$ is called an eigenvector of A corresponding to the eigenvalue λ. If λ is negative, then the corresponding eigenvector gives the direction of the stable manifold, and if λ is positive, then the eigenvector gives the direction of the unstable manifold.

When $\lambda_1 \neq \lambda_2$, it is known from elementary linear algebra that the eigenvectors $\mathbf{e_1}$ and $\mathbf{e_2}$, corresponding to the eigenvalues λ_1 and λ_2, are linearly independent. Therefore, the general solution to the differential equations given by (8.1) is given by

$$\mathbf{x}(t) = C_1 \exp(\lambda_1 t)\mathbf{e_1} + C_2 \exp(\lambda_2 t)\mathbf{e_2},$$

where C_1, C_2 are constants. In fact, for any given initial condition, this solution is unique by the existence and uniqueness theorem.

8. Planar Systems

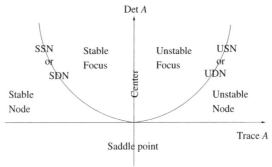

SSN - Stable Singular Node
SDN - Stable Degenerate Node
USN - Unstable Singular Node
UDN - Unstable Degenerate Node

Figure 8.5: Classification of critical points for system (8.2). The parabola has equation $T^2 - 4D = 0$, where $D = \det(A)$ and $T = \operatorname{trace}(A)$.

Definition 5. Suppose that $\mathbf{0} \in \Re^2$ is a critical point of the linear system (8.2). Then the stable and unstable manifolds of the critical point $\mathbf{0}$ are defined by

$$E_S(\mathbf{0}) = \{\mathbf{x} : \Lambda^+(\mathbf{x}) = \mathbf{0}\}, \quad E_U(\mathbf{0}) = \{\mathbf{x} : \Lambda^-(\mathbf{x}) = \mathbf{0}\},$$

where Λ^+ is the positive limit set and Λ^- is the negative limit set (see Definition 4 in Chapter 10).

Consider the following two simple illustrations.

Example 2. Determine the stable and unstable manifolds for the linear system

$$\dot{x} = 2x + y, \qquad \dot{y} = x + 2y.$$

Solution. The system can be written as $\dot{\mathbf{x}} = A\mathbf{x}$, where

$$A = \begin{pmatrix} 2 & 1 \\ 1 & 2 \end{pmatrix}.$$

The characteristic equation for matrix A is given by $\det(A - \lambda I) = 0$, or in this case,

$$\begin{vmatrix} 2 - \lambda & 1 \\ 1 & 2 - \lambda \end{vmatrix} = 0.$$

Therefore, the characteristic equation is $\lambda^2 - 4\lambda + 3 = 0$, which has roots $\lambda_1 = 1$ and $\lambda_2 = 3$. Since both eigenvalues are real and positive, the critical point at the origin is an unstable node. The manifolds are determined from the eigenvectors

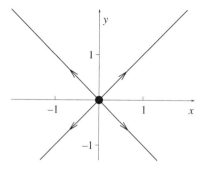

Figure 8.6: The unstable manifolds for Example 2.

corresponding to these eigenvalues. The eigenvector for λ_1 is $(1, -1)^T$ and the eigenvector for λ_2 is $(1, 1)^T$. The manifolds are shown in Figure 8.6.

For the sake of completeness, the general solution in this case is given by

$$\mathbf{x}(t) = C_1 \exp(t)(1, -1)^T + C_2 \exp(3t)(1, 1)^T.$$

Example 3. Determine the stable and unstable manifolds for the linear system

$$\dot{\mathbf{x}} = \begin{pmatrix} -3 & 4 \\ -2 & 3 \end{pmatrix} \mathbf{x}.$$

Solution. The characteristic equation for matrix A is given by

$$\begin{vmatrix} -3 - \lambda & 4 \\ -2 & 3 - \lambda \end{vmatrix} = 0.$$

Therefore, the characteristic equation is $\lambda^2 - 1 = 0$, which has roots $\lambda_1 = 1$ and $\lambda_2 = -1$. Since one eigenvalue is real and positive and the other is real and negative, the critical point at the origin is a saddle point. The manifolds are derived from the eigenvectors corresponding to these eigenvalues. The eigenvector for λ_1 is $(1, 1)^T$ and the eigenvector for λ_2 is $(2, 1)^T$. The manifolds are shown in Figure 8.7.

For the sake of completeness, the general solution in this case is given by

$$\mathbf{x}(t) = C_1 \exp(t)(1, 1)^T + C_2 \exp(-t)(2, 1)^T.$$

Notation. The stable and unstable manifolds of linear systems will be denoted by E_S and E_U, respectively. Center manifolds (where the eigenvalues have zero real part) will be discussed in Chapter 13.

8.3 Phase Portraits of Linear Systems in the Plane

Definition 6. Two systems of first-order autonomous differential equations are said to be *qualitatively* (or *topologically*) *equivalent* if there exists an invertible

8. Planar Systems

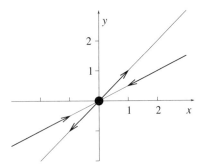

Figure 8.7: The stable and unstable manifolds for Example 3. The trajectories lying on the stable manifold tend to the origin as $t \to \infty$ but never reach it.

mapping that maps one phase portrait onto the other while preserving the orientation of the trajectories.

Phase portraits can be constructed using isoclines, vector fields, and eigenvectors (for real eigenvalues).

Example 4. Consider the system

$$\dot{x} = \begin{pmatrix} 2 & 1 \\ 1 & 2 \end{pmatrix} x.$$

Find (a) the eigenvalues and corresponding eigenvectors of A; (b) a nonsingular matrix P such that $J = P^{-1}AP$ is diagonal; and (c) new coordinates (u, v) such that substituting $x = x(u, v)$, $y = y(u, v)$, converts the linear dynamical system

$$\dot{x} = 2x + y, \qquad \dot{y} = x + 2y \quad \text{into} \quad \dot{u} = \lambda_1 u, \qquad \dot{v} = \lambda_2 v$$

for suitable λ_1, λ_2; (d) sketch phase portraits for these qualitatively equivalent systems.

Solutions. The origin is a unique critical point.

(a) From Example 2, the eigenvalues and corresponding eigenvectors are given by $\lambda_1 = 1$, $(1, -1)^T$ and $\lambda_2 = 3$, $(1, 1)^T$; the critical point is an unstable node.

(b) Using elementary linear algebra, the columns of matrix P are these eigenvectors and so

$$P = \begin{pmatrix} 1 & 1 \\ -1 & 1 \end{pmatrix}$$

and

$$J = P^{-1}AP = \begin{pmatrix} 1 & 0 \\ 0 & 3 \end{pmatrix}.$$

(c) Take the linear transformation $\mathbf{x} = P\mathbf{u}$ to obtain the system $\dot{u} = u$, $\dot{v} = 3v$.

(d) Consider the *isoclines*. In the xy plane, the flow is horizontal on the line where $\dot{y} = 0$ and hence on the line $y = -\frac{x}{2}$. On this line, $\dot{x} = \frac{3x}{2}$; thus $\dot{x} > 0$ if $x > 0$ and $\dot{x} < 0$ if $x < 0$. The flow is vertical on the line $y = -2x$. On this line, $\dot{y} < 0$ if $x > 0$ and $\dot{y} > 0$ if $x < 0$.

Vector fields: The directions of the vector fields can be determined from $\dot{x}$ and $\dot{y}$ at points (x, y) in the plane.

Consider the *slope of the trajectories*. If $x + 2y > 0$ and $2x + y > 0$, then $\frac{dy}{dx} > 0$; if $x + 2y < 0$ and $2x + y > 0$, then $\frac{dy}{dx} < 0$; if $x + 2y > 0$ and $2x + y < 0$, then $\frac{dy}{dx} < 0$; and if $x + 2y < 0$ and $2x + y < 0$, then $\frac{dy}{dx} > 0$.

Manifolds: From the eigenvectors, both manifolds are unstable. One passes through $(0, 0)$ and $(1, 1)$ and the other through $(0, 0)$ and $(1, -1)$.

Putting all of this information together gives the phase portrait in Figure 8.8(a). The canonical phase portrait is shown in Figure 8.8(b).

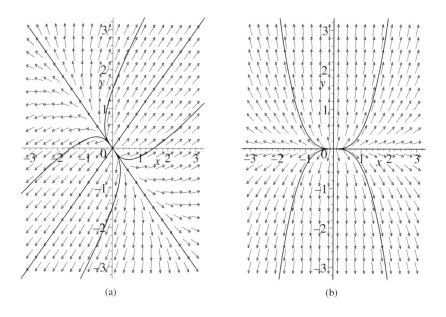

(a) (b)

Figure 8.8: Qualitatively equivalent phase portraits for Example 4.

Example 5. Sketch a phase portrait for the system

$$\dot{x} = -x - y, \qquad \dot{y} = x - y.$$

Solution. The origin is the only critical point. The characteristic equation is given by

$$|A - \lambda I| = \lambda^2 + 2\lambda + 2 = 0,$$

196   8. Planar Systems

which has complex solutions $\lambda_{1,2} = -1 \pm i$. The critical point at the origin is a stable focus.

Consider the *isoclines*. In the xy plane, the flow is horizontal on the line where $\dot{y} = 0$ and hence on the line $y = x$. On this line, $\dot{x} = -2x$; thus $\dot{x} < 0$ if $x > 0$ and $\dot{x} > 0$ if $x < 0$. The flow is vertical on the line where $\dot{x} = 0$ and hence on the line $y = -x$. On this line $\dot{y} < 0$ if $x > 0$ and $\dot{y} > 0$ if $x < 0$.

Vector fields: The directions of the vector fields can be determined from $\dot{x}$ and $\dot{y}$ at points (x, y) in the plane.

Consider the *slope of the trajectories*. If $y > x$ and $y > -x$, then $\frac{dy}{dx} > 0$; if $y > x$ and $y < -x$, then $\frac{dy}{dx} < 0$; if $y < x$ and $y > -x$, then $\frac{dy}{dx} < 0$; and if $y < x$ and $y < -x$, then $\frac{dy}{dx} > 0$.

Manifolds: The eigenvectors are complex and there are no real manifolds.

Converting to polar coordinates gives $\dot{r} = -r, \dot{\theta} = 1$. Putting all of this information together gives the phase portrait in Figure 8.9.

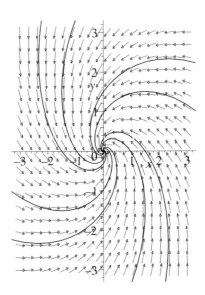

Figure 8.9: Some trajectories for Example 5. The critical point is a stable focus.

Example 6. Sketch a phase portrait for the system

$$\dot{x} = -2x, \qquad \dot{y} = -4x - 2y.$$

Solution. The origin is the only critical point. The characteristic equation is given by

$$|A - \lambda I| = \lambda^2 - 4\lambda + 4 = 0,$$

which has repeated roots $\lambda_{1,2} = -2$.

Consider the *isoclines*. In the xy plane, the flow is horizontal on the line where $\dot{y} = 0$ and hence on the line $y = -2x$. Trajectories that start on the y-axis remain there forever.

Vector fields: The directions of the vector fields can be determined from $\dot{x}$ and $\dot{y}$ at points (x, y) in the plane.

Consider the *slope of the trajectories*. The slopes are given by $\frac{dy}{dx}$ at each point (x, y) in the plane.

Manifolds: There is one linearly independent eigenvector, $(0, 1)^T$. Therefore, the critical point is a stable degenerate node. The stable manifold E_S is the y-axis.

Putting all of this together gives the phase portrait in Figure 8.10.

Figure 8.10: Some trajectories for Example 6. The critical point is a stable degenerate node.

Phase portraits of nonlinear planar autonomous systems will be considered in the following sections, where stable and unstable manifolds do not necessarily lie on straight lines. However, all is not lost as the manifolds for certain critical points are tangent to the eigenvectors of the linearized system at that point.

Manifolds in three-dimensional systems will be discussed in Chapter 13.

8.4 Linearization and Hartman's Theorem

Suppose that the nonlinear autonomous system

(8.4) $$\dot{x} = P(x, y), \qquad \dot{y} = Q(x, y)$$

8. Planar Systems

has a critical point at (u, v), where P and Q are at least quadratic in x and y. Take a linear transformation that moves the critical point to the origin. Let $X = x - u$ and $Y = y - v$. Then system (8.4) becomes

$$\dot{X} = P(X+u, Y+v) = P(u,v) + X\frac{\partial P}{\partial x}\bigg|_{x=u,y=v} + Y\frac{\partial P}{\partial y}\bigg|_{x=u,y=v} + R(X,Y),$$

$$\dot{Y} = Q(X+u, Y+v) = Q(u,v) + X\frac{\partial Q}{\partial x}\bigg|_{x=u,y=v} + Y\frac{\partial Q}{\partial y}\bigg|_{x=u,y=v} + S(X,Y)$$

after a Taylor series expansion. The nonlinear terms R and S satisfy the conditions $\frac{R}{r} \to 0$ and $\frac{S}{r} \to 0$ as $r = \sqrt{X^2 + Y^2} \to 0$. The functions R and S are said to be "big Oh of r^2," or in mathematical notation, $R = O(r^2)$ and $S = O(r^2)$. Discard the nonlinear terms in the system and note that $P(u,v) = Q(u,v) = 0$ since (u,v) is a critical point of system (8.4). The *linearized* system is then of the form

(8.5)
$$\dot{X} = X\frac{\partial P}{\partial x}\bigg|_{x=u,y=v} + Y\frac{\partial P}{\partial y}\bigg|_{x=u,y=v},$$

$$\dot{Y} = X\frac{\partial Q}{\partial x}\bigg|_{x=u,y=v} + Y\frac{\partial Q}{\partial y}\bigg|_{x=u,y=v},$$

and the Jacobian matrix is given by

$$J(u,v) = \begin{pmatrix} \frac{\partial P}{\partial x} & \frac{\partial P}{\partial y} \\ \frac{\partial Q}{\partial x} & \frac{\partial Q}{\partial y} \end{pmatrix}\bigg|_{x=u,y=v}.$$

Definition 7. A critical point is called *hyperbolic* if the real part of the eigenvalues of the Jacobian matrix $J(u, v)$ are nonzero. If the real part of either of the eigenvalues of the Jacobian are equal to zero, then the critical point is called *nonhyperbolic*.

Hartman's Theorem. *Suppose that (u, v) is a hyperbolic critical point of system (8.4). Then there is a neighborhood of this critical point on which the phase portrait for the nonlinear system resembles that of the linearized system (8.5). In other words, there is a curvilinear continuous change of coordinates taking one phase portrait to the other, and in a small region around the critical point, the portraits are qualitatively equivalent.*

A proof of this theorem may be found in Hartman's book [7]. Note that the stable and unstable manifolds of the nonlinear system will be tangent to the manifolds of the linearized system near the relevant critical point. These trajectories diverge as one moves away from the critical point; this is illustrated in Examples 7 and 8.

Notation. Stable and unstable manifolds of a nonlinear system are labeled W_S and W_U, respectively.

Hartman's theorem implies that W_S and W_U are tangent to E_S and E_U at the relevant critical point. If any of the critical points are nonhyperbolic, then other methods must be used to sketch a phase portrait, and numerical solvers may be required.

8.5 Constructing Phase Plane Diagrams

The method for plotting phase portraits for nonlinear planar systems having hyperbolic critical points may be broken down into three distinct steps:

- Locate all of the critical points.
- Linearize and classify each critical point according to Hartman's theorem.
- Determine the isoclines and use $\frac{dy}{dx}$ to obtain slopes of trajectories.

The method can be illustrated with some simple examples. Examples 10–12 illustrate possible approaches when a critical point is not hyperbolic.

Example 7. Sketch a phase portrait for the nonlinear system

$$\dot{x} = x, \qquad \dot{y} = x^2 + y^2 - 1.$$

Solution. Locate the critical points by solving the equations $\dot{x} = \dot{y} = 0$. Hence $\dot{x} = 0$ if $x = 0$ and $\dot{y} = 0$ if $x^2 + y^2 = 1$. If $x = 0$, then $\dot{y} = 0$ if $y^2 = 1$, which has solutions $y = 1$ and $y = -1$. Therefore, there are two critical points, $(0, 1)$ and $(0, -1)$.

Linearize by finding the Jacobian matrix; hence

$$J = \begin{pmatrix} \frac{\partial P}{\partial x} & \frac{\partial P}{\partial y} \\ \frac{\partial Q}{\partial x} & \frac{\partial Q}{\partial y} \end{pmatrix} = \begin{pmatrix} 1 & 0 \\ 2x & 2y \end{pmatrix}.$$

Linearize at each critical point; hence

$$J_{(0,1)} = \begin{pmatrix} 1 & 0 \\ 0 & 2 \end{pmatrix}.$$

The matrix is in diagonal form. There are two distinct positive eigenvalues and hence the critical point is an unstable node.

For the other critical point,

$$J_{(0,-1)} = \begin{pmatrix} 1 & 0 \\ 0 & -2 \end{pmatrix}.$$

There is one positive and one negative eigenvalue, and so this critical point is a saddle point or col.

8. Planar Systems

Note that the matrices $J_{(0,1)}$ and $J_{(0,-1)}$ are in diagonal form. The eigenvectors for both critical points are $(1, 0)^T$ and $(0, 1)^T$. Thus in a small neighborhood around each critical point, the stable and unstable manifolds are tangent to the lines generated by the eigenvectors through each critical point. Therefore, near each critical point the manifolds are horizontal and vertical. Note that the manifolds of the nonlinear system W_S and W_U need not be straight lines but are tangent to E_S and E_U at the relevant critical point.

Consider the isoclines. Now $\dot{x} = 0$ on $x = 0$, and on this line $\dot{y} = y^2 - 1$. Thus if $|y| < 1$, then $\dot{y} < 0$, and if $|y| > 1$, then $\dot{y} > 0$. Also, $\dot{y} = 0$ on the circle $x^2 + y^2 = 1$, and on this curve $\dot{x} = x$. Thus if $x > 0$, then $\dot{x} > 0$, and if $x < 0$, then $\dot{x} < 0$. The slope of the trajectories is given by

$$\frac{dy}{dx} = \frac{x^2 + y^2 - 1}{x}.$$

Putting all of this information together gives a phase portrait as depicted in Figure 8.11.

Figure 8.11: A phase portrait for Example 7. The stable and unstable manifolds (W_S, W_U) are tangent to horizontal or vertical lines (E_S, E_U) in a small neighborhood of each critical point.

Example 8. Sketch a phase portrait for the nonlinear system

$$\dot{x} = y, \qquad \dot{y} = x(1 - x^2) + y.$$

Solution. Locate the critical points by solving the equations $\dot{x} = \dot{y} = 0$. Hence $\dot{x} = 0$ if $y = 0$ and $\dot{y} = 0$ if $x(1 - x^2) + y = 0$. If $y = 0$, then $\dot{y} = 0$ if $x(1 - x^2) = 0$, which has solutions $x = 0$, $x = 1$, and $x = -1$. Therefore, there are three critical points, $(0, 0)$, $(1, 0)$, and $(-1, 0)$.

Linearize by finding the Jacobian matrix; hence

$$J = \begin{pmatrix} \frac{\partial P}{\partial x} & \frac{\partial P}{\partial y} \\ \frac{\partial Q}{\partial x} & \frac{\partial Q}{\partial y} \end{pmatrix} = \begin{pmatrix} 0 & 1 \\ 1 - 3x^2 & 1 \end{pmatrix}.$$

Linearize at each critical point; hence

$$J_{(0,0)} = \begin{pmatrix} 0 & 1 \\ 1 & 1 \end{pmatrix}.$$

The eigenvalues are

$$\lambda_1 = \frac{1 + \sqrt{5}}{2} \quad \text{and} \quad \lambda_2 = \frac{1 - \sqrt{5}}{2}.$$

The corresponding eigenvectors are $(1\ \lambda_1)^T$ and $(1\ \lambda_2)^T$. Thus the critical point at the origin is a saddle point or col.

For the other critical points,

$$J_{(1,0)} = J_{(-1,0)} = \begin{pmatrix} 0 & 1 \\ -2 & 1 \end{pmatrix}.$$

The eigenvalues are

$$\lambda = \frac{1 \pm i\sqrt{7}}{2},$$

and so both critical points are unstable foci.

Consider the isoclines. Now $\dot{x} = 0$ on $y = 0$, and on this line $\dot{y} = x(1 - x^2)$. Thus if $0 < x < 1$, then $\dot{y} > 0$; if $x > 1$, then $\dot{y} < 0$; if $-1 < x < 0$, then $\dot{y} < 0$, and if $x < -1$, then $\dot{y} > 0$. Also, $\dot{y} = 0$ on the curve $y = x - x^3$, and on this curve, $\dot{x} = y$. Thus if $y > 0$, then $\dot{x} > 0$, and if $y < 0$, then $\dot{x} < 0$. The slope of the trajectories is given by

$$\frac{dy}{dx} = \frac{x - x^3 + y}{y}.$$

Note that on $x = 0$ and $x = \pm 1$, $\frac{dy}{dx} = 1$. Putting all of this information together gives a phase portrait as depicted in Figure 8.12.

8. Planar Systems

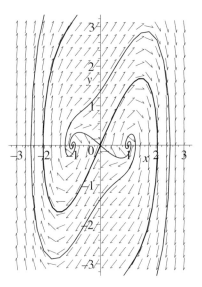

Figure 8.12: A phase portrait for Example 8. Note that in a small neighborhood of the origin, the unstable manifold (W_U) is tangent to the line E_U given by $y = \lambda_1 x$, and the stable manifold (W_S) is tangent to the line E_S given by $y = \lambda_2 x$. The equations of both lines are derived from the respective eigenvectors.

Example 9. Plot a phase portrait for the system

$$\dot{x} = x\left(1 - \frac{x}{2} - y\right), \qquad \dot{y} = y\left(x - 1 - \frac{y}{2}\right).$$

Solution. Locate the critical points by solving the equations $\dot{x} = \dot{y} = 0$. Hence $\dot{x} = 0$ if either $x = 0$ or $y = 1 - \frac{x}{2}$. Suppose that $x = 0$. Then $\dot{y} = 0$ if $y(-1 - \frac{y}{2}) = 0$, which has solutions $y = 0$ or $y = -2$. Suppose that $y = 1 - \frac{x}{2}$. Then $\dot{y} = 0$ if either $1 - \frac{x}{2} = 0$ or $1 - \frac{x}{2} = 2x - 2$, which has solutions $x = 2$ or $x = \frac{6}{5}$. Thus there are four critical points at $(0, 0)$, $(2, 0)$, $(0, -2)$, and $(\frac{6}{5}, \frac{2}{5})$. Notice that $\dot{x} = 0$ when $x = 0$, which means that the flow is vertical on the y-axis. Similarly, $\dot{y} = 0$ when $y = 0$, and the flow is horizontal along the x-axis. In this case, the axes are invariant.

Linearize by finding the Jacobian matrix; hence

$$J = \begin{pmatrix} \frac{\partial P}{\partial x} & \frac{\partial P}{\partial y} \\ \frac{\partial Q}{\partial x} & \frac{\partial Q}{\partial y} \end{pmatrix} = \begin{pmatrix} 1 - x - y & -x \\ y & x - 1 - y \end{pmatrix}.$$

Linearize around each of the critical points and apply Hartman's theorem. Consider the critical point at $(0, 0)$. The eigenvalues are $\lambda = \pm 1$ and the critical point is a saddle point or col. Next, consider the critical point at $(2,0)$; now the eigenvalues are $\lambda_1 = 1$ and $\lambda_2 = -1$. The corresponding eigenvectors are $(-1, 1)^T$ and $(1, 0)^T$, respectively. This critical point is also a saddle point or col. Consider the critical point at $(0, -2)$. Now the eigenvalues are $\lambda_1 = 3$ and $\lambda_2 = 1$; the corresponding eigenvectors are $(1, -1)^T$ and $(0, 1)^T$, respectively. The critical point at $(0, -2)$ is therefore an unstable node. Finally, consider the critical point at $(\frac{6}{5}, \frac{2}{5})$. The eigenvalues in this case are

$$\lambda = \frac{-2 \pm i\sqrt{11}}{5},$$

and the critical point is a stable focus. There is no need to find the eigenvectors; they are complex in this case.

Consider the isoclines. Now $\dot{x} = 0$ on $x = 0$ or on $y = 1 - \frac{x}{2}$, and $\dot{y} = 0$ on $y = 0$ or on $y = 2x - 2$. The directions of the flow can be found by considering $\dot{y}$ and $\dot{x}$ on these curves.

The slope of the trajectories is given by

$$\frac{dy}{dx} = \frac{y\left(x - 1 - \frac{y}{2}\right)}{x\left(1 - \frac{x}{2} - y\right)}.$$

A phase portrait indicating the stable and unstable manifolds of the critical points is shown in Figure 8.13.

Figure 8.13: A phase portrait for Example 9. The axes are invariant.

8. Planar Systems

Example 10. Sketch a phase portrait for the nonlinear system

$$\dot{x} = y^2, \qquad \dot{y} = x.$$

Solution. Locate the critical points by solving the equations $\dot{x} = \dot{y} = 0$. Therefore, $\dot{x} = 0$ if $y = 0$ and $\dot{y} = 0$ if $x = 0$. Thus the origin is the only critical point.

Attempt to linearize by finding the Jacobian matrix; hence

$$J = \begin{pmatrix} \frac{\partial P}{\partial x} & \frac{\partial P}{\partial y} \\ \frac{\partial Q}{\partial x} & \frac{\partial Q}{\partial y} \end{pmatrix} = \begin{pmatrix} 0 & 2y \\ 1 & 0 \end{pmatrix}.$$

Linearize at the origin to obtain

$$J_{(0,0)} = \begin{pmatrix} 0 & 0 \\ 1 & 0 \end{pmatrix}.$$

The eigenvalues are both zero and so the origin is a nonhyberbolic critical point. To sketch a phase portrait, solve the differential equation

$$\frac{dy}{dx} = \frac{\dot{y}}{\dot{x}} = \frac{x}{y^2}.$$

This differential equation was solved in Chapter 7 and the solution curves were given in Figure 7.2.

Consider the isoclines. Now $\dot{x} = 0$ on $y = 0$, and on this line $\dot{y} = x$. Thus if $x > 0$, then $\dot{y} > 0$, and if $x < 0$, then $\dot{y} < 0$. Also, $\dot{y} = 0$ on $x = 0$, and on this line $\dot{x} = y^2$. Thus $\dot{x} > 0$ for all y. The slope of the trajectories is given by $\frac{dy}{dx} = \frac{x}{y^2}$. Putting all of this information together gives a phase portrait as depicted in Figure 8.14.

Example 11. A simple model for the spread of an *epidemic* in a city is given by

$$\dot{S} = -\tau SI, \qquad \dot{I} = \tau SI - rI,$$

where $S(t)$ and $I(t)$ represent the numbers of susceptible and infected individuals scaled by 1000, respectively; τ is a constant measuring how quickly the disease is transmitted; r measures the rate of recovery (assume that those who recover become immune); and t is measured in days. Determine a value for S at which the infected population is a maximum.

Given that $\tau = 0.003$ and $r = 0.5$, sketch a phase portrait showing three trajectories whose initial points are at $(1000, 1)$, $(700, 1)$, and $(500, 1)$. Give a physical interpretation in each case.

Solution. The maximum number of infected individuals occurs when $\frac{dI}{dS} = 0$. Now

$$\frac{dI}{dS} = \frac{\dot{I}}{\dot{S}} = \frac{\tau S - r}{-\tau S}.$$

8.5. Constructing Phase Plane Diagrams

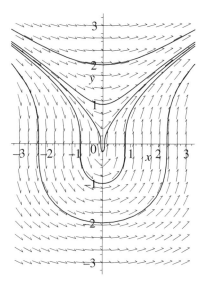

Figure 8.14: A phase portrait for Example 10 that has a nonhyperbolic critical point at the origin. There is a *cusp* at the origin.

Therefore, $\frac{dI}{dS} = 0$ when $S = \frac{r}{\tau}$. The number $\frac{r}{\tau}$ is called the *threshold value*.

The critical points for this system are found by solving the equations $\dot{S} = \dot{I} = 0$. Therefore, there are an infinite number of critical points lying along the x-axis. A phase portrait is plotted in Figure 8.15.

In each case, the population of susceptibles decreases to a constant value and the population of infected individuals increases and then decreases to zero. Note that in each case, the maximum number of infected individuals occurs at $S = \frac{r}{\tau} \approx 167,000$.

Example 12. Chemical kinetics involving the derivation of one differential equation were introduced in Chapter 7. This example will consider a system of two differential equations. Consider the isothermal chemical reaction

$$A + B \rightleftharpoons C,$$

in which one molecule of A combines with one molecule of B to form one molecule of C. In the reverse reaction, one molecule of C returns to A + B. Suppose that the rate of the forward reaction is k_f and the rate of the backward reaction is k_r. Let the concentrations of A, B, and C be a, b, and c, respectively. Assume that the concentration of A is much larger than the concentrations of B and C and can therefore be thought of as constant. From the law of mass action, the equations for the kinetics of b and c are

$$\dot{b} = k_r c - k_f ab, \qquad \dot{c} = k_f ab - k_r c.$$

206   8. Planar Systems

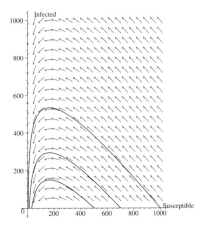

Figure 8.15: A phase portrait showing three trajectories for Example 11. The axes are scaled by 10^3 in each case. Trajectories are only plotted in the first quadrant since populations cannot be negative.

Find the critical points and sketch a typical trajectory for this system. Interpret the results in physical terms.

Solution. The critical points are found by determining where $\dot{b} = \dot{c} = 0$. Clearly, there are an infinite number of critical points along the line $c = \frac{k_f a}{k_r} b$. The slope of the trajectories is given by

$$\frac{dc}{db} = \frac{\dot{c}}{\dot{b}} = -1.$$

If $c < \frac{k_f a}{k_r} b$, then $\dot{b} < 0$ and $\dot{c} > 0$. Similarly, if $c > \frac{k_f a}{k_r} b$, then $\dot{b} > 0$ and $\dot{c} < 0$. Two typical solution curves are plotted in Figure 8.16.

Thus the final concentrations of B and C depend upon the initial concentrations of these chemicals. Two trajectories starting from the initial points at $(b_0, 0)$ and (b_0, c_0) are plotted in Figure 8.16. Note that the chemical reaction obeys the law of *conservation of mass*; this explains why the trajectories lie along the lines $b + c =$ constant.

Example 13. Suppose that H is a population of healthy rabbits and I is the subpopulation of infected rabbits that never recover once infected, both measured in millions. The following differential equations can be used to model the dynamics of the system:

$$\dot{H} = (b - d)H - \delta I, \qquad \dot{I} = \tau I(H - I) - (\delta + d)I,$$

where b is the birth rate, d is the natural death rate, δ is the rate of death of the diseased rabbits, and τ is the rate at which the disease is transmitted.

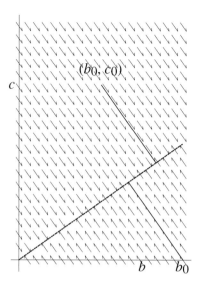

Figure 8.16: Two solution curves for the chemical kinetic equation in Example 12, where a is assumed to be constant. The dotted line represents the critical points lying on the line $c = \frac{k_f a}{k_r} b$.

Given that $b = 4, d = 1, \delta = 6$, and $\tau = 1$ and given an initial population of $(H_0, I_0) = (2, 2)$, plot a phase portrait and explain what happens to the rabbits in real-world terms.

Solution. There are two critical points in the first quadrant at $0 = (0, 0)$ and $P = (14, 7)$. The Jacobian matrix is given by

$$J = \begin{pmatrix} (b-d) & -\delta \\ \tau I & \tau H - 2\tau I - (\delta + b) \end{pmatrix}.$$

The critical point at the origin is a col with eigenvalues and corresponding eigenvectors given by $\lambda_1 = 3, (1, 0)^T$ and $\lambda_2 = -7, (3, 5)^T$. The critical point at $P = (14, 7)$ has eigenvalues $\lambda = -2 \pm i\sqrt{17}$ and is therefore a stable focus. A phase portrait is plotted in Figure 8.17.

Either the population of rabbits stabilizes to the values at P or they become extinct, depending on the initial populations. For example, plot a solution curve for the trajectory starting at $(H_0, I_0) = (7, 14)$.

Models of interacting species will be considered in Chapter 9.

8. Planar Systems

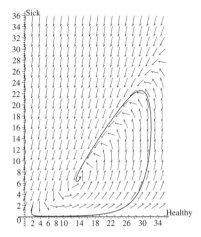

Figure 8.17: A trajectory starting from the initial point (2, 2). The population stabilizes to 14 million healthy rabbits and 7 million infected rabbits.

8.6 MATLAB Commands

The first M-file is a program for plotting vector fields and should be saved as vectorfield.m. All of the phase portraits in this chapter have the vector fields superimposed. If you have problems with the numerics (for example, if the trajectories are not smooth), you can increase the accuracy using the **RelTol** and **AbsTol** commands, as in Program 9b in Chapter 9.

```
% Program_8a - The vector field (see phase portraits in
% book).
% IMPORTANT - Save M-file as vectorfield.m.

function vectorfield(deqns,xval,yval,t)
if nargin==3;
    t=0;
end
m=length(xval);
n=length(yval);
x1=zeros(n,m);
y1=zeros(n,m);
for a=1:m
  for b=1:n
    pts = feval(deqns,t,[xval(a);yval(b)]);
    x1(b,a) = pts(1);
    y1(b,a) = pts(2);
  end
```

```
end
arrow=sqrt(x1.^2+y1.^2);
quiver(xval,yval,x1./arrow,y1./arrow,.5,'r');
axis tight;
% End of Program_8a.
```

```
% Program_8b - Phase portrait of a linear system (Figure
% 8.8(a)).
% Remember to save M-file Program8a.m as vectorfield.m.

clear
sys=inline('[2*x(1)+x(2);x(1)+2*x(2)]','t', 'x');
warning off MATLAB:divideByZero
vectorfield(sys,-3:.25:3,-3:.25:3)
     hold on
     for x0=-3:1.5:3
         for y0=-3:1.5:3
            [ts,xs] = ode45(sys,[0 5],[x0 y0]);
            plot(xs(:,1),xs(:,2))
         end
     end
     for x0=-3:1.5:3
         for y0=-3:1.5:3
            [ts,xs] = ode45(sys,[0 -5],[x0 y0]);
            plot(xs(:,1),xs(:,2))
         end
     end
     hold off
axis([-3 3 -3 3])
fsize=15;
set(gca,'xtick',[-3:1:3],'FontSize',fsize)
set(gca,'ytick',[-3:1:3],'FontSize',fsize)
xlabel('x(t)','FontSize',fsize)
ylabel('y(t)','FontSize',fsize)
hold off
% End of Program_8b.
```

```
% Program_8c - Phase portrait of a nonlinear system (Figure
% 8.12).
% Remember to save M-file Program8a.m as vectorfield.m.

clear
sys=inline('[x(2);x(1)*(1-(x(1))^2)+x(2)]','t', 'x');
warning off MATLAB:divideByZero
vectorfield(sys,-3:.25:3,-3:.25:3);
     hold on
```

```
        sep=.5;
        for x0=-3:sep:3
            for y0=-3:sep:3
                [ts,xs] = ode45(sys,[0 5],[x0 y0]);
                plot(xs(:,1),xs(:,2))
            end
        end
        for x0=-3:sep:3
            for y0=-3:sep:3
                [ts,xs] = ode45(sys,[0 -5],[x0 y0]);
                plot(xs(:,1),xs(:,2))
            end
        end
        hold off
axis([-3 3 -3 3])
fsize=15;
set(gca,'xtick',[-3:1:3],'FontSize',fsize)
set(gca,'ytick',[-3:1:3],'FontSize',fsize)
xlabel('x(t)','FontSize',fsize)
ylabel('y(t)','FontSize',fsize)
hold off
% End of Program_8c.
```

8.7 Exercises

1. (a) Find the eigenvalues and eigenvectors of the matrix

$$B = \begin{pmatrix} -7 & 6 \\ 2 & -6 \end{pmatrix}.$$

Sketch a phase portrait for the system $\dot{\mathbf{x}} = B\mathbf{x}$ and its corresponding canonical form.

(b) Carry out the same procedures as in part (a) for the system

$$\dot{x} = -4x - 8y, \qquad \dot{y} = -2y.$$

2. Sketch phase portraits for the following linear systems:

(a) $\dot{x} = 0, \dot{y} = x + 2y$;
(b) $\dot{x} = x + 2y, \dot{y} = 0$;
(c) $\dot{x} = 3x + 4y, \dot{y} = 4x - 3y$;
(d) $\dot{x} = 3x + y, \dot{y} = -x + 3y$;
(e) $\dot{x} = y, \dot{y} = -x - 2y$;
(f) $\dot{x} = x - y, \dot{y} = y - x$.

3. A very simple mechanical oscillator can be modeled using the second-order differential equation

$$\frac{d^2x}{dt^2} + \mu \frac{dx}{dt} + 25x = 0,$$

where x measures displacement from equilibrium.

   (a) Rewrite this equation as a linear first-order system by setting $\dot{x} = y$.
   (b) Sketch phase portraits when (i) $\mu = -8$, (ii) $\mu = 0$, (iii) $\mu = 8$, and (iv) $\mu = 26$.
   (c) Describe the dynamical behavior in each case given that $x(0) = 1$ and $\dot{x}(0) = 0$.

   Plot the corresponding solutions in the tx plane.

4. Plot phase portraits for the following systems:

   (a) $\dot{x} = y, \dot{y} = x - y + x^3$;
   (b) $\dot{x} = -2x - y + 2, \dot{y} = xy$;
   (c) $\dot{x} = x^2 - y^2, \dot{y} = xy - 1$;
   (d) $\dot{x} = 2 - x - y^2, \dot{y} = -y(x^2 + y^2 - 3x + 1)$;
   (e) $\dot{x} = y^2, \dot{y} = x^2$;
   (f) $\dot{x} = x^2, \dot{y} = y^2$;
   (g) $\dot{x} = y, \dot{y} = x^3$;
   (h) $\dot{x} = x, \dot{y} = \mu - y^2$ for $\mu < 0, \mu = 0$, and $\mu > 0$.

5. Construct a nonlinear system that has four critical points: two cols, one stable focus, and one unstable focus.

6. A nonlinear capacitor–resistor electrical circuit can be modeled using the differential equations

$$\dot{x} = y, \qquad \dot{y} = -x + x^3 - (a_0 + x)y,$$

where a_0 is a nonzero constant and $x(t)$ represents the current in the circuit at time t. Sketch phase portraits when $a_0 > 0$ and $a_0 < 0$ and give a physical interpretation of the results.

7. An age-dependent population can be modeled by the differential equations

$$\dot{p} = \beta + p(a - bp), \qquad \dot{\beta} = \beta(c + (a - bp)),$$

where p is the population, β is the birth rate, and a, b, and c are all positive constants. Find the critical points of this system and determine the long-term solution.

8. Planar Systems

8. The power, say, P, generated by a water wheel of velocity V can be modeled by the system

$$\dot{P} = -\alpha P + PV, \qquad \dot{V} = 1 - \beta V - P^2,$$

where α and β are both positive. Describe the qualitative behavior of this system as α and β vary and give physical interpretations of the results.

9. A very simple model for the economy is given by

$$\dot{I} = I - KS, \qquad \dot{S} = I - CS - G_0,$$

where I represents income, S is the rate of spending, G_0 denotes constant government spending, and C and K are positive constants.

   (a) Plot possible solution curves when $C = 1$ and interpret the solutions in economic terms. What happens when $C \neq 1$?

   (b) Plot the solution curve when $K = 4$, $C = 2$, $G_0 = 4$, $I(0) = 15$, and $S(0) = 5$. What happens for other initial conditions?

10. Given that

$$\frac{d^3\eta}{d\tau^3} = -\eta \frac{d^2\eta}{d\tau^2}$$

and

$$x = \frac{\eta \frac{d\eta}{d\tau}}{\frac{d^2\eta}{d\tau^2}}, \qquad y = \frac{\left(\frac{d\eta}{d\tau}\right)^2}{\eta \frac{d^2\eta}{d\tau^2}}, \qquad \text{and} \quad t = \log\left|\frac{d\eta}{d\tau}\right|,$$

prove that

$$\dot{x} = x(1 + x + y), \qquad \dot{y} = y(2 + x - y).$$

Plot a phase portrait in the xy plane.

Recommended Textbooks

[1] J. H. Liu, *A First Course in the Qualitative Theory of Differential Equations*, Prentice–Hall, Upper Saddle River, NJ, 2002.

[2] D. W. Jordan and P. Smith, *Nonlinear Ordinary Differential Equations*, 3rd ed., Oxford University Press, London, 1999.

[3] B. West, S. Strogatz, J. M. McDill, J. Cantwell, and H. Holn, *Interactive Differential Equations*, Version 2.0, Addison–Wesley, Reading, MA, 1997.

[4] E. J. Kostelich and D. Armbruster, *Introductory Differential Equations*, Addison–Wesley, 1997.

[5] J. Cronin, *Differential Equations: Introduction and Qualitative Theory*, 2nd ed., Marcel Dekker, New York, 1994.

[6] D. K. Arrowsmith and C. M. Place, *Dynamical Systems*, Chapman and Hall, London, 1992.

[7] P. Hartman, *Ordinary Differential Equations*, Wiley, New York, 1964.

9
Interacting Species

Aims and Objectives

- To apply the theory of planar systems to modeling interacting species.

On completion of this chapter the reader should be able to

- plot solution curves to modeling problems for planar systems;
- interpret the results in terms of species behavior.

The theory of planar ODEs is applied to the study of interacting species. The models are restricted in that only two species are considered and external factors such as pollution, environment, refuge, age classes, and other species interactions, for example, are ignored. However, even these restricted systems give useful results. These simple models can be applied to species living in our oceans and to both animal and insect populations on land. Note that the continuous differential equations used in this chapter are only relevant if the species populations under consideration are large, typically scaled by 10^4, 10^5, or 10^6 in applications.

9.1 Competing Species

Suppose that there are two species in competition with one another in an environment where the common food supply is limited. For example, sea lions and penguins, red and grey squirrels, and ants and termites are all species that fall into this category. There are two particular types of outcome that are often observed in

9. Interacting Species

the real world. In the first case, there is *coexistence*, in which the two species live in harmony. (In nature, this is the most likely outcome; otherwise, one of the species would be extinct.) In the second case, there is *mutual exclusion*, in which one of the species becomes extinct. (For example, American grey squirrels imported into the United Kingdom are causing the extinction of the smaller native red squirrels.)

Both coexistence and mutual exclusion can be observed when plotting solution curves on a phase plane diagram. Consider the following general model for two competing species.

Example 1. Sketch possible phase plane diagrams for the following system:

(9.1) $$\dot{x} = x(\beta - \delta x - \gamma y), \qquad \dot{y} = y(b - dy - cx),$$

where $\beta, \delta, \gamma, a, b,$ and c are all positive constants with $x(t)$ and $y(t)$—both positive—representing the two species populations measured in tens or hundreds of thousands.

Solution. The terms appearing in the right-hand sides of equation (9.1) have a physical meaning as follows:

- The terms $\beta x - \delta x^2$ and $by - dy^2$ represent the usual logistic growth of one species (Verhulst's equation).

- Both species suffer as a result of competition over a limited food supply, hence the terms $-\gamma xy$ and $-cxy$ in $\dot{x}$ and $\dot{y}$.

Construct a phase plane diagram in the usual way. Find the critical points, linearize around each one, determine the isoclines, and plot the phase plane portrait.

Locate the critical points by solving the equations $\dot{x} = \dot{y} = 0$. There are four critical points at

$$O = (0,0), \qquad P = \left(0, \frac{b}{d}\right), \qquad Q = \left(\frac{\beta}{\delta}, 0\right),$$

and

$$R = \left(\frac{\gamma b - \beta d}{\gamma c - \delta d}, \frac{\beta c - \delta b}{\gamma c - \delta d}\right).$$

Suppose that $C_1 = \gamma c - \delta d$, $C_2 = \gamma b - \beta d$, and $C_3 = \beta c - \delta b$. For the critical point to lie in the first quadrant, one of the following conditions must hold: Either

(i) $C_1, C_2,$ and C_3 are all negative, or

(ii) $C_1, C_2,$ and C_3 are all positive.

Linearize by finding the Jacobian matrix. Therefore,

$$J = \begin{pmatrix} \beta - 2\delta x - \gamma y & -\gamma x \\ -cy & b - 2dy - cx \end{pmatrix}.$$

Linearize at each critical point. Thus

$$J_O = \begin{pmatrix} \beta & 0 \\ 0 & b \end{pmatrix}.$$

For the critical point at P,

$$J_P = \begin{pmatrix} \beta - \gamma b/d & 0 \\ -bc/d & -b \end{pmatrix}.$$

For the critical point at Q,

$$J_Q = \begin{pmatrix} -\beta & -\gamma\beta/\delta \\ 0 & b - \beta c/\delta \end{pmatrix}.$$

Finally, for the critical point at R,

$$J_R = \frac{1}{C_1}\begin{pmatrix} \delta C_2 & \gamma C_2 \\ c C_3 & d C_3 \end{pmatrix}.$$

Consider case (i) first. The fixed points are all simple and it is not difficult to show that O is an unstable node, P and Q are cols, and for certain parameter values R is a stable fixed point. A phase portrait is plotted in Figure 9.1(a), where eight of an infinite number of solution curves are plotted. Each trajectory is plotted numerically for both positive and negative time steps; in this way, critical points are easily identified in the phase plane. For the parameter values chosen here, the two species coexist and the populations stabilize to constant values after long time periods. The arrows in Figure 9.1(a) show the vector field plot and define the direction of the trajectories for system (9.1). The slope of each arrow is given by $\frac{dy}{dx}$ at the point, and the direction of the arrows is determined from $\dot{x}$ and $\dot{y}$. There is a stable node lying wholly in the first quadrant at R, and the nonzero populations $x(t)$ and $y(t)$ tend to this critical point with increasing time no matter what the initial populations are. The domain of stability for the critical point at R is therefore $S_R = \{(x, y) \in \Re^2 : x > 0, y > 0\}$.

Now consider case (ii). The fixed points are all simple, and it is not difficult to show that O is an unstable node, P and Q are stable or improper nodes, and R is a col. A phase portrait is shown in Figure 9.1(b), where nine of an infinite number of solution curves are plotted. Once more the trajectories are plotted for both positive and negative time iterations. In this case, one of the species becomes extinct.

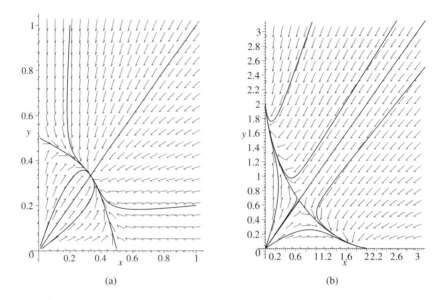

Figure 9.1: (a) A possible phase portrait showing coexistence and (b) a possible phase portrait depicting mutual exclusion. Note that the axes are invariant in both cases.

In Figure 9.1(b), the critical point lying wholly in the first quadrant is a saddle point or col, which is unstable. The long-term behavior of the system is divided along the diagonal in the first quadrant. Trajectories starting to the right of the diagonal will tend to the critical point at $Q = (2, 0)$, which implies that species y becomes extinct. Trajectories starting to the left of the diagonal will tend to the critical point at $P = (0, 2)$, which means that species x will become extinct. Numerically, the trajectories lying on the stable manifold of the saddle point in the first quadrant will tend towards the critical point at R. However, in the real world, populations cannot remain exactly on the stable manifold, and trajectories will be diverted from this critical point leading to extinction of one of the species. The domain of stability for the critical point at $P = (0, 2)$ is given by, $S_P = \{(x, y) \in \Re^2 : x > 0, y > 0, y > x\}$. The domain of stability for the critical point at $Q = (2, 0)$ is given by $S_Q = \{(x, y) \in \Re^2 : x > 0, y > 0, y < x\}$.

9.2 Predator–Prey Models

Consider a two-species predator–prey model in which one species preys on another. Examples in the natural world include sharks and fish, lynx and snow-shoe hares, and ladybirds and aphids. A very simple differential equation—first used by Volterra in 1928 [5,6] and known as the *Lotka–Volterra model*—is given in Example 2.

9.2. Predator–Prey Models

Example 2. Sketch a phase portrait for the system

(9.2) $$\dot{x} = x(\alpha - cy), \qquad \dot{y} = y(\gamma x - \delta),$$

where α, c, γ, and δ are all positive constants, with $x(t)$ and $y(t)$ representing the scaled population of prey and predator, respectively, and t is measured in years.

Solution. The terms appearing in the right-hand sides of equation (9.2) have a physical meaning as follows:

- The term αx represents the growth of the population of prey in the absence of any predators. This is obviously a crude model; the population of a species cannot increase forever.

- The terms $-cxy$ and $+\gamma xy$ represent species interaction. The population of prey suffers and predators gain from the interaction.

- The term $-\delta y$ represents the extinction of predators in the absence of prey.

Attempt to construct a phase plane diagram in the usual way. Find the critical points, linearize around each one, determine the isoclines, and plot the phase plane portrait.

The critical points are found by solving the equations $\dot{x} = \dot{y} = 0$. There are two critical points, one at $O = (0, 0)$ and the other at $P = (\frac{\delta}{\gamma}, \frac{\alpha}{c})$.

Linearize to obtain

$$J = \begin{pmatrix} \alpha - cy & -cx \\ \gamma y & -\delta + \gamma x \end{pmatrix}.$$

The critical point at the origin is a saddle point, and the stable and unstable manifolds lie along the axes. The stable manifold lies on the positive y-axis and the unstable manifold lies on the x-axis. The critical point at P is not hyperbolic, and so Hartman's theorem cannot be applied. System (9.2) has solution curves (the differential equation is separable) given by $x^\delta y^\alpha e^{-\gamma x} e^{-cy} = K$, where K is a constant. These solution curves may be plotted in the phase plane. The isoclines are given by $x = 0$, $y = \frac{\alpha}{c}$, where the flow is vertical, and $y = 0$, $x = \frac{\delta}{\gamma}$, where the flow is horizontal. The vector fields are found by considering $\dot{x}$, $\dot{y}$, and $\frac{dy}{dx}$. A phase portrait is shown in Figure 9.2.

The population fluctuations can also be represented in the tx and ty planes. The graphs shown in Figure 9.3 show how the populations of predator and prey oscillate.

Note that the oscillations are dependent on the initial conditions. In Figure 9.3, the period of both cycles is about 10 years. Different sets of initial conditions can give solutions with different amplitudes. For example, plot the solution curves in the tx and ty planes for the initial conditions $x(0) = 1$ and $y(0) = 3$.

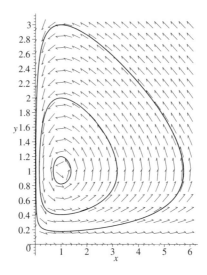

Figure 9.2: A phase portrait for the Lotka–Volterra model.

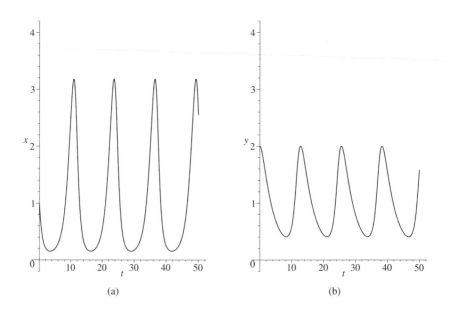

Figure 9.3: (a) Periodic behavior of the prey and (b) periodic behavior of the predators for one set of initial conditions, namely, $x(0) = 1$, $y(0) = 2$.

How can this system be interpreted in terms of species behavior? Consider the trajectory passing through the point (1, 3) in Figure 9.2. At this point the ratio of predators to prey is high; as a result the population of predators drops. The ratio of predators to prey drops, and so the population of prey increases. Once there are lots of prey, the predator numbers will again start to increase. The resulting cyclic behavior is repeated over and over and is shown as the largest closed trajectory in Figure 9.2.

If small perturbations are introduced into system (9.2)—to model other factors, for example—then the qualitative behavior changes. The periodic cycles can be destroyed by adding small terms into the right-hand sides of system (9.2). The system is said to be *structurally unstable* (or *not robust*).

Many predator–prey interactions have been modeled in the natural world. For example, there are data dating back over 150 years for the populations of lynx and snowshoe hares from the Hudson Bay Company in Canada. The data clearly shows that the populations periodically rise and fall (with a period of about 10 years) and that the maximum and minimum values (amplitudes) are relatively constant. This is not true for the Lotka–Volterra model (see Figure 9.2). Different initial conditions can give solutions with different amplitudes. In 1975, Holling and Tanner constructed a system of differential equations whose solutions have the same amplitudes in the long term, no matter what the initial populations. Two particular examples of the Holling–Tanner model for predator–prey interactions are given in Example 3.

The reader is encouraged to compare the terms (and their physical meaning) appearing in the right-hand sides of the differential equations in Examples 1–3.

Example 3. Consider the specific Holling–Tanner model

(9.3) $$\dot{x} = x\left(1 - \frac{x}{7}\right) - \frac{6xy}{(7+7x)}, \qquad \dot{y} = 0.2y\left(1 - \frac{Ny}{x}\right),$$

where N is a constant with $x(t) \neq 0$ and $y(t)$ representing the populations of prey and predators, respectively. Sketch phase portraits when (i) $N = 2.5$ and (ii) $N = 0.5$.

Solution. The terms appearing in the right-hand sides of equation (9.3) have a physical meaning as follows:

- The term $x(1 - \frac{x}{7})$ represents the usual logistic growth in the absence of predators.

- The term $-\frac{6xy}{(7+7x)}$ represents the effect of predators subject to a maximum predation rate.

- The term $0.2y(1 - \frac{Ny}{x})$ denotes the predator growth rate when a maximum of x/N predators is supported by x prey.

Construct a phase plane diagram in the usual way. Find the critical points, linearize around each one, determine the isoclines, and plot a phase plane portrait.

Consider case (i). The critical points are found by solving the equations $\dot{x} = \dot{y} = 0$. There are two critical points in the first quadrant, $A = (5, 2)$ and $B = (7, 0)$. The Jacobian matrices are given by

$$J_A = \begin{pmatrix} -1 & -3/4 \\ 0 & 1/5 \end{pmatrix}$$

and

$$J_B = \begin{pmatrix} -10/21 & -5/7 \\ 2/25 & -1/5 \end{pmatrix}.$$

The eigenvalues and eigenvectors of J_A are given by $\lambda_1 = -1$, $(1, 0)^T$ and $\lambda_2 = \frac{1}{5}$, $(-\frac{5}{8}, 1)^T$. Therefore, this critical point is a saddle point or col with the stable manifold lying along the x-axis and the unstable manifold tangent to the line with slope $-\frac{8}{5}$ in a small neighborhood around the critical point. The eigenvalues of J_B are given by $\lambda \approx -0.338 \pm 0.195i$. Therefore, the critical point at B is a stable focus.

A phase portrait showing four trajectories and the vector field is shown in Figure 9.4(a).

The populations eventually settle down to constant values. If there are any natural disasters or diseases, for example, the populations would both decrease but eventually return to the stable values. This is, of course, assuming that neither species becomes extinct. There is no periodic behavior in this model.

Consider case (ii). The critical points are found by solving the equations $\dot{x} = \dot{y} = 0$. There are two critical points in the first quadrant, $A = (1, 2)$ and $B = (7, 0)$. The Jacobian matrices are given by

$$J_A = \begin{pmatrix} -1 & -3/4 \\ 0 & 1/5 \end{pmatrix}$$

and

$$J_B = \begin{pmatrix} 2/7 & -3/7 \\ 2/5 & -1/5 \end{pmatrix}.$$

The eigenvalues and eigenvectors of J_A are given by $\lambda_1 = -1$, $(1, 0)^T$ and $\lambda_2 = \frac{1}{5}$, $(-\frac{5}{8}, 1)^T$. Therefore, this critical point is a saddle point or col with the stable manifold lying along the x-axis and the unstable manifold tangent to the line with slope $-\frac{8}{5}$ near to the critical point. The eigenvalues of J_B are given by $\lambda \approx 0.043 \pm 0.335i$. Therefore, the critical point at B is an unstable focus.

All trajectories lying in the first quadrant are drawn to the closed periodic cycle shown in Figure 9.5(a). Therefore, no matter what the initial values of $x(t)$

9.2. Predator–Prey Models

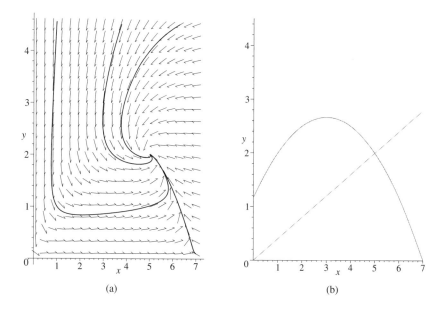

Figure 9.4: (a) A phase portrait for system (9.3) when $N = 2.5$. (b) Intersection of the isoclines.

and $y(t)$, the populations eventually rise and fall periodically. This isolated periodic trajectory is known as a *stable limit cycle*. In the long term, all trajectories in the first quadrant are drawn to this periodic cycle, and once there, remain there forever. Definitions and the theory of limit cycles will be introduced in Chapter 10. The isoclines are plotted in Figure 9.5(b); these curves show where the flow is horizontal or vertical, in this case.

The limit cycle persists if small terms are added to the right-hand sides of the differential equations in system (9.3). The system is *structurally stable* (or *robust*) since small perturbations do not affect the qualitative behavior. Again the populations of both predator and prey oscillate in a similar manner to the Lotka–Volterra model with another major exception. The final steady-state solution for the Holling–Tanner model is independent of the initial conditions. Use MATLAB to plot time series plots for the solutions plotted in Figure 9.5(a). A program is listed in Section 9.4. The period of the limit cycle can be easily established from the time series plot. This model appears to match very well with what happens for many predator–prey species in the natural world—for example, house sparrows and sparrow hawks in Europe, muskrat and mink in Central North America, and white-tailed deer and wolf in Ontario.

From the time series plot, the period, say, T, of the limit cycle is approximately 19 units of time. Thus if t is measured in six-month intervals, then this would be

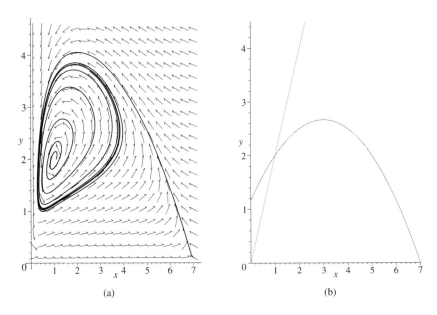

Figure 9.5: (a) A phase portrait for system (9.3) when $N = 0.5$. (b) Intersection of the isoclines.

a good model for the lynx and snowshoe hare populations, which have a natural period of about 10 years. Periodicity of limit cycles will be discussed in the next chapter.

9.3 Other Characteristics Affecting Interacting Species

A simple model of one species infected with a disease was considered in Chapter 8. The models considered thus far for interacting species have been limited to only two populations, and external factors have been ignored. There are a great many research papers published every year on interacting species, and the author hopes that this chapter will inspire the reader to investigate further. To conclude Chapter 9, some other characteristics ignored here will be listed. Of course, the differential equations will become more complicated and are beyond the scope of this chapter.

- Age classes—for example, young, mature, and old; time lags need to be introduced into the differential equations (see Chapter 1).

- Diseases—epidemics affecting one or more species (see Chapter 8).

- Environmental effects.

- Enrichment of prey—this can lead to extinction of predators.

- Harvesting and culling policies (see Chapter 1).
- Pollution—persistence and extinction.
- Refuge—for example, animals in Africa find refuge in the bush.
- Seasonal effects—for example, some animals hibernate in winter.
- Three or more species interactions (see the exercises in Section 9.5).

One interesting example is discussed by Lenbury et al. [2], where predator–prey interaction is coupled to parasitic infection. One or both of the species can become infected, and this can lead to mathematical problems involving four systems of differential equations. The dynamics become far more complicated, and more interesting behavior is possible. Higher-dimensional systems will be discussed later in the book.

9.4 MATLAB Commands

```
% Program_9a - The Holling-Tanner Model (Figure 9.5(a)).
% Phase portrait of a predator/prey model.

clear
hold on
sys=inline('[x(1)*(1-x(1)/7)-6*x(1)*x(2)/(7+7*x(1));
0.2*x(2)*(1-.5*x(2)/x(1))]','t','x')
options = odeset('RelTol',1e-4,'AbsTol',1e-4);
[t,xa]=ode45(sys,[0 100],[7.1 .1],options);
plot(xa(:,1),xa(:,2))
axis([0 8 0 5])
fsize=15;
set(gca,'xtick',[0:2:8],'FontSize',fsize)
set(gca,'ytick',[0:2.5:5],'FontSize',fsize)
xlabel('x(t)','FontSize',fsize)
ylabel('y(t)','FontSize',fsize)
hold off
% End of Program_9a.
```

```
% Program_9b - The Holling-Tanner Model (Time series).
% Time series of a predator/prey model.

clear
hold on
sys=inline('[x(1)*(1-x(1)/7)-6*x(1)*x(2)/(7+7*x(1));
0.2*x(2)*(1-.5*x(2)/x(1))]','t','x')
options = odeset('RelTol',1e-4,'AbsTol',1e-4);
```

```
[t,xa]=ode45(sys,[0 200],[7.1 .1],options);
plot(t,xa(:,1),'r')
plot(t,xa(:,2),'b')
legend('prey','predator')
axis([0 200 0 8])
fsize=15;
set(gca,'xtick',[0:50:200],'FontSize',fsize)
set(gca,'ytick',[0:2:8],'FontSize',fsize)
xlabel('time','FontSize',fsize)
ylabel('population','FontSize',fsize)
hold off
% End of Program_9b.
```

9.5 Exercises

1. Plot a phase portrait for the following competing species model

$$\dot{x} = 2x - x^2 - xy, \qquad \dot{y} = 3y - y^2 - 2xy$$

and describe what happens in terms of species behavior.

2. Plot a phase plane diagram for the following predator–prey system and interpret the solutions in terms of species behavior:

$$\dot{x} = 2x - xy, \qquad \dot{y} = -3y + xy.$$

3. Plot a phase portrait for the following system and describe what happens to the population for different initial conditions:

$$\dot{x} = 2x - x^2 - xy, \qquad \dot{y} = -y - y^2 + xy.$$

4. The differential equations used to model a competing species are given by

$$\dot{x} = x(2 - x - y), \qquad \dot{y} = y\left(\mu - y - \mu^2 x\right),$$

where μ is a constant. Describe the qualitative behavior of this system as the parameter μ varies.

5. (a) Sketch a phase portrait for the system

$$\dot{x} = x(4 - y - x), \qquad \dot{y} = y(3x - 1 - y), \qquad x \geq 0, \quad y \geq 0$$

given that the critical points occur at $O = (0, 0)$, $A = (4, 0)$, and $B = (\frac{5}{4}, \frac{11}{4})$.

(b) Sketch a phase portrait for the system

$$\dot{x} = x(2 - y - x), \qquad \dot{y} = y(3 - 2x - y), \qquad x \geq 0, \qquad y \geq 0,$$

given that the critical points occur at $O = (0, 0)$, $C = (0, 3)$, $D = (2, 0)$, and $E = (1, 1)$.

One of the systems can be used to model predator–prey interactions and the other competing species. Describe which system applies to which model and interpret the results in terms of species behavior.

6. A predator–prey system may be modeled using the differential equations

$$\dot{x} = x(1 - y - \epsilon x), \qquad \dot{y} = y(-1 + x - \epsilon y),$$

where $x(t)$ is the population of prey and $y(t)$ is the predator population size at time t, respectively. Classify the critical points for $\epsilon \geq 0$ and plot phase portraits for the different types of qualitative behavior. Interpret the results in physical terms.

7. A predator–prey model is given by

$$\dot{x} = x(x - x^2 - y), \qquad \dot{y} = y(x - 0.6).$$

Sketch a phase portrait and interpret the results in physical terms.

8. Use MATLAB to plot a trajectory for the predator–prey system

$$\dot{x} = x(x - x^2 - y), \qquad \dot{y} = y(x - 0.48)$$

using the initial condition $(0.6, 0.1)$. What can you deduce about the long-term populations?

9. Suppose that there are three species of insect X, Y, and Z, say. Give rough sketches to illustrate the possible ways in which these species can interact with one another. You should include the possibility of a species being cannibalistic. Three-dimensional systems will be discussed later.

10. The following three differential equations are used to model a combined predator–prey and competing species system:

$$\dot{x} = x(a_{10} - a_{11}x + a_{12}y - a_{13}z),$$
$$\dot{y} = y(a_{20} - a_{21}x - a_{22}y - a_{23}z),$$
(9.4) $\qquad \dot{z} = z(a_{30} + a_{31}x - a_{32}y - a_{33}z),$

where a_{ij} are positive constants. Give a physical interpretation for the terms appearing in the right-hand sides of these differential equations.

Recommended Reading

[1] F. Brauer and C. Castillo-Chavez, *Mathematical Models in Population Biology and Epidemiology*, Springer-Verlag, New York, 2001.

[2] Y. Lenbury, S. Rattanamongkonkul, N. Tumravsin, and S. Amornsamankul, Predator-prey interaction coupled by parasitic infection: Limit cycles and chaotic behavior, *Math. Comput. Model.*, **30**-9/10 (1999), 131–146.

[3] R. E. Kooij and A. Zegeling, Qualitative properties of two-dimensional predator-prey systems, *Nonlinear Anal. Theory Meth. Appl.*, **29** (1997), 693–715.

[4] E. C. Pielou, *Mathematical Ecology*, Wiley, New York, 1977.

[5] V. Volterra, Variazioni e fluttuazioni del numero d'individui in specie animali conviventi, *Mem. R. Accad. Naz. Lincei*, **2**-3 (1926), 30–113.

[6] A. J. Lotka, *Elements of Physical Biology*, William and Wilkins, Baltimore, 1925.

10
Limit Cycles

Aims and Objectives
- To give a brief historical background.
- To define features of phase plane portraits.
- To introduce the theory of planar limit cycles.

On completion of this chapter the reader should be able to
- prove existence and uniqueness of a limit cycle;
- prove that certain systems have no limit cycles;
- interpret limit cycle behavior in physical terms.

Limit cycles, or isolated periodic solutions, are the most common form of solution observed when modeling physical systems in the plane. Early investigations were concerned with mechanical and electronic systems, but periodic behavior is evident in all branches of science. The chapter begins with a historical introduction, and then the theory of planar limit cycles is introduced.

10.1 Historical Background

Definition 1. A *limit cycle* is an isolated periodic solution.

10. Limit Cycles

Limit cycles in planar differential systems commonly occur when modeling both the technological and natural sciences. Most of the early history in the theory of limit cycles in the plane was stimulated by practical problems. For example, the differential equation derived by Rayleigh in 1877 [4], related to the oscillation of a violin string, is given by

$$\ddot{x} + \epsilon \left(\frac{1}{3}(\dot{x})^2 - 1 \right) \dot{x} + x = 0,$$

where $\ddot{x} = \frac{d^2 x}{dt^2}$ and $\dot{x} = \frac{dx}{dt}$. Let $\dot{x} = y$. Then this differential equation can be written as a system of first-order autonomous differential equations in the plane

(10.1) $$\dot{x} = y, \qquad \dot{y} = -x - \epsilon \left(\frac{y^2}{3} - 1 \right) y.$$

A phase portrait is shown in Figure 10.1.

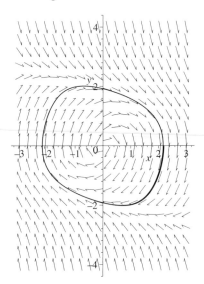

Figure 10.1: Periodic behavior in the Rayleigh system (10.1) when $\epsilon = 1.0$.

Following the invention of the triode vacuum tube, which was able to produce stable self-excited oscillations of constant amplitude, van der Pol [6] obtained the following differential equation to describe this phenomenon

$$\ddot{x} + \epsilon(x^2 - 1)\dot{x} + x = 0,$$

which can be written as a planar system of the form

(10.2) $$\dot{x} = y, \qquad \dot{y} = -x - \epsilon(x^2 - 1)y.$$

10.1. Historical Background

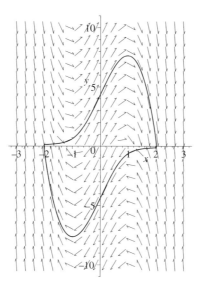

Figure 10.2: Periodic behavior for system (10.2) when $\epsilon = 5.0$.

A phase portrait is shown in Figure 10.2.

Perhaps the most famous class of differential equations that generalize (10.2) are those first investigated by Liénard in 1928 [5],

$$\ddot{x} + f(x)\dot{x} + g(x) = 0,$$

or in the phase plane

(10.3) $\qquad \dot{x} = y, \qquad \dot{y} = -g(x) - f(x)y.$

This system can be used to model mechanical systems, where $f(x)$ is known as the *damping* term and $g(x)$ is called the *restoring force* or *stiffness*. Equation (10.3) is also used to model resistor–inductor–capacitor circuits (see Chapter 7) with nonlinear circuit elements. Limit cycles of Liénard systems will be discussed in some detail in Chapters 15 and 16.

Possible physical interpretations for limit cycle behavior of certain dynamical systems are listed below:

- For predator–prey and epidemic models, the populations oscillate in phase with one another and the systems are robust (see Example 3 in Chapter 9 and Example 10 in Chapter 13).

- For mechanical systems, examples include the motion of simple nonlinear pendula, wing rock oscillations in aircraft flight dynamics, and surge oscillations in axial flow compressors.

- For periodic chemical reactions, examples include the Landolt clock reaction and the Belousov–Zhabotinski reaction (see Chapter 13).

- For electrical or electronic circuits, it is possible to construct simple electronic oscillators using a nonlinear circuit element; a limit cycle can be observed if the circuit is connected to an oscilloscope.

Limit cycles are common solutions for all types of dynamical systems. Sometimes it becomes necessary to prove the existence and uniqueness of a limit cycle, as described in the next section.

10.2 Existence and Uniqueness of Limit Cycles in the Plane

To understand the existence and uniqueness theorem, it is necessary to define some features of phase plane portraits. Assume that the existence and uniqueness theorem from Chapter 7 holds for all solutions considered here.

Definition 2. A *flow* on $\Re^2$ is a mapping $\pi : \Re^2 \to \Re^2$ such that

1. π is continuous;
2. $\pi(\mathbf{x}, 0) = \mathbf{x}$ for all $\mathbf{x} \in \Re^2$;
3. $\pi(\pi(\mathbf{x}, t_1), t_2) = \pi(\mathbf{x}, t_1 + t_2)$.

Definition 3. Suppose that $I_\mathbf{x}$ is the maximal interval of existence. The *trajectory* (or *orbit*) through $\mathbf{x}$ is defined as $\gamma(\mathbf{x}) = \{\pi(\mathbf{x}, t) : t \in I_\mathbf{x}\}$.
The *positive semiorbit* is defined as $\gamma^+(\mathbf{x}) = \{\pi(\mathbf{x}, t) : t > 0\}$.
The *negative semiorbit* is defined as $\gamma^-(\mathbf{x}) = \{\pi(\mathbf{x}, t) : t < 0\}$.

Definition 4. The *positive limit set* of a point $\mathbf{x}$ is defined as

$$\Lambda^+(\mathbf{x}) = \{\mathbf{y} : \text{there exists a sequence } t_n \to \infty \text{ such that } \pi(\mathbf{x}, t) \to \mathbf{y}\}.$$

The *negative limit set* of a point $\mathbf{x}$ is defined as

$$\Lambda^-(\mathbf{x}) = \{\mathbf{y} : \text{there exists a sequence } t_n \to -\infty \text{ such that } \pi(\mathbf{x}, t) \to \mathbf{y}\}.$$

In the phase plane, trajectories tend to a critical point, a closed orbit, or infinity.

Definition 5. A set S is *invariant* with respect to a flow if $\mathbf{x} \in S$ implies that $\gamma(\mathbf{x}) \subset S$.
A set S is *positively invariant* with respect to a flow if $\mathbf{x} \in S$ implies that $\gamma^+(\mathbf{x}) \subset S$.
A set S is *negatively invariant* with respect to a flow if $\mathbf{x} \in S$ implies that $\gamma^-(\mathbf{x}) \subset S$.
A general trajectory can be labeled γ for simplicity.

10.2. Existence and Uniqueness of Limit Cycles in the Plane

Definition 6. A limit cycle, say, Γ, is

- a *stable limit cycle* if $\Lambda^+(\mathbf{x}) = \Gamma$ for all $\mathbf{x}$ in some neighborhood; this implies that nearby trajectories are attracted to the limit cycle;

- an *unstable limit cycle* if $\Lambda^-(\mathbf{x}) = \Gamma$ for all $\mathbf{x}$ in some neighborhood; this implies that nearby trajectories are repelled away from the limit cycle;

- a *semistable limit cycle* if it is attracting on one side and repelling on the other.

The stability of limit cycles can also be deduced analytically using the Poincaré map (see Chapter 14). The following example will be used to illustrate each of the Definitions 1–6 above and 7 below.

Definition 7. The period, say, T, of a limit cycle is given by $\mathbf{x}(t) = \mathbf{x}(t + T)$, where T is the minimum period. The period can be found by plotting a time series plot of the limit cycle (see Program 9b).

Example 1. Describe some of the features for the following set of polar differential equations in terms of Definitions 1–7:

$$(10.4) \qquad \dot{r} = r(1-r)(2-r)(3-r), \qquad \dot{\theta} = -1.$$

Solution. A phase portrait is shown in Figure 10.3. There is a unique critical point at the origin since $\dot{\theta}$ is nonzero. There are three limit cycles that may be determined from the equation $\dot{r} = 0$. They are the circles of radii one, two, and three, all centered at the origin. Let Γ_i denote the limit cycle of radius $r = i$.

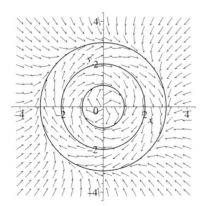

Figure 10.3: Three limit cycles for system (10.4).

There is one critical point at the origin. If a trajectory starts at this point, it remains there forever. A trajectory starting at $(1, 0)$ will reach the point $(-1, 0)$

when $t_1 = \pi$ and the motion is clockwise. Continuing on this path for another time interval $t_2 = \pi$, the orbit returns to $(1, 0)$. Using part 3 of Definition 2, one can write $\pi(\pi((1, 0), t_1), t_2) = \pi((1, 0), 2\pi)$ since the limit cycle is of period 2π (see below). On the limit cycle Γ_1, both the positive and negative semiorbits lie on Γ_1.

Suppose that $P = (\frac{1}{2}, 0)$ and $Q = (4, 0)$ are two points in the plane. The limit sets are given by $\Lambda^+(P) = \Gamma_1$, $\Lambda^-(P) = (0, 0)$, $\Lambda^+(Q) = \Gamma_3$, and $\Lambda^-(Q) = \infty$.

The annulus $A_1 = \{r \in \Re^2 : 0 < r < 1\}$ is positively invariant, and the annulus $A_2 = \{r \in \Re^2 : 1 < r < 2\}$ is negatively invariant.

If $0 < r < 1$, then $\dot{r} > 0$ and the critical point at the origin is unstable. If $1 < r < 2$, then $\dot{r} < 0$ and Γ_1 is a stable limit cycle. If $2 < r < 3$, then $\dot{r} > 0$ and Γ_2 is an unstable limit cycle. Finally, if $r > 3$, then $\dot{r} < 0$ and Γ_3 is a stable limit cycle.

Integrate both sides of $\dot{\theta} = -1$ with respect to time to show that the period of all of the limit cycles is 2π.

The Poincaré–Bendixson Theorem. *Suppose that γ^+ is contained in a bounded region in which there are finitely many critical points. Then $\Lambda^+(\gamma)$ is either*

- *a single critical point;*
- *a single closed orbit;*
- *a graphic—critical points joined by heteroclinic orbits (see Chapter 11).*

Corollary. *Let D be a bounded closed set containing no critical points and suppose that D is positively invariant. Then there exists a limit cycle contained in D.*

A proof of this theorem involves topological arguments and can be found in [1], for example.

Example 2. *By considering the flow across the rectangle with corners at $(-1, 2)$, $(1, 2)$, $(1, -2)$, and $(-1, -2)$, prove that the following system has at least one limit cycle:*

(10.5) $$\dot{x} = y - 8x^3, \qquad \dot{y} = 2y - 4x - 2y^3.$$

Solution. The critical points are found by solving the equations $\dot{x} = \dot{y} = 0$. Set $y = 8x^3$. Then $\dot{y} = 0$ if $x(1 - 4x^2 + 256x^8) = 0$. The graph of the function $y = 1 - 4x^2 + 256x^8$ is given in Figure 10.4(a). The graph has no roots and the origin is the only critical point.

Linearize at the origin in the usual way. It is not difficult to show that the origin is an unstable focus.

Consider the flow on the sides of the given rectangle:

- On $y = 2$, $|x| \leq 1$, $\dot{y} = -4x - 12 < 0$.
- On $y = -2$, $|x| \leq 1$, $\dot{y} = -4x + 12 > 0$.

10.2. Existence and Uniqueness of Limit Cycles in the Plane

- On $x = 1$, $|y| \leq 2$, $\dot{x} = y - 8 < 0$.
- On $x = -1$, $|y| \leq 2$, $\dot{y} = y + 8 > 0$.

The flow is depicted in Figure 10.4(b). The rectangle is positively invariant and there are no critical points other than the origin, which is unstable. Consider a small deleted neighborhood, say, N_ϵ, around this critical point. For example, the boundary of N_ϵ could be a small ellipse. On this ellipse, all trajectories will cross outwards. Therefore, there exists a stable limit cycle lying inside the rectangular region and outside of N_ϵ by the corollary to the Poincaré–Bendixson theorem.

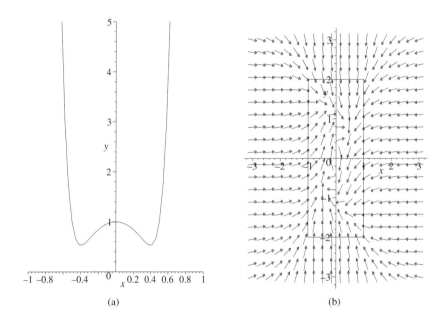

Figure 10.4: (a) Polynomial of degree 8. (b) Flow across the rectangle for system (10.5).

Definition 8. A planar simple closed curve is called a *Jordan curve*.

Consider the system

(10.6) $$\dot{x} = P(x, y), \qquad \dot{y} = Q(x, y),$$

where P and Q have continuous first-order partial derivatives. Let the vector field be denoted by $\mathbf{X}$ and let ψ be a weighting factor that is continuously differentiable. Recall Green's theorem, which will be required to prove the following two theorems.

10. Limit Cycles

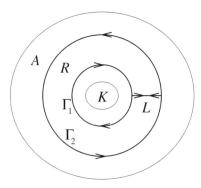

Figure 10.5: Two limit cycles encircling the region K.

Green's Theorem. *Let J be a Jordan curve of finite length. Suppose that P and Q are two continuously differentiable functions defined on the interior of J, say, D. Then*

$$\iint_D \left[\frac{\partial P}{\partial x} + \frac{\partial Q}{\partial y} \right] dx\, dy = \oint_J P\, dy - Q\, dx.$$

Dulac's Criteria. *Consider an annular region, say, A, contained in an open set E. If*

$$\nabla \cdot (\psi \mathbf{X}) = \operatorname{div}(\psi \mathbf{X}) = \frac{\partial}{\partial x}(\psi P) + \frac{\partial}{\partial y}(\psi Q)$$

does not change sign in A, then there is at most one limit cycle entirely contained in A.

Proof. Suppose that Γ_1 and Γ_2 are limit cycles encircling K, as depicted in Figure 10.5, of periods T_1 and T_2, respectively. Apply Green's theorem to the region R shown in Figure 10.5:

$$\iint_R \left[\frac{\partial (\psi P)}{\partial x} + \frac{\partial (\psi Q)}{\partial y} \right] dx\, dy$$

$$= \oint_{\Gamma_2} \psi P\, dy - \psi Q\, dx + \int_L \psi P\, dy - \psi Q\, dx$$

$$- \oint_{\Gamma_1} \psi P\, dy - \psi Q\, dx - \int_L \psi P\, dy - \psi Q\, dx.$$

Now on Γ_1 and Γ_2, $\dot{x} = P$ and $\dot{y} = Q$, so

$$\iint_R \left[\frac{\partial (\psi P)}{\partial x} + \frac{\partial (\psi Q)}{\partial y} \right] dx\, dy$$

$$= \int_0^{T_2} (\psi PQ - \psi QP)dt - \int_0^{T_1} (\psi PQ - \psi QP)dt,$$

which is zero and contradicts the hypothesis that $\text{div}(\psi X) \neq 0$ in A. Therefore, there is at most one limit cycle entirely contained in the annulus A. □

Example 3. Use Dulac's criteria to prove that the system

(10.7) $\quad \dot{x} = -y + x(1 - 2x^2 - 3y^2), \qquad \dot{y} = x + y(1 - 2x^2 - 3y^2)$

has a unique limit cycle in an annulus.

Solution. Convert to polar coordinates using the transformations

$$r\dot{r} = x\dot{x} + y\dot{y}, \qquad r^2\dot{\theta} = x\dot{y} - y\dot{x}.$$

Therefore, system (10.7) becomes

$$\dot{r} = r(1 - 2r^2 - r^2 \sin^2 \theta), \qquad \dot{\theta} = 1.$$

Since $\dot{\theta} = 1$, the origin is the only critical point. On the circle $r = \frac{1}{2}$, $\dot{r} = \frac{1}{2}(\frac{1}{2} - \frac{1}{4}\sin^2 \theta)$. Hence $\dot{r} > 0$ on this circle. On the circle $r = 1$, $\dot{r} = -1 - \sin^2 \theta$. Hence $\dot{r} < 0$ on this circle. If $r \geq 1$, then $\dot{r} < 0$, and if $0 < r \leq \frac{1}{2}$, then $\dot{r} > 0$. Therefore, there exists a limit cycle in the annulus $A = \{r : \frac{1}{2} < r < 1\}$ by the corollary to the Poincaré–Bendixson theorem.

Consider the annulus A. Now $\text{div}(X) = 2(1 - 4r^2 - 2r^2 \sin^2 \theta)$. If $\frac{1}{2} < r < 1$, then $\text{div}(X) < 0$. Since the divergence of the vector field does not change sign in the annulus A, there is at most one limit cycle in A by Dulac's criteria.

A phase portrait is given in Figure 10.6.

Example 4. Plot a phase portrait for the Liénard system

$$\dot{x} = y, \qquad \dot{y} = -x - y(a_2 x^2 + a_4 x^4 + a_6 x^6 + a_8 x^8 + a_{10} x^{10} + a_{12} x^{12} + a_{14} x^{14}).$$

where $a_2 = 90$, $a_4 = -882$, $a_6 = 2598.4$, $a_8 = -3359.997$, $a_{10} = 2133.34$, $a_{12} = -651.638$, and $a_{14} = 76.38$.

Solution. Not all limit cycles are convex closed curves, as Figure 10.7 demonstrates.

10.3 Nonexistence of Limit Cycles in the Plane

Bendixson's Criteria. *Consider system (10.6) and suppose that D is a simply connected domain (no holes in D) and that*

$$\nabla \cdot (\psi X) = \text{div}(\psi X) = \frac{\partial}{\partial x}(\psi P) + \frac{\partial}{\partial y}(\psi Q) \neq 0$$

in D. Then there are no limit cycles entirely contained in D.

238   **10. Limit Cycles**

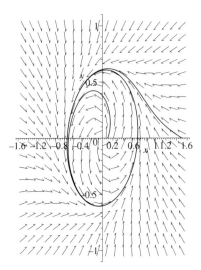

Figure 10.6: A phase portrait for system (10.7) showing the unique limit cycle.

Figure 10.7: A phase portrait for Example 4. The limit cycle is a nonconvex closed curve.

10.3. Nonexistence of Limit Cycles in the Plane

Proof. Suppose that D contains a limit cycle Γ of period T. Then from Green's theorem

$$\iint_D \left[\frac{\partial(\psi P)}{\partial x} + \frac{\partial(\psi Q)}{\partial y} \right] dxdy = \oint_\Gamma (\psi P dy - \psi Q dx)$$

$$= \int_0^T \left(\psi P \frac{dy}{dt} - \psi Q \frac{dx}{dt} \right) dt = 0$$

since on Γ, $\dot{x} = P$ and $\dot{y} = Q$. This contradicts the hypothesis that $\operatorname{div}(\psi \mathbf{X}) \neq 0$, and therefore D contains no limit cycles entirely. □

Definition 9. Suppose there is a compass on a Jordan curve C and that the needle points in the direction of the vector field. The compass is moved in a counterclockwise direction around the Jordan curve by 2π radians. When it returns to its initial position, the needle will have moved through an angle, say, Θ. The *index*, say, $I_\mathbf{X}(C)$, is defined as

$$I_\mathbf{X}(C) = \frac{\Delta \Theta}{2\pi},$$

where $\Delta \Theta$ is the overall change in the angle Θ.

The above definition can be applied to isolated critical points. For example, the index of a node, focus, or center is $+1$ and the index of a col is -1. The following result is clear.

Theorem 1. *The sum of the indices of the critical points contained entirely within a limit cycle is $+1$.*

The next theorem then follows.

Theorem 2. *A limit cycle contains at least one critical point.*

When proving that a system has no limit cycles, the following items should be considered:

1. Bendixson's criteria;
2. indices;
3. invariant lines;
4. critical points.

Example 5. Prove that none of the following systems have any limit cycles:

(a) $\dot{x} = 1 + y^2 - e^{xy}$, $\dot{y} = xy + \cos^2 y$.

(b) $\dot{x} = y^2 - x$, $\dot{y} = y + x^2 + yx^3$.

(c) $\dot{x} = y + x^3$, $\dot{y} = x + y + y^3$.

10. Limit Cycles

(d) $\dot{x} = 2xy - 2y^4$, $\dot{y} = x^2 - y^2 - xy^3$.

(e) $\dot{x} = x(2 - y - x)$, $\dot{y} = y(4x - x^2 - 3)$ given $\psi = \frac{1}{xy}$.

Solutions.

(a) The system has no critical points and hence no limit cycles by Theorem 2.

(b) The origin is the only critical point and it is a saddle point or col. Since the index of a col is -1, there are no limit cycles from Theorem 1.

(c) Find the divergence, div $\mathbf{X} = \frac{\partial P}{\partial x} + \frac{\partial Q}{\partial y} = 3x^2 + 3y^2 + 1 \neq 0$. Hence there are no limit cycles by Bendixson's criteria.

(d) Find the divergence, div $\mathbf{X} = \frac{\partial P}{\partial x} + \frac{\partial Q}{\partial y} = -3x^2 y$. Now div $\mathbf{X} = 0$ if either $x = 0$ or $y = 0$. However, on the line $x = 0$, $\dot{x} = -2y^4 \leq 0$, and on the line $y = 0$, $\dot{y} = x^2 \geq 0$. Therefore, a limit cycle must lie wholly in one of the four quadrants. This is not possible since div $\mathbf{X}$ is nonzero here. Hence there are no limit cycles by Bendixson's criteria.

(e) The axes are invariant since $\dot{x} = 0$ if $x = 0$ and $\dot{y} = 0$ if $y = 0$. The weighted divergence is given by div$(\psi \mathbf{X}) = \frac{\partial}{\partial x}(\psi P) + \frac{\partial}{\partial y}(\psi Q) = -\frac{1}{y}$. Therefore, there are no limit cycles contained entirely in any of the quadrants, and since the axes are invariant, there are no limit cycles in the whole plane.

Example 6. Prove that the system

$$\dot{x} = x(1 - 4x + y), \qquad \dot{y} = y(2 + 3x - 2y)$$

has no limit cycles by applying Bendixson's criteria with $\psi = x^m y^n$.

Solution. The axes are invariant since $\dot{x} = 0$ on $x = 0$ and $\dot{y} = 0$ on $y = 0$. Now

$$\text{div}(\psi \mathbf{X}) = \frac{\partial}{\partial x}(x^{m+1} y^n - 4x^{m+2} y^n + x^{m+1} y^{n+1})$$

$$+ \frac{\partial}{\partial y}(2x^m y^{n+1} + 3x^{m+1} y^{n+1} - 2x^m y^{n+2}),$$

which simplifies to

$$\text{div}(\psi \mathbf{X}) = (m + 2n + 2)x^m y^n + (-4m + 3n - 5)x^{m+1} y^n + (m - 2n - 3)x^m y^{n+1}.$$

Select $m = \frac{1}{2}$ and $n = -\frac{5}{4}$. Then

$$\text{div}(\psi \mathbf{X}) = -\frac{43}{4} x^{\frac{3}{2}} y^{-\frac{5}{4}}.$$

Therefore, there are no limit cycles contained entirely in any of the four quadrants, and since the axes are invariant, there are no limit cycles at all.

MATLAB program files for plotting phase portraits may be found in Chapters 8 and 9.

10.4 Exercises

1. Prove that the system

$$\dot{x} = y + x\left(\frac{1}{2} - x^2 - y^2\right), \qquad \dot{y} = -x + y(1 - x^2 - y^2)$$

   has a stable limit cycle.

2. Plot the limit cycle for Exercise 1.

3. By considering the flow across the square with coordinates $(1, 1)$, $(1, -1)$, $(-1, -1)$, $(-1, 1)$, centered at the origin, prove that the system

$$\dot{x} = -y + x\cos(\pi x), \qquad \dot{y} = x - y^3$$

   has a stable limit cycle.

4. Plot the vector field, limit cycle, and square in Exercise 3.

5. Prove that the system

$$\dot{x} = x - y - x^3, \qquad \dot{y} = x + y - y^3$$

   has a unique limit cycle.

6. Prove that the system.

$$\dot{x} = y + x(\alpha - x^2 - y^2), \qquad \dot{y} = -x + y(1 - x^2 - y^2),$$

   where $0 < \alpha < 1$, has a limit cycle and determine its stability.

7. For which parameter values does the Holling–Tanner model

$$\dot{x} = x\beta\left(1 - \frac{x}{k}\right) - \frac{rxy}{(a+ax)}, \qquad \dot{y} = by\left(1 - \frac{Ny}{x}\right)$$

   have a limit cycle?

8. Plot phase portraits for the Liénard system

$$\dot{x} = y - \mu(-x + x^3), \qquad \dot{y} = -x,$$

   when (a) $\mu = 0.01$ and (b) $\mu = 10$.

9. Prove that none of the following systems have limit cycles:

   (a) $\dot{x} = y$, $\dot{y} = -x - (1 + x^2 + x^4)y$;
   (b) $\dot{x} = x - x^2 + 2y^2$, $\dot{y} = y(x + 1)$;

(c) $\dot{x} = y^2 - 2x, \dot{y} = 3 - 4y - 2x^2 y$;
(d) $\dot{x} = -x + y^3 - y^4, \dot{y} = 1 - 2y - x^2 y + x^4$;
(e) $\dot{x} = x^2 - y - 1, \dot{y} = y(x - 2)$;
(f) $\dot{x} = x - y^2(1 + x^3), \dot{y} = x^5 - y$;
(g) $\dot{x} = 4x - 2x^2 - y^2, \dot{y} = x(1 + xy)$.

10. Prove that neither of the following systems have limit cycles using the given multipliers:

(a) $\dot{x} = x(4 + 5x + 2y), \dot{y} = y(-2 + 7x + 3y), \psi = \frac{1}{xy^2}$;

(b) $\dot{x} = x(\beta - \delta x - \gamma y), \dot{y} = y(b - dy - cx), \psi = \frac{1}{xy}$.

In case (b), prove that there are no limit cycles in the first quadrant only. These differential equations were used as a general model for competing species in Chapter 9.

Recommended Reading

[1] L. Perko, *Differential Equations and Dynamical Systems*, 3rd ed., Springer-Verlag, New York, 2001.

[2] J. A. Yorke (contributor), K. Alligood (ed.), and T. Sauer (ed.), *Chaos: An Introduction to Dynamical Systems*, Springer-Verlag, New York, 1996.

[3] Ye Yan Qian, *Theory of Limit Cycles*, Translations of Mathematical Monographs 66, American Mathematical Society, Providence, 1986.

[4] J. Rayleigh, *The Theory of Sound*, Dover, New York, 1945.

[5] A. Liénard, Étude des oscillations entrenues, *Rev. Gén. Électricité*, **23** (1928), 946–954.

[6] B. van der Pol, On relaxation oscillations, *Philos. Magazine*, **7** (1926), 901–912, 946–954.

11
Hamiltonian Systems, Lyapunov Functions, and Stability

Aims and Objectives

- To study Hamiltonian systems in the plane.
- To investigate stability using Lyapunov functions.

On completion of this chapter the reader should be able to

- prove whether or not a system is Hamiltonian;
- sketch phase portraits of Hamiltonian systems;
- use Lyapunov functions to determine the stability of a critical point;
- distinguish between stability and asymptotic stability.

The theory of Hamiltonian (or conservative) systems in the plane is introduced. The differential equations are used to model dynamical systems in which there is no energy loss. Hamiltonian systems are also used extensively when bifurcating limit cycles in the plane (see Chapters 15 and 16).

Sometimes it is not possible to apply the linearization techniques to determine the stability of a critical point or invariant set. In certain cases, the flow across level curves, defined by Lyapunov functions, can be used to determine the stability.

11.1 Hamiltonian Systems in the Plane

Definition 1. A system of differential equations on $\Re^2$ is said to be *Hamiltonian* with one degree of freedom if it can be expressed in the form

$$\text{(11.1)} \qquad \frac{dx}{dt} = \frac{\partial H}{\partial y}, \qquad \frac{dy}{dt} = -\frac{\partial H}{\partial x},$$

where $H(x, y)$ is a twice-continuously differentiable function. The system is said to be *conservative* and there is no dissipation. In applications, the Hamiltonian is defined by

$$H(x, y) = K(x, y) + V(x, y),$$

where K is the kinetic energy and V is the potential energy. Hamiltonian systems with two degrees of freedom will be discussed in Chapter 14.

Theorem 1 (Conservation of Energy). *The total energy $H(x, y)$ is a first integral and a constant of the motion.*

Proof. The total derivative along a trajectory is given by

$$\frac{dH}{dt} = \frac{\partial H}{\partial x}\frac{dx}{dt} + \frac{\partial H}{\partial y}\frac{dy}{dt} = 0$$

from the chain rule and (11.1). Therefore, $H(x, y)$ is constant along the solution curves of (11.1), and the trajectories lie on the contours defined by $H(x, y) = C$, where C is a constant. □

Consider a simple mechanical system that is Hamiltonian in the plane.

The Simple Nonlinear Pendulum. The differential equation used to model the motion of a pendulum in the plane may be derived using Newton's law of motion:

$$\text{(11.2)} \qquad \frac{d^2\theta}{dt^2} + \frac{g}{l}\sin\theta = 0,$$

where θ is the angular displacement from the vertical, l is the length of the arm of the pendulum, which swings in the plane, and g is the acceleration due to gravity.

This model does not take into account any resistive forces, so once the pendulum is set into motion, it will swing periodically forever, thus obeying the conservation of energy. The system is called conservative since no energy is lost. A periodically forced pendulum will be discussed in Chapter 14.

Let $\dot\theta = \phi$. Then system (11.2) can be written as a planar system in the form

$$\text{(11.3)} \qquad \dot\theta = \phi, \qquad \dot\phi = -\frac{g}{l}\sin\theta.$$

The critical points occur at $(n\pi, 0)$ in the (θ, ϕ) plane, where n is an integer. It is not difficult to show that the critical points are hyperbolic if n is odd and

11.1. Hamiltonian Systems in the Plane

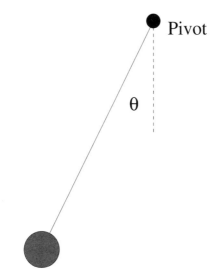

Figure 11.1: A simple nonlinear pendulum.

nonhyperbolic if n is even. Therefore, Hartman's theorem cannot be applied when n is even. However, system (11.3) is a Hamiltonian system with $H(\theta, \phi) = \frac{\phi^2}{2} - \frac{g}{l}\cos\theta$ (kinetic energy + potential energy), and therefore the solution curves may be plotted. The direction field may be constructed by considering $\frac{d\phi}{d\theta}$, $\dot\theta$, and $\dot\phi$. Solution curves and direction fields are given in Figure 11.2(a).

The axes of Figure 11.2(a) are the angular displacement (θ) and angular velocity ($\dot\theta$). The closed curves surrounding the critical points $(2n\pi, 0)$ represent periodic oscillations, and the wavy lines for large angular velocities correspond to motions in which the pendulum spins around its pivotal point. The closed curves correspond to local minima on the surface $z = H(\theta, \phi)$, and the unstable critical points correspond to local maxima on the same surface.

Definition 2. A critical point of the system

(11.4) $$\dot{\mathbf{x}} = \mathbf{f}(\mathbf{x}), \quad \mathbf{x} \in \Re^2,$$

at which the Jacobian matrix has no zero eigenvalues is called a *nondegenerate critical point*; otherwise, it is called a *degenerate critical point*.

Theorem 2. *Any nondegenerate critical point of an analytic Hamiltonian system is either a saddle point or a center.*

11. Hamiltonian Systems, Lyapunov Functions, and Stability

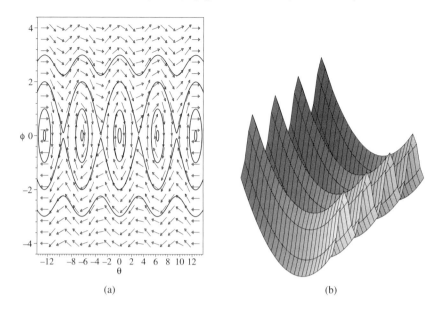

(a) (b)

Figure 11.2: (a) A phase portrait for system (11.3) when $-4\pi \leq \theta \leq 4\pi$. (b) The surface $z = H(\theta, \phi)$.

Proof. Assume that the critical point is at the origin. The Jacobian matrix is equal to

$$J_O = \begin{pmatrix} \frac{\partial^2 H}{\partial x \partial y}(0,0) & \frac{\partial^2 H}{\partial y^2}(0,0) \\ -\frac{\partial^2 H}{\partial x^2}(0,0) & -\frac{\partial^2 H}{\partial y \partial x}(0,0) \end{pmatrix}.$$

Now trace(J_0) = 0 and

$$\det(J_0) = \frac{\partial^2 H}{\partial x^2}(0,0) \frac{\partial^2 H}{\partial y^2}(0,0) - \left(\frac{\partial^2 H}{\partial x \partial y}(0,0) \right)^2.$$

The origin is a saddle point if $\det(J_0) < 0$. If $\det(J_0) > 0$, then the origin is either a center or a focus. Note that the critical points of system (11.1) correspond to the stationary points on the surface $z = H(x, y)$. If the origin is a focus, then the origin is not a strict local maximum or minimum of the Hamiltonian function. Suppose that the origin is a stable focus, for instance. Then

$$H(x_0, y_0) = \lim_{t \to \infty} H(x(t, x_0, y_0), y(t, x_0, y_0)) = H(0, 0),$$

for all $(x_0, y_0) \in N_\epsilon(0, 0)$, where N_ϵ denotes a small deleted neighborhood of the origin. However, $H(x, y) > H(0, 0)$ at a local minimum and $H(x, y) < H(0, 0)$ at a local maximum, a contradiction. A similar argument can be applied when the origin is an unstable focus.

Therefore, a nondegenerate critical point of a Hamiltonian is either a saddle point or a center. □

Example 1. Find the Hamiltonian for each of the following systems and sketch the phase portraits:

(a) $\dot{x} = y$, $\dot{y} = x + x^2$;

(b) $\dot{x} = y + x^2 - y^2$, $\dot{y} = -x - 2xy$.

Solution. (a) Integration gives $H(x, y) = \frac{y^2}{2} - \frac{x^2}{2} - \frac{x^3}{3}$; the solution curves are given by $H(x, y) = C$. There are two critical points at $(0, 0)$ and $(-1, 0)$, which are both nondegenerate. The critical point at the origin is a saddle point or col from linearization, and the eigenvectors are $(1, -1)^T$ and $(1, 1)^T$. The critical point at $(-1, 0)$ is a center from Theorem 1. If $y > 0$, then $\dot{x} > 0$, and if $y < 0$, then $\dot{x} < 0$. A phase portrait is given in Figure 11.3.

Figure 11.3: A phase portrait for Example 1(a).

(b) Integration gives $H(x, y) = \frac{x^2}{2} + \frac{y^2}{2} + x^2 y - \frac{y^3}{3}$; the solution curves are given by $H(x, y) = C$. There are four critical points at $O = (0, 0)$, $A = (0, 1)$, $B = (\frac{\sqrt{3}}{2}, -\frac{1}{2})$, and $C = (-\frac{\sqrt{3}}{2}, -\frac{1}{2})$, which are all nondegenerate. The critical point at the origin is a center by Theorem 1, and the critical points at A, B, and C are saddle points or cols from linearization. The eigenvectors determine the stable and unstable manifolds of the cols. The eigenvectors for point A are $(1, \sqrt{3})^T$ and $(1, -\sqrt{3})^T$; the eigenvectors for B are $(1, -\sqrt{3})^T$ and $(1, 0)^T$; and the eigenvectors

11. Hamiltonian Systems, Lyapunov Functions, and Stability

Figure 11.4: A phase portrait for Example 1(b). The lines $y = -\frac{1}{2}$, $y = -\sqrt{3}x+1$, and $y = \sqrt{3}x + 1$ are invariant.

for C are $(1, 0)^T$ and $(1, \sqrt{3})^T$. The solution curves and direction fields are shown in Figure 11.4.

Definition 3. Suppose that $\mathbf{x_0}$ is a critical point of system (11.4). If $\Lambda^+(\gamma) = \Lambda^-(\gamma) = \mathbf{x_0}$, then γ is a *homoclinic orbit*.

An example of a homoclinic orbit is given in Figure 11.3. The unstable and stable manifolds from the origin form a homoclinic loop around the critical point at $(-1, 0)$. A homoclinic orbit connects a critical point to itself and takes an infinite amount of time to make the connection.

Definition 4. Suppose that $\mathbf{x_0}$ and $\mathbf{y_0}$ are distinct critical points. If $\Lambda^+(\gamma) = \mathbf{x_0}$ and $\Lambda^-(\gamma) = \mathbf{y_0}$, then γ is called a *heteroclinic orbit*.

Examples of heteroclinic orbits are given in Figure 11.4. They are the three orbits lying on the line segments $\{y = -\frac{1}{2}, -\frac{\sqrt{3}}{2} < x < \frac{\sqrt{3}}{2}\}$, $\{y = -\sqrt{3}x + 1, -\frac{\sqrt{3}}{2} < x < \frac{\sqrt{3}}{2}\}$, and $\{y = \sqrt{3}x + 1, -\frac{\sqrt{3}}{2} < x < \frac{\sqrt{3}}{2}\}$.

Definition 5. A *separatrix* is an orbit that divides the phase plane into two distinctly different types of qualitative behavior. The homoclinic and heteroclinic orbits are examples of separatrix cycles.

For example, in Figure 11.3, orbits are bounded inside the homoclinic orbit surrounding the point $(-1, 0)$ and unbounded outside it.

11.2 Lyapunov Functions and Stability

Consider nonlinear systems of the form (11.4). The stability of hyperbolic critical points may be determined from the eigenvalues of the Jacobian matrix. The critical point is stable if the real part of all of the eigenvalues is negative and unstable otherwise. If a critical point is nonhyberbolic, then a method due to Lyapunov may sometimes be used to determine the stability of the critical point.

Imagine a system defined by the potential function $V(x, y)$, where

$$\dot{x} = -\frac{\partial V}{\partial x}, \qquad \dot{y} = -\frac{\partial V}{\partial y}.$$

The negative signs arise from the analogies with potential energy from physics. Now

$$\frac{dV}{dt} = \frac{\partial V}{\partial x}\frac{dx}{dt} + \frac{\partial V}{\partial y}\frac{dy}{dt} = -\left(\frac{\partial V}{\partial x}\right)^2 - \left(\frac{\partial V}{\partial y}\right)^2 \leq 0.$$

This implies that $V(t)$ decreases along trajectories and the motion is always toward lower potentials. Now $\dot{x} = \dot{y} = 0$ when $\frac{\partial V}{\partial x} = \frac{\partial V}{\partial y} = 0$, corresponding to local maxima, minima, or saddle points on $V(x, y)$. Local maxima correspond to unstable critical points, and local minima correspond to stable critical points.

Example 2. Plot a phase portrait for the system $\dot{x} = x - x^3$, $\dot{y} = -y$ and plot the potential function for this system.

Solution. There are three critical points at $O = (0, 0)$, $A = (-1, 0)$, and $B = (1, 0)$. The origin is unstable and the critical points A and B are stable, as seen in Figure 11.5(a). The function $z = V(x, y) = -\frac{x^2}{2} + \frac{x^4}{4} + \frac{y^2}{2}$ is plotted in Figure 11.5(b) is known as the *double-well potential*. The system is *multistable* since it has two stable critical points.

The local minima in Figure 11.5(b) correspond to the stable critical points at A and B. The local maximum at the origin corresponds to the saddle point in Figure 11.5(a).

Definition 6. A critical point, say, $\mathbf{x_0}$, of system (11.4) is called *stable* if given $\epsilon > 0$ there is a $\delta > 0$ such that for all $t \geq t_0$, $\| \mathbf{x}(t) - \mathbf{x_0}(t) \| < \epsilon$ whenever $\| \mathbf{x}(t_0) - \mathbf{x_0}(t_0) \| < \delta$, where $\mathbf{x}(t)$ is a solution of (11.4).

Definition 7. A critical point, say, $\mathbf{x_0}$, of system (11.4) is called *asymptotically stable* if it is stable and there is an $\eta > 0$ such that

$$\lim_{t \to \infty} \| \mathbf{x}(t) - \mathbf{x_0}(t) \| = 0,$$

whenever $\| \mathbf{x}(t_0) - \mathbf{x_0}(t_0) \| < \eta$.

A trajectory near a stable critical point will remain close to that point, whereas a trajectory near an asymptotically stable critical point will move closer and closer to the critical point as $t \to \infty$.

11. Hamiltonian Systems, Lyapunov Functions, and Stability

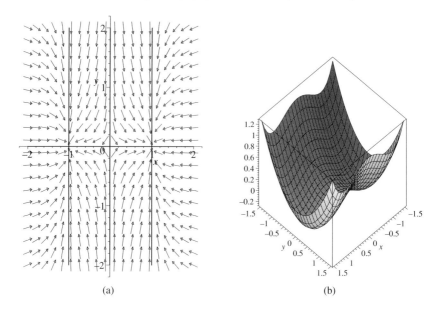

Figure 11.5: (a) A phase portrait for Example 2. (b) The double-well potential.

The following theorem holds for system (11.4) when $\mathbf{x} \in \Re^n$. Examples in $\Re^3$ are given in Chapter 13.

The Lyapunov Stability Theorem. *Let E be an open subset of $\Re^n$ containing an isolated critical point $\mathbf{x_0}$. Suppose that $\mathbf{f}$ is continuously differentiable and that there exists a continuously differentiable function, say, $V(\mathbf{x})$, which satisfies the conditions:*

- $V(\mathbf{x_0}) = 0$;
- $V(\mathbf{x}) > 0$ if $\mathbf{x} \neq \mathbf{x_0}$,

where $\mathbf{x} \in \Re^n$. Then

1. *if $\dot{V}(\mathbf{x}) \leq 0$ for all $\mathbf{x} \in E$, $\mathbf{x_0}$ is stable;*
2. *if $\dot{V}(\mathbf{x}) < 0$ for all $\mathbf{x} \in E$, $\mathbf{x_0}$ is asymptotically stable;*
3. *if $\dot{V}(\mathbf{x}) > 0$ for all $\mathbf{x} \in E$, $\mathbf{x_0}$ is unstable.*

Proof. 1. Choose a small neighborhood N_ϵ surrounding the critical point $\mathbf{x_0}$. In this neighborhood, $\dot{V}(\mathbf{x}) \leq 0$, so a positive semiorbit starting inside N_ϵ remains there forever. The same conclusion is drawn no matter how small ϵ is chosen to be. The critical point is therefore stable.

11.2. Lyapunov Functions and Stability

2. Since $\dot{V}(\mathbf{x}) < 0$, the Lyapunov function must decrease monotonically on every positive semiorbit $\mathbf{x}(t)$. Let ϕ_t define the flow on (11.4). Then either $V(\phi_t) \to \mathbf{x_0}$ as $t \to \infty$ or there is a positive semiorbit $\mathbf{x}(t)$ such that

(11.5) $$V(\phi_t) \geq n > 0 \quad \text{for all } t \geq t_0,$$

for some $n > 0$. Since $\mathbf{x_0}$ is stable, there is an annular region A, defined by $n \leq V(\mathbf{x}) \leq c$, containing this semiorbit. Suppose that $\dot{V}$ attains its upper bound in A, say, $-N$, so

$$\dot{V}(\mathbf{x}) \leq -N < 0, \quad \mathbf{x} \in A, \quad N > 0.$$

Integration gives

$$V(\mathbf{x}(t)) - V(\mathbf{x}(t_0)) \leq -N(t - t_0),$$

where $t > t_0$. This contradicts (11.5), and therefore no path fails to approach the critical point at $\mathbf{x_0}$. The critical point is asymptotically stable.

3. Since $\dot{V}(\mathbf{x}) > 0$, $V(\mathbf{x})$ is strictly increasing along trajectories of (11.4). If ϕ_t is the flow of (11.4), then

$$V(\phi_t) > V(\mathbf{x_0}) > 0$$

for $t > 0$ in a small neighborhood of $\mathbf{x_0}$, N_ϵ. Therefore,

$$V(\phi_t) - V(\mathbf{x_0}) \geq kt$$

for some constant k and $t \geq 0$. Hence for sufficiently large t,

$$V(\phi_t) > kt > K,$$

where K is the maximum of the continuous function $V(\mathbf{x})$ on the compact set $\overline{N_\epsilon}$. Therefore, ϕ_t lies outside the closed set N_ϵ and $\mathbf{x_0}$ is unstable. □

Definition 8. The function $V(\mathbf{x})$ is called a *Lyapunov function*.

Unfortunately, there is no systematic way to construct a Lyapunov function. The Lyapunov functions required for specific examples will be given in this book. Note that if $\dot{V}(\mathbf{x}) = 0$, then all trajectories lie on the curves (surfaces in $\Re^n$) defined by $V(\mathbf{x}) = C$, where C is a constant. The quantity $\dot{V}$ gives the rate of change of V along trajectories, or in other words, $\dot{V}$ gives the direction that trajectories cross the level curves $V(\mathbf{x}) = C$.

Example 3. Determine the stability of the origin for the system

$$\dot{x} = -y^3, \qquad \dot{y} = x^3.$$

Solution. The eigenvalues are both zero and the origin is a degenerate critical point. A Lyapunov function for this system is given by $V(x, y) = x^4 + y^4$, and furthermore

$$\frac{dV}{dt} = \frac{\partial V}{\partial x}\frac{dx}{dt} + \frac{\partial V}{\partial y}\frac{dy}{dt} = 4x^3(-y^3) + 4y^3(x^3) = 0.$$

Hence the solution curves lie on the closed curves given by $x^4 + y^4 = C$. The origin is thus stable but not asymptotically stable. The trajectories that start near to the origin remain there but do not approach the origin asymptotically. If $y > 0$, then $\dot{x} < 0$, and if $y < 0$, then $\dot{x} > 0$. The level curves and direction fields are given in Figure 11.6.

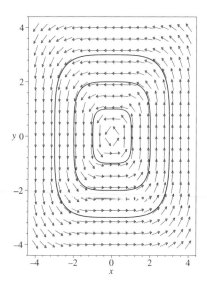

Figure 11.6: A phase portrait for Example 3.

Example 4. Investigate the stability of the origin for the system

$$\dot{x} = y, \qquad \dot{y} = -x - y(1 - x^2)$$

using the Lyapunov function $V(x, y) = x^2 + y^2$.

Solution. Now

$$\frac{dV}{dt} = \frac{\partial V}{\partial x}\frac{dx}{dt} + \frac{\partial V}{\partial y}\frac{dy}{dt} = 2x(y) + 2y(-x - y + yx^2),$$

so

$$\frac{dV}{dt} = 2y^2(x^2 - 1)$$

and $\dot{V} \leq 0$ if $|x| \leq 1$. Therefore, $\dot{V} = 0$, if either $y = 0$ or $x = \pm 1$. When $y = 0$, $\dot{x} = 0$ and $\dot{y} = -x$, which means that a trajectory will move off the line $y = 0$ when $x \neq 0$. Hence if a trajectory starts inside the circle of radius one centered at the origin, then it will approach the origin asymptotically. The origin is asymptotically stable.

Example 5. Prove that the origin of the system

$$\dot{x} = -8x - xy^2 - 3y^3, \qquad \dot{y} = 2x^2y + 2xy^2$$

is asymptotically stable using the Lyapunov function $V(x, y) = 2x^2 + 3y^2$. Determine the *Lyapunov domain of stability* based on $V(x, y)$.

Solution. Now

$$\dot{V} = 4x(-8x - xy^2 - 3y^3) + 6y(2x^2y + 2xy^2) = 8x^2(y^2 - 4)$$

and $\dot{V} \leq 0$ if $|y| \leq 2$. Therefore, $\dot{V} = 0$ if either $x = 0$ or $y = \pm 2$. When $x = 0$, $\dot{x} = -3y^3$ and $\dot{y} = 0$, which means that a trajectory will move off the line $x = 0$ when $y \neq 0$. Now $\dot{V} < 0$ if $|y| < 2$. This implies that $\dot{V} < 0$ as long as $V(x, y) = 2x^2 + 3y^2 < 12$. This region defines the domain of Lyapunov stability. Therefore, if a trajectory lies wholly inside the ellipse $2x^2 + 3y^2 = 12$, it will move to the origin asymptotically. Hence the origin is asymptotically stable.

An approximation of the true domain of stability for the origin of the system in Example 5 is indicated in Figure 11.7(a). Notice that it is larger than the Lyapunov domain of stability (Figure 11.7(b)) and that the x-axis is invariant.

The MATLAB commands for plotting surfaces and contour plots may be found in Chapter 0. The MATLAB program files for plotting phase plane portraits may be found in Chapters 8 and 9.

11.3 Exercises

1. Find the Hamiltonian of the system

$$\dot{x} = y, \qquad \dot{y} = x - x^3$$

and sketch a phase portrait.

2. Given the Hamiltonian function $H(x, y) = \frac{y^2}{2} + \frac{x^2}{2} - \frac{x^4}{4}$, sketch a phase portrait for the Hamiltonian system.

3. Plot a phase portrait for the damped pendulum equation

$$\ddot{\theta} + 0.15\dot{\theta} + \sin\theta = 0$$

and describe what happens physically.

11. Hamiltonian Systems, Lyapunov Functions, and Stability

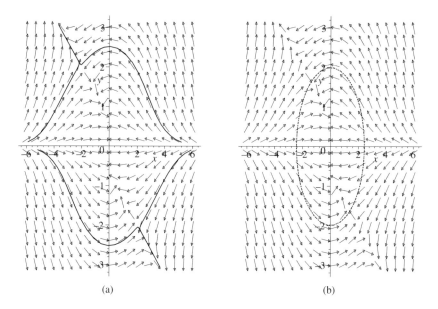

Figure 11.7: (a) A phase portrait for Example 5. (b) The domain of Lyapunov stability.

4. Plot a phase portrait of the system

$$\dot{x} = y(y^2 - 1), \qquad \dot{y} = x(1 - x^2).$$

5. Investigate the stability of the critical points at the origin for the following systems:

   (a) $\dot{x} = -y - x^3$, $\dot{y} = x - y^3$ using the Lyapunov function $V(x, y) = x^2 + y^2$;

   (b) $\dot{x} = x(x - \alpha)$, $\dot{y} = y(y - \beta)$ using the Lyapunov function $V(x, y) = (\frac{x}{\alpha})^2 + (\frac{y}{\beta})^2$;

   (c) $\dot{x} = y$, $\dot{y} = y - x^3$ using the Lyapunov function $V(x, y) = ax^4 + bx^2 + cxy + dy^2$.

6. Prove that the origin is a unique critical point of the system

$$\dot{x} = -\frac{1}{2}y(1+x) + x(1 - 4x^2 - y^2), \qquad \dot{y} = 2x(1+x) + y(1 - 4x^2 - y^2).$$

Determine the stability of the origin using the Lyapunov function $V(x, y) = (1 - 4x^2 - y^2)^2$. Find $\Lambda^+(p)$ for each $p \in \Re^2$. Plot a phase portrait.

7. Determine the values of a for which $V(x, y) = x^2 + ay^2$ is a Lyapunov function for the system

$$\dot{x} = -x + y - x^2 - y^2 + xy^2, \qquad \dot{y} = -y + xy - y^2 - x^2 y.$$

8. Determine the basin of attraction of the origin for the system

$$\dot{x} = x(x^2 + y^2 - 4) - y, \qquad \dot{y} = x + y(x^2 + y^2 - 4)$$

using the Lyapunov function $V(x, y) = x^2 + y^2$.

9. Plot a phase portrait for the system in Exercise 8.

10. Use the Lyapunov function $V(x, y) = (x^4 + 2y^2 - 10)^2$ to investigate the invariant sets of the system

$$\dot{x} = y - x(x^4 + 2y^2 - 10), \qquad \dot{y} = -x^3 - 3y^5(x^4 + 2y^2 - 10).$$

Recommended Reading

[1] G. M. Zaslavsky, *Physics of Chaos in Hamiltonian Systems*, World Scientific, Singapore, 1998.

[2] J. Cronin, *Differential Equations: Introduction and Qualitative Theory*, Marcel Dekker, New York, 1994.

[3] J. Moser, Recent developments in the theory of Hamiltonian systems, *SIAM Rev.*, **28**-4 (1986), 459–485.

[4] M. W. Hirsch and S. Smale, *Differential Equations, Dynamical Systems and Linear Algebra*, Academic Press, New York, 1974.

[5] V. V. Nemitskii and V. V. Stepanov, *Qualitative Theory of Differential Equations*, Princeton University Press, Princeton, NJ, 1960.

12
Bifurcation Theory

Aims and Objectives

- To introduce bifurcation theory in the plane.

- To introduce the notion of steady-state solution and investigate multistability and bistability.

On completion of this chapter the reader should be able to

- describe how a phase portrait changes as a parameter changes;

- plot bifurcation diagrams;

- interpret the bifurcation diagrams in terms of physical behavior.

If the behavior of a dynamical system changes suddenly as a parameter is varied, then it is said to have undergone a bifurcation. At a point of bifurcation, stability may be gained or lost. The study of bifurcations to chaos will be discussed in later chapters.

It may be possible for a nonlinear system to have more than one steady-state solution. For example, different initial conditions can lead to different stable solutions. A system of this form is said to be multistable. Bifurcations of so-called large-amplitude limit cycles are discussed. By introducing a feedback mechanism into the system it is possible to obtain hysteresis, or bistable behavior.

12.1 Bifurcations of Nonlinear Systems in the Plane

Definition 1. A vector field $\mathbf{f} \in \Re^2$, which is continuously differentiable, is called *structurally stable* if small perturbations in the system $\dot{\mathbf{x}} = \mathbf{f}(\mathbf{x})$ leave the qualitative behavior unchanged. If small perturbations cause a change in the qualitative behavior of the system, then $\mathbf{f}$ is called *structurally unstable*.

For example, the Lotka–Volterra model (Example 2, Chapter 9) is structurally unstable, while the Holling–Tanner model (Example 3, Chapter 9) is structurally stable.

Peixoto's Theorem in the Plane. *Let the vector field $\mathbf{f}$ be continuously differentiable on a compact set, say, D. Then $\mathbf{f}$ is structurally stable on D if and only if*

- the number of critical points and limit cycles is finite and each is hyperbolic (see Theorem 1 in Chapter 14);
- there are no trajectories connecting saddle points to saddle points.

Consider systems of the form

$$(12.1) \qquad \dot{\mathbf{x}} = \mathbf{f}(\mathbf{x}, \mu),$$

where $\mathbf{x} \in \Re^2$ and $\mu \in \Re$. A value, say, μ_0, for which the vector field $\mathbf{f}(\mathbf{x}, \mu_0)$ is not structurally stable is called a *bifurcation value*.

Four simple types of bifurcation, all at nonhyperbolic critical points, will be given in order to illustrate how the qualitative behavior of a structurally unstable system of differential equations can change with respect to a parameter value. Certain bifurcations can be classified by so-called *normal forms*. By finding suitable transformations it is possible to reduce systems to a normal form. Four normal form bifurcations are illustrated below, and more bifurcation diagrams of varying normal forms can be found in [2].

12.1.I A Saddle-Node Bifurcation.
Consider the system

$$(12.2) \qquad \dot{x} = \mu - x^2, \qquad \dot{y} = -y.$$

The critical points are found by solving the equations $\dot{x} = \dot{y} = 0$. There are (i) zero, (ii) one, or (iii) two critical points, depending on the value of μ. Consider the three cases separately.

Case (i). When $\mu < 0$, there are no critical points in the plane and the flow is from right to left since $\dot{x} < 0$. If $y > 0$, then $\dot{y} < 0$ and if $y < 0$, then $\dot{y} > 0$. A plot of the vector field is given in Figure 12.1(a). Note that the flow is invariant on the x-axis.

Case (ii). When $\mu = 0$, there is one critical point at the origin and it is nonhyperbolic. The solution curves may be found by solving the differential equation

$$\frac{dy}{dx} = \frac{\dot{y}}{\dot{x}} = -\frac{y}{x^2}.$$

12.1. Bifurcations of Nonlinear Systems in the Plane

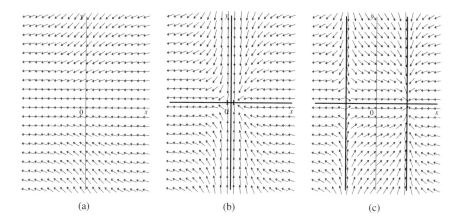

Figure 12.1: Vector field plots and manifolds when (a) $\mu < 0$, (b) $\mu = 0$, and (c) $\mu > 0$. There are no manifolds when $\mu < 0$.

This is a separable differential equation (see Chapter 7) and the solution is given by $|y| = Ke^{-\frac{1}{x}}$, where K is a constant. Note that $\dot{x} < 0$ for all x. The vector field is plotted in Figure 12.1(b). Note that the flow is invariant along both the x-axis and the y-axis.

Case (iii). When $\mu > 0$, there are two critical points at $A = (\sqrt{\mu}, 0)$ and $B = (-\sqrt{\mu}, 0)$. Linearize in the usual way. The Jacobian matrix is given by

$$J = \begin{pmatrix} \frac{\partial P}{\partial x} & \frac{\partial P}{\partial y} \\ \frac{\partial Q}{\partial x} & \frac{\partial Q}{\partial y} \end{pmatrix} = \begin{pmatrix} -2x & 0 \\ 0 & -1 \end{pmatrix},$$

where $\dot{x} = P(x, y)$ and $\dot{y} = Q(x, y)$. Therefore,

$$J_A = \begin{pmatrix} -2\sqrt{\mu} & 0 \\ 0 & -1 \end{pmatrix}$$

and the eigenvalues and eigenvectors are given by $\lambda_1 = -2\sqrt{\mu}$, $(1, 0)^T$ and $\lambda_2 = -1$, $(0, 1)^T$. The critical point at A is thus a stable node and the stable manifolds are orthogonal to one another.

The Jordan matrix for the critical point at B is

$$J_B = \begin{pmatrix} 2\sqrt{\mu} & 0 \\ 0 & -1 \end{pmatrix}$$

and the eigenvalues and eigenvectors are $\lambda_1 = 2\sqrt{\mu}$, $(1, 0)^T$ and $\lambda_2 = -1$, $(0, 1)^T$. This critical point is a saddle point. The vector field and orthogonal stable and unstable manifolds are plotted in Figure 12.1(c).

12. Bifurcation Theory

In summary, there are no critical points if μ is negative; there is one nonhyperbolic critical point at the origin if $\mu = 0$; and there are two critical points—one a saddle and the other a node—when μ is positive. The qualitative behavior of the system changes as the parameter μ passes through the bifurcation value $\mu_0 = 0$. The behavior of the critical points can be summarized on a *bifurcation diagram* as depicted in Figure 12.2.

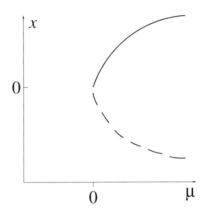

Figure 12.2: A bifurcation diagram for system (12.2) showing a saddle-node bifurcation. The solid curves depict stable behavior and the dashed curves depict unstable behavior.

When $\mu < 0$, there are no critical points, and as μ passes through zero the qualitative behavior changes and two critical points bifurcate from the origin. As μ increases, the critical points move farther and farther apart. Note that the critical points satisfy the equation $\mu = x^2$, hence the parabolic form of the bifurcation curve. More examples of saddle-node bifurcations are given in Section 12.2.

12.1.II A Transcritical Bifurcation. Consider the system

(12.3) $$\dot{x} = \mu x - x^2, \qquad \dot{y} = -y.$$

The critical points are found by solving the equations $\dot{x} = \dot{y} = 0$. There are either one or two critical points depending on the value of the parameter μ. The bifurcation value is again $\mu_0 = 0$. Consider the cases (i) $\mu < 0$, (ii) $\mu = 0$, and (iii) $\mu > 0$ separately.

Case (i). When $\mu < 0$, there are two critical points, one at $O = (0, 0)$ and the other at $A = (\mu, 0)$. The origin is a stable node and A is a saddle point. A vector field and manifolds are plotted in Figure 12.3(a).

12.1. Bifurcations of Nonlinear Systems in the Plane

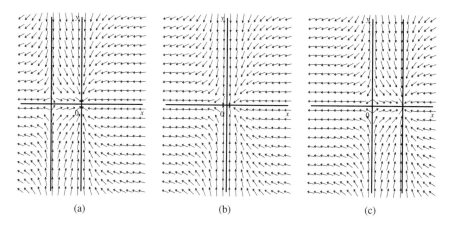

Figure 12.3: Vector field plots and manifolds when (a) $\mu < 0$, (b) $\mu = 0$, and (c) $\mu > 0$.

Case (ii). When $\mu = 0$, there is one nonhyperbolic critical point at the origin. The solution curves satisfy the differential equation

$$\frac{dy}{dx} = \frac{y}{x^2}$$

which has solutions $|y| = Ke^{-\frac{1}{x}}$, where K is a constant. A vector field and the manifolds through the origin are shown in Figure 12.3(b).

Case (iii). When $\mu > 0$, there are two critical points, one at $O = (0, 0)$ and the other at $B = (\mu, 0)$. The origin is now a saddle point and B is a stable node. A vector field and manifolds are plotted in Figure 12.3(c).

The behavior of the critical points can be summarized on a bifurcation diagram as depicted in Figure 12.4.

12.1.III A Pitchfork Bifurcation. Consider the system

(12.4) $$\dot{x} = \mu x - x^3, \qquad \dot{y} = -y.$$

The critical points are found by solving the equations $\dot{x} = \dot{y} = 0$. There are either one or three critical points depending on the value of the parameter μ. The bifurcation value is again $\mu_0 = 0$. Consider the cases (i) $\mu < 0$, (ii) $\mu = 0$, and (iii) $\mu > 0$ separately.

Case (i). When $\mu < 0$, there is one critical point at $O = (0, 0)$. The origin is a stable node. A vector field and the manifolds at the origin are shown in Figure 12.5(a).

262   12. Bifurcation Theory

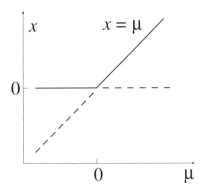

Figure 12.4: A bifurcation diagram for system (12.3) showing a transcritical bifurcation. The solid lines depict stable behavior and the dashed lines depict unstable behavior.

Case (ii). When $\mu = 0$, there is one nonhyperbolic critical point at the origin. The solution curves satisfy the differential equation

$$\frac{dy}{dx} = \frac{y}{x^3}$$

which has solutions $|y| = Ke^{-\frac{1}{2x^2}}$, where K is a constant. A vector field is plotted in Figure 12.5(a).

Case (iii). When $\mu > 0$, there are three critical points at $O = (0, 0)$, $A = (\sqrt{\mu}, 0)$, and $B = (-\sqrt{\mu}, 0)$. The origin is now a saddle point and A and B are both stable nodes. A vector field and all of the stable and unstable manifolds are plotted in Figure 12.5(b).

The behavior of the critical points can be summarized on a bifurcation diagram as depicted in Figure 12.6.

12.1.IV A Hopf Bifurcation.
Consider the system

(12.5) $\qquad \dot{r} = r(\mu - r^2), \qquad \dot{\theta} = -1.$

The origin is the only critical point since $\dot{\theta} \neq 0$. There are no limit cycles if (i) $\mu \leq 0$ and one if (ii) $\mu > 0$. Consider the two cases separately.

Case (i). When $\mu \leq 0$, the origin is a stable focus. Since $\dot{\theta} < 0$, the flow is clockwise. A phase portrait and vector field is shown in Figure 12.7(a).

Case (ii). When $\mu > 0$, there is an unstable focus at the origin and a stable limit cycle at $r = \sqrt{\mu}$ since $\dot{r} > 0$ if $0 < r < \sqrt{\mu}$ and $\dot{r} < 0$ if $r > \sqrt{\mu}$. A phase portrait is shown in Figure 12.7(b).

12.1. Bifurcations of Nonlinear Systems in the Plane   263

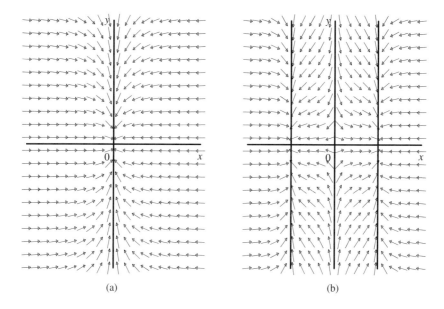

Figure 12.5: Vector field plots and manifolds when (a) $\mu \leq 0$ and (b) $\mu > 0$.

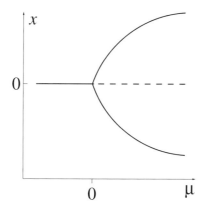

Figure 12.6: A bifurcation diagram for system (12.4) showing a pitchfork bifurcation. The solid curves depict stable behavior and the dashed curves depict unstable behavior. Note the resemblance of the stable branches to a pitchfork.

The qualitative behavior can be summarized on a bifurcation diagram as shown in Figure 12.8. As the parameter μ passes through the bifurcation value $\mu_0 = 0$, a limit cycle bifurcates from the origin. The amplitude of the limit cycle grows as μ increases. Think of the origin blowing a smoke ring.

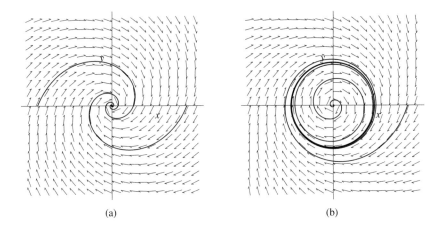

Figure 12.7: Phase portraits when (a) $\mu \leq 0$ and (b) $\mu > 0$.

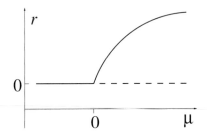

Figure 12.8: A bifurcation diagram for system (12.5) showing a Hopf bifurcation. The solid curves depict stable behavior and the dashed curves depict unstable behavior.

12.2 Multistability and Bistability

There are two types of Hopf bifurcation, one in which stable limit cycles are created about an unstable critical point, called the *supercritical Hopf bifurcation* (see Figure 12.8), and the other in which an unstable limit cycle is created about a stable critical point, called the *subcritical Hopf bifurcation* (see Figure 12.9).

In the engineering literature, supercritical bifurcations are sometimes called *soft* (or *safe*). The amplitude of the limit cycles build up gradually as the parameter, μ in this case, is moved away from the bifurcation point. In contrast, subcritical bifurcations are *hard* (or *dangerous*). A steady state, say, at the origin, could become unstable as a parameter varies and the nonzero solutions could tend to infinity. An example of this type of behavior can be found in Figure 12.9. As μ passes through zero from positive to negative values, the steady-state solution at the origin becomes

12.2. Multistability and Bistability

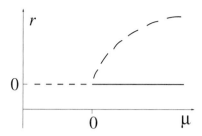

Figure 12.9: A bifurcation diagram showing a subcritical Hopf bifurcation. The solid curves depict stable behavior and the dashed curves depict unstable behavior.

unstable and trajectories starting anywhere other than the origin would tend to infinity.

It is also possible for limit cycles of finite amplitude to suddenly appear as the parameter μ is varied. These limit cycles are known as *large-amplitude limit cycles*. Examples of this type of behavior include surge oscillations in axial flow compressors and wing rock oscillations in aircraft flight dynamics; see [5] for examples. Unstable limit cycles are not observed in physical applications, so it is only the stable large-amplitude limit cycles that are of interest. These limit cycles can appear in one of two ways. Either there is a jump from a stable critical point to a stable large-amplitude limit cycle or there is a jump from one stable limit cycle to another of larger amplitude. These bifurcations are illustrated in the following examples.

Large-Amplitude Limit Cycle Bifurcations. Consider the system

(12.6) $$\dot{r} = r(\mu + r^2 - r^4), \qquad \dot{\theta} = -1.$$

The origin is the only critical point since $\dot{\theta} \neq 0$. This critical point is stable if $\mu < 0$ and unstable if $\mu > 0$. The system undergoes a subcritical Hopf bifurcation at $\mu = 0$ as in Figure 12.9. However, the new feature here is the stable large-amplitude limit cycle that exists for, say, $\mu > \mu_S$. In the range $\mu_S < \mu < 0$, there exist two different steady-state solutions; hence system (12.6) is multistable in this range. The choice of initial conditions determines which steady state will be approached as $t \to \infty$.

Definition 2. A dynamical system, say, (12.1), is said to be *multistable* if there is more than one possible steady-state solution for a fixed value of the parameter μ. The steady state obtained depends on the initial conditions.

The existence of multistable solutions allows for the possibility of *bistability* (or *hysteresis*) as a parameter is varied. The two essential ingredients for bistable behavior are nonlinearity and *feedback*. To create a bistable region there must

12. Bifurcation Theory

be some history in the system (see Chapter 4). Suppose that the parameter μ is increased from some value less than μ_S. The steady state remains at $r = 0$ until $\mu = 0$, where the origin loses stability. There is a sudden jump (a subcritical Hopf bifurcation) to the large-amplitude limit cycle, and the steady state remains on this cycle as μ is increased further. If the parameter μ is now decreased, then the steady state remains on the large-amplitude limit cycle until $\mu = \mu_S$, where the steady state suddenly jumps back to the origin (a saddle-node bifurcation) and remains there as μ is decreased further. In this way a bistable region is obtained as depicted in Figure 12.10.

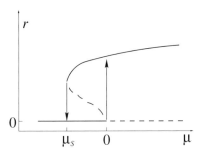

Figure 12.10: A bifurcation diagram depicting bistable behavior for system (12.6).

Definition 3. A dynamical system, say, (12.1), has a *bistable* solution if there are two steady states for a fixed parameter μ and the steady state obtained depends on the history of the system.

Now consider the system

(12.7) $$\dot{r} = r(\mu - 0.28r^6 + r^4 - r^2), \qquad \dot{\theta} = -1.$$

A bistable region may be obtained by increasing and then decreasing the parameter μ as in the example above. A possible bifurcation diagram is given in Figure 12.11. In this case, there is a supercritical Hopf bifurcation at $\mu = 0$ and saddle-node bifurcations at μ_B and μ_A, respectively.

Jumps between different steady states have been observed in mechanical systems. Parameters need to be chosen that avoid such large-amplitude limit cycle bifurcations, and research is currently underway in this respect.

Bistability also has many positive applications in the real world; for example, *nonlinear bistable optical resonators* were investigated in Chapter 4. The author is also currently investigating multistability and bistability in a wide range of disciplines.

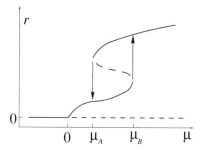

Figure 12.11: A bifurcation diagram depicting bistable behavior for system (12.7).

12.3 MATLAB Commands

```
% Programs_12a - Animation of a simple curve.
% The curve y=mu-x^2, for mu from -4 to +4.

axis tight
% Record the movie
x=-4:.1:4;
for n = 1:9
    plot(x,(n-5)-x.^2,x,0);
    M(n) = getframe;
end
% Use the movieviewer to watch the animation.
movieview(M)
% End of Programs_12a.
```

```
% Programs_12b - Finding critical points.
% Critical points for y=mu*x-x^3.
syms x y
[x,y]=solve('mu*x-x^3','-y')

% Plot a simple bifurcation diagram (Figure 12.11).
r=0:.01:2;
mu=.28*r.^6-r.^4+r.^2;
plot(mu,r)
fsize=15;
xlabel('\mu','Fontsize',fsize);
ylabel('r','Fontsize',fsize);
% End of Programs_12b.
```

12.4 Exercises

1. Consider the following one-parameter families of first-order differential equations defined on $\Re$:

   (a) $\dot{x} = \mu - x - e^{-x}$;
   (b) $\dot{x} = x(\mu + e^x)$;
   (c) $\dot{x} = x - \frac{\mu x}{1+x^2}$.

   Determine the critical points and the bifurcation values, plot vector fields on the line, and draw a bifurcation diagram in each case.

2. Use the animate command in MATLAB to show how $\dot{x}$ varies as μ increases from -4 to $+4$, for each of the differential equations in Exercise 1.

3. Construct first-order ODEs having the following:

   (i) three critical points (one stable and two unstable) when $\mu < 0$, one critical point when $\mu = 0$, and three critical points (one unstable and two stable) when $\mu > 0$;

   (ii) two critical points (one stable and one unstable) for $\mu \neq 0$ and one critical point when $\mu = 0$;

   (iii) one critical point if $|\mu| \geq 1$ and three critical points if $|\mu| < 1$.

   Draw a bifurcation diagram in each case.

4. A certain species of fish in a large lake is harvested. The differential equation used to model the population, $x(t)$ in hundreds of thousands, is given by

   $$\frac{dx}{dt} = x\left(1 - \frac{x}{5}\right) - \frac{hx}{0.2 + x}.$$

   Determine and classify the critical points and plot a bifurcation diagram. How can the model be interpreted in physical terms?

5. Consider the following one-parameter systems of differential equations:

   (a) $\dot{x} = x$, $\dot{y} = \mu - y^4$;
   (b) $\dot{x} = x^2 - x\mu^2$, $\dot{y} = -y$;
   (c) $\dot{x} = -x^4 + 5\mu x^2 - 4\mu^2$, $\dot{y} = -y$.

   Find the critical points, plot phase portraits, and sketch a bifurcation diagram in each case.

6. Consider the following one-parameter systems of differential equations in polar form:

(a) $\dot{r} = \mu r(r+\mu)^2$, $\dot{\theta} = 1$;
(b) $\dot{r} = r(\mu - r)(\mu - 2r)$, $\dot{\theta} = -1$;
(c) $\dot{r} = r(\mu^2 - r^2)$, $\dot{\theta} = 1$.

Plot phase portraits for $\mu < 0$, $\mu = 0$, and $\mu > 0$ in each case. Sketch the corresponding bifurcation diagrams.

7. Plot a bifurcation diagram for the planar system

$$\dot{r} = r(\mu - 0.2r^6 + r^4 - r^2), \qquad \dot{\theta} = -1$$

and indicate the regions where the system is multistable and/or possibly bistable.

8. Plot a bifurcation diagram for the system

$$\dot{r} = r((r-1)^2 - \mu r), \qquad \dot{\theta} = 1.$$

Give a possible explanation as to why this type of bifurcation should be known as a *fold bifurcation*.

9. Show that the one-parameter system

$$\dot{x} = y + \mu x - xy^2, \qquad \dot{y} = \mu y - x - y^3$$

undergoes a Hopf bifurcation at $\mu_0 = 0$. Plot phase portraits and sketch a bifurcation diagram.

10. Thus far, the analysis has been restricted to bifurcations involving only one parameter, and these are known as *codimension-1 bifurcations*. This example illustrates what can happen when two parameters are varied, allowing so-called *codimension-2 bifurcations*.

The following two-parameter system of differential equations may be used to model a simple laser:

$$\dot{x} = x(y-1), \qquad \dot{y} = \alpha + \beta y - xy.$$

Find and classify the critical points and sketch the phase portraits. Illustrate the different types of behavior in the (α, β) plane and determine whether or not any bifurcations occur.

Recommended Reading

[1] K. Ikeda and K. Murota, *Imperfect Bifurcations in Structures and Materials*, Springer-Verlag, New York, 2002.

12. Bifurcation Theory

[2] J. M. T. Thompson and H. B. Stewart, *Nonlinear Dynamics and Chaos*, 2nd ed., Wiley, New York, 2002.

[3] S. H. Strogatz, *Nonlinear Dynamics and Chaos with Applications to Physics, Biology, Chemistry and Engineering*, Perseus Books, New York, 2001.

[4] M. Demazure and D. Chillingworth (translator), *Bifurcations and Catastrophes: Geometry of Solutions to Nonlinear Problems*, Springer-Verlag, New York, 2000.

[5] S. Lynch and C. J. Christopher, Limit cycles in highly nonlinear differential equations, *J. Sound Vibration*, **224**-3 (1999), 505–517.

[6] G. Iooss and D. D. Joseph, *Elementary Stability and Bifurcation Theory*, Springer-Verlag, New York, 1997.

[7] R. Seydel, *Practical Bifurcation and Stability Analysis, from Equilibrium to Chaos*, Springer-Verlag, New York, 1994.

13
Three-Dimensional Autonomous Systems and Chaos

Aims and Objectives

- To introduce first-order ODEs in three variables.
- To discuss chaos in more depth.

On completion of this chapter the reader should be able to

- construct phase portraits for linear systems in three dimensions;
- use the MATLAB package to plot phase portraits and time series for nonlinear systems;
- identify chaotic solutions;
- interpret the solutions to modeling problems taken from various scientific disciplines, and in particular, chemical kinetics, electric circuits, and meteorology.

Three-dimensional autonomous systems of differential equations are considered. Critical points and stability are discussed and the concept of chaos is expanded upon. Examples include the Lorenz equations, used as a simple meteorological model and in the theory of lasers; Chua's circuit, used in nonlinear electronics and radiophysics; and the Belousov–Zhabotinski reaction, used in chemistry and biophysics. All of these systems can display highly complex behavior that can be interpreted from phase portrait analysis or Poincaré maps (see Chapter 14).

13. Three-Dimensional Autonomous Systems and Chaos

Basic concepts are explained by means of example rather than mathematical rigor. Strange or chaotic attractors are constructed using the MATLAB package, and the reader is encouraged to investigate these systems through the exercises at the end of the chapter. Chaos is discussed in other chapters of the book.

13.1 Linear Systems and Canonical Forms

Consider linear three-dimensional autonomous systems of the form

(13.1)
$$\dot{x} = a_{11}x + a_{12}y + a_{13}z,$$
$$\dot{y} = a_{21}x + a_{22}y + a_{23}z,$$
$$\dot{z} = a_{31}x + a_{32}y + a_{33}z,$$

where the a_{ij} are constants. The existence and uniqueness theorem (see Section 7.4) holds, which means that trajectories do not cross in three-dimensional space. The real canonical forms for 3×3 matrices are

$$J_1 = \begin{pmatrix} \lambda_1 & 0 & 0 \\ 0 & \lambda_2 & 0 \\ 0 & 0 & \lambda_3 \end{pmatrix}, \quad J_2 = \begin{pmatrix} \alpha & -\beta & 0 \\ \beta & \alpha & 0 \\ 0 & 0 & \lambda_3 \end{pmatrix},$$

$$J_3 = \begin{pmatrix} \lambda_1 & 1 & 0 \\ 0 & \lambda_1 & 0 \\ 0 & 0 & \lambda_2 \end{pmatrix}, \quad J_4 = \begin{pmatrix} \lambda_1 & 1 & 0 \\ 0 & \lambda_1 & 1 \\ 0 & 0 & \lambda_1 \end{pmatrix}.$$

Matrix J_1 has three real eigenvalues; matrix J_2 has a pair of complex eigenvalues; and matrices J_3 and J_4 have repeated eigenvalues. The type of phase portrait is determined from each of these canonical forms.

Definition 1. Suppose that $\mathbf{p} \in \mathbb{R}^3$ is a critical point of the system $\dot{\mathbf{x}} = \mathbf{f}(\mathbf{x})$, where $\mathbf{x} \in \mathbb{R}^3$. Then the stable and unstable manifolds of the critical point $\mathbf{p}$ are defined by

$$W_S(\mathbf{p}) = \{\mathbf{x} : \Lambda^+(\mathbf{x}) = \mathbf{p}\}, \qquad W_U(\mathbf{p}) = \{\mathbf{x} : \Lambda^-(\mathbf{x}) = \mathbf{p}\}.$$

As for two-dimensional systems, three-dimensional systems can have stable and unstable manifolds. These manifolds can be convoluted surfaces in three-dimensional space.

Example 1. Solve the following system of differential equations, sketch a phase portrait, and define the manifolds:

(13.2) $$\dot{x} = x, \qquad \dot{y} = y, \qquad \dot{z} = -z.$$

Solution. There is one critical point at the origin. Each differential equation is integrable with solutions given by $x(t) = C_1 e^t$, $y(t) = C_2 e^t$, and $z(t) = C_3 e^{-t}$. The eigenvalues and corresponding eigenvectors are $\lambda_{1,2} = 1$, $(0, 1, 0)^T$, $(1, 0, 0)^T$ and $\lambda_3 = -1$, $(0, 0, 1)^T$. System (13.2) may be uncoupled in any of the xy, xz, or yz planes. Planar analysis gives an unstable singular node in the xy plane and cols in each of the xz and yz planes. The phase plane portraits for two of the uncoupled systems are given in Figure 13.1. If $z > 0$, $\dot{z} < 0$, and if $z < 0$, $\dot{z} > 0$. The z-axis is a one-dimensional stable manifold since trajectories on this line are attracted to the origin as $t \to +\infty$. The xy plane is a two-dimensional unstable manifold since all trajectories in this plane are attracted to the origin as $t \to -\infty$.

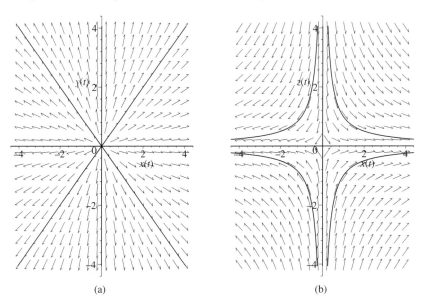

Figure 13.1: Phase plane portraits in the (a) xy and (b) xz planes. Note that (a) is an unstable planar manifold.

Putting all of this together, any trajectories not lying on the manifolds flow along "lampshades" in three-dimensional space, as depicted in Figure 13.2.

Example 2. Given the linear transformations $x = x_1 - 2y_1$, $y = -y_1$, and $z = -y_1 + z_1$, show that the system

$$\dot{x}_1 = -3x_1 + 10y_1, \qquad \dot{y}_1 = -2x_1 + 5y_1, \qquad \dot{z}_1 = -2x_1 + 2y_1 + 3z_1$$

can be transformed into

(13.3) $\qquad \dot{x} = x - 2y, \qquad \dot{y} = 2x + y, \qquad \dot{z} = 3z.$

Make a sketch of some trajectories in xyz space.

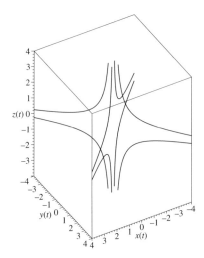

Figure 13.2: Phase portrait for system (13.2). The manifolds are not shown here.

Solution. The origin is the only critical point. Consider the transformations. Then

$\dot{x} = \dot{x}_1 - 2\dot{y}_1 = (-3x_1 + 10y_1) - 2(-2x_1 + 5y_1) = x_1 = x - 2y,$
$\dot{y} = -\dot{y}_1 = -(-2x_1 + 5y_1) = 2x_1 - 5y_1 = 2x + y,$
$\dot{z} = -\dot{y}_1 + \dot{z}_1 = -(-2x_1 + 5y_1) + (-2x_1 + 2y_1 + 3z_1) - 3(-y_1 + z_1) = 3z.$

System (13.3) is already in canonical form, and the eigenvalues are $\lambda_{1,2} = 1 \pm i$ and $\lambda_3 = 3$; hence the critical point is hyperbolic. The system can be uncoupled; the critical point at the origin in the xy plane is an unstable focus. A phase plane portrait is given in Figure 13.3.

Note that all trajectories spiral away from the origin, as depicted in Figure 13.4. Since all trajectories tend to the origin as $t \to -\infty$, the whole phase space forms an unstable manifold.

Example 3. Solve the following initial value problem

(13.4) $\qquad \dot{x} = z - x, \qquad \dot{y} = -y, \qquad \dot{z} = z - 17x + 16$

with $x(0) = y(0) = z(0) = 0.8$, and plot the solution curve in three-dimensional space.

Solution. System (13.4) can be uncoupled. The differential equation $\dot{y} = -y$ has general solution $y(t) = y_0 e^{-t}$, and substituting $y_0 = 0.8$ gives $y(t) = 0.8e^{-t}$. Now $z = \dot{x} + x$, and therefore the equation $\dot{z} = z - 17x + 16$ becomes

$$(\ddot{x} + \dot{x}) = (\dot{x} + x) - 17x + 16,$$

13.1. Linear Systems and Canonical Forms

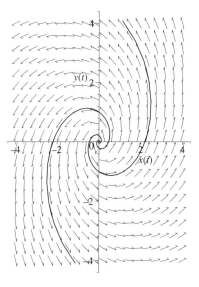

Figure 13.3: Some trajectories in the xy plane.

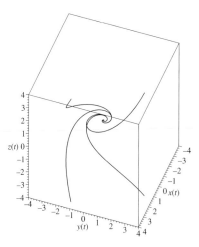

Figure 13.4: Phase portrait for system (13.3).

which simplifies to

$$\ddot{x} + 16x = 16.$$

Take Laplace transforms of both sides and insert the initial conditions to obtain

$$\bar{x}(s) = \frac{1}{s} - \frac{0.2s}{s^2 + 16}.$$

Take inverse transforms to get

$$x(t) = 1 - 0.2\cos(4t),$$

and therefore

$$z(t) = 1 + 0.8\sin(4t) - 0.2\cos(4t).$$

The solution curve is plotted in Figure 13.5.

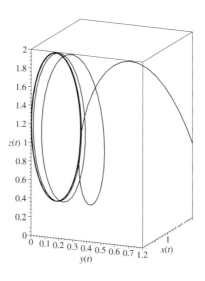

Figure 13.5: The solution curve for the initial value problem in Example 3. The trajectory ends up on an ellipse in the $y = 0$ plane.

13.2 Nonlinear Systems and Stability

If the critical point of a three-dimensional autonomous system is hyperbolic, then the linearization methods of Hartman can be applied (see [7, Chapter 8]). If the critical point is not hyperbolic, then other methods must be used.

Theorem 1. *Consider the differential equation*

$$\dot{\mathbf{x}} = \mathbf{f}(\mathbf{x}), \quad \mathbf{x} \in \Re^n,$$

where $\mathbf{f} \in C^1(E)$ *and* E *is an open subset of* $\Re^n$ *containing the origin. Suppose that* $\mathbf{f}(\mathbf{0}) = \mathbf{0}$ *and that the Jacobian matrix has n eigenvalues with nonzero real part. Then in a small neighborhood of* $\mathbf{x} = \mathbf{0}$, *there exist stable and unstable*

13.2. Nonlinear Systems and Stability

manifolds W_S and W_U with the same dimensions n_S and n_U as the stable and unstable manifolds (E_S, E_U) of the linearized system

$$\dot{\mathbf{x}} = J\mathbf{x},$$

where W_S and W_U are tangent to E_S and E_U at $\mathbf{x} = \mathbf{0}$.

A proof of this theorem can be found in Hartman's book (see [7, Chapter 8]).
If the Jacobian matrix J has eigenvalues with zero real part, then there exists a so-called *center manifold*, which in general is not unique.

Theorem 2 (The Center Manifold Theorem). *Let* $\mathbf{f} \in C^r(E)$, $(r \geq 1)$, *where E is an open subset of $\Re^n$ containing the origin. If* $\mathbf{f}(\mathbf{0}) = \mathbf{0}$ *and the Jacobian matrix has n_S eigenvalues with negative real part, n_U eigenvalues with positive real part, and $n_C = n - n_S - n_U$ purely imaginary eigenvalues, then there exists an n_C-dimensional center manifold W_C of class C^r that is tangent to the center manifold E_C of the linearized system.*

To find out more about center manifolds, see Wiggins [2].

Example 4. Determine the stable, unstable, and center manifolds of the nonlinear system

$$\dot{x} = x^2, \qquad \dot{y} = -y, \qquad \dot{z} = -2z.$$

Solution. There is a unique critical point at the origin. This system is easily solved, and it is not difficult to plot phase portraits for each of the uncoupled systems. The solutions are $x(t) = \frac{1}{C_1 - t}$, $y(t) = C_2 e^{-t}$, and $z(t) = C_3 e^{-2t}$. The eigenvalues and corresponding eigenvectors of the Jacobian matrix are $\lambda_1 = 0$, $(1, 0, 0)^T$, $\lambda_2 = -1$, $(0, 1, 0)^T$, and $\lambda_3 = -2$, $(0, 0, 1)^T$. In this case, $W_C = E_C$, the x-axis, and the yz plane forms a two-dimensional stable manifold, where $W_S = E_S$. Note that the center manifold is unique in this case, but it is not in general.

Example 5. Solve the following nonlinear differential system

$$\dot{x} = -x, \qquad \dot{y} = -y + x^2, \qquad \dot{z} = z + x^2,$$

and determine the stable and unstable manifolds.

Solution. The point $O = (0, 0, 0)$ is a unique critical point. Linearize by finding the Jacobian matrix. Hence

$$J = \begin{pmatrix} \frac{\partial P}{\partial x} & \frac{\partial P}{\partial y} & \frac{\partial P}{\partial z} \\ \frac{\partial Q}{\partial x} & \frac{\partial Q}{\partial y} & \frac{\partial Q}{\partial z} \\ \frac{\partial R}{\partial x} & \frac{\partial R}{\partial y} & \frac{\partial R}{\partial z} \end{pmatrix},$$

278   13. Three-Dimensional Autonomous Systems and Chaos

where $\dot{x} = P(x, y, z)$, $\dot{y} = Q(x, y, z)$, and $\dot{z} = R(x, y, z)$. Therefore,

$$J_O = \begin{pmatrix} -1 & 0 & 0 \\ 0 & -1 & 0 \\ 0 & 0 & 1 \end{pmatrix},$$

and the origin is an unstable critical point. Note that two of the eigenvalues are negative. These give a two-dimensional stable manifold, which will now be defined.

The differential equation $\dot{x} = -x$ is integrable and has solution $x(t) = C_1 e^{-t}$. The other two differential equations are linear and have solutions $y(t) = C_2 e^{-t} + C_1^2 (e^{-t} - e^{-2t})$ and $z(t) = C_3 e^{t} + \frac{C_1^2}{3}(e^{t} - e^{-2t})$. Now $\Lambda^{+}(\mathbf{x}) = \mathbf{0}$ if and only if $C_3 + \frac{C_1^2}{3} = 0$, where $\mathbf{x} \in \Re^3$, $C_1 = x(0)$, $C_2 = y(0)$, and $C_3 = z(0)$. Therefore, the stable manifold is given by

$$W_S = \left\{ \mathbf{x} \in \Re^3 : z = -\frac{x^2}{3} \right\}.$$

Using similar arguments, $\Lambda^{-}(\mathbf{x}) = \mathbf{0}$ if and only if $C_1 = C_2 = 0$. Hence the unstable manifold is given by

$$W_U = \{ \mathbf{x} \in \Re^3 : x = y = 0 \}.$$

Note that the surface W_S is tangent to the xy plane at the origin.

Example 6. Sketch a phase portrait for the system

(13.5) $\dot{x} = x + y - x(x^2 + y^2), \qquad \dot{y} = -x + y - y(x^2 + y^2), \qquad \dot{z} = -z.$

Solution. Convert to cylindrical polar coordinates by setting $x = r\cos\theta$ and $y = r\sin\theta$. System (13.5) then becomes

$$\dot{r} = r(1 - r^2), \qquad \dot{\theta} = -1, \qquad \dot{z} = -z.$$

The origin is the only critical point. The system uncouples; in the xy plane, the flow is clockwise and the origin is an unstable focus. If $z > 0$, then $\dot{z} < 0$, and if $z < 0$, then $\dot{z} > 0$. If $r = 1$, then $\dot{r} = 0$. Trajectories spiral towards the xy plane and onto the limit cycle, say, Γ_1, of radius 1 centered at the origin. Hence $\Lambda^{+}(\mathbf{x}) = \Gamma_1$ if $\mathbf{x} \neq \mathbf{0}$ and Γ_1 is a stable limit cycle. A phase portrait is shown in Figure 13.6.

Lyapunov functions were introduced in Chapter 11 and were used to determine the stability of critical points for certain planar systems. The theory is easily extended to the three-dimensional case as the following examples demonstrate. Once again, there is no systematic way to determine the Lyapunov functions, and they are given in the question.

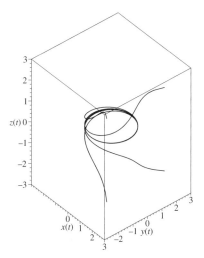

Figure 13.6: Trajectories are attracted to a stable limit cycle in the xy plane.

Example 7. Prove that the origin of the system

$$\dot{x} = -2y + yz, \qquad \dot{y} = x(1-z), \qquad \dot{z} = xy$$

is stable but not asymptotically stable by using the Lyapunov function $V(x, y, z) = ax^2 + by^2 + cz^2$.

Solution. Now

$$\frac{dV}{dt} = \frac{\partial V}{\partial x}\frac{dx}{dt} + \frac{\partial V}{\partial y}\frac{dy}{dt} + \frac{\partial V}{\partial z}\frac{dz}{dt} = 2(a - b + c)xyz + 2(b - 2a)xy.$$

If $b = 2a$ and $a = c > 0$, then $V(x, y, z) > 0$ for all $\mathbf{x} \neq 0$ and $\frac{dV}{dt} = 0$. Thus the trajectories lie on the ellipsoids defined by $x^2 + 2y^2 + z^2 = r^2$. The origin is thus stable but not asymptotically stable.

Example 8. Prove that the origin of the system

$$\dot{x} = -y - xy^2 + z^2 - x^3, \qquad \dot{y} = x + z^3 - y^3, \qquad \dot{z} = -xz - x^2z - yz^2 - z^5$$

is asymptotically stable by using the Lyapunov function $V(x, y, z) = x^2 + y^2 + z^2$.

Solution. Now

$$\frac{dV}{dt} = \frac{\partial V}{\partial x}\frac{dx}{dt} + \frac{\partial V}{\partial y}\frac{dy}{dt} + \frac{\partial V}{\partial z}\frac{dz}{dt} = -2(x^4 + y^4 + x^2z^2 + x^2y^2 + z^6).$$

Since $\frac{dV}{dt} < 0$ for $x, y, z \neq 0$, the origin is asymptotically stable. In fact, the origin is *globally asymptotically stable* since $\Lambda^+(\mathbf{x}) = (0, 0, 0)$ for all $\mathbf{x} \in \Re^3$.

13.3 The Rössler System and Chaos

13.3.I The Rössler Attractor. In 1976, Otto E. Rössler [13] constructed the following three-dimensional system of differential equations:

(13.6) $\quad \dot{x} = -(y+z), \quad \dot{y} = x + ay, \quad \dot{z} = b + xz - cz,$

where a, b, and c are all constants. Note that the only nonlinear term appears in the $\dot{z}$ equation and is quadratic. As the parameters vary, this simple system can display a wide range of behavior. Set $a = b = 0.2$, for example, and vary the parameter c. The dynamics of the system can be investigated using the MATLAB package. Four examples are considered here. Transitional trajectories have been omitted to avoid confusion. The initial conditions are $x(0) = y(0) = z(0) = 1$ in all cases.

Definition 2. A limit cycle is of period one if $\mathbf{x}(t) = \mathbf{x}(t+T)$ for some minimum constant T called the period. The periodic orbit is called a *period-one cycle*.

When $c = 2.3$, there is a period-one limit cycle that can be plotted in three-dimensional space. Figure 13.7(a) shows the limit cycle in phase space, and the periodic behavior with respect to $x(t)$ is shown in Figure 13.7(b).

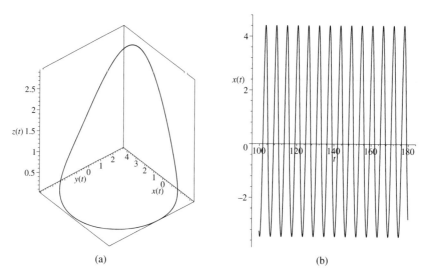

Figure 13.7: (a) A limit cycle for system (13.6) when $c = 2.3$. (b) Period-one behavior for $x(t)$.

When $c = 3.3$, there is period-two behavior. Figure 13.8(a) shows the closed orbit in phase space, and the periodic behavior is shown in Figure 13.8(b). Notice that there are two distinct amplitudes in Figure 13.8(b). This periodic behavior can be easily detected using Poincaré maps (see Chapter 14).

13.3. The Rössler System and Chaos   281

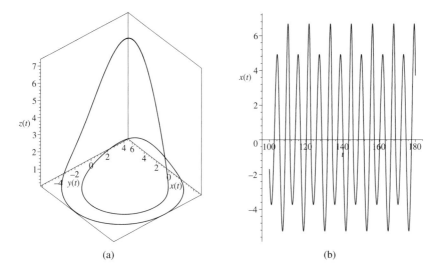

Figure 13.8: (a) A period-two limit cycle for system (13.6) when $c = 3.3$. (b) Period-two behavior for $x(t)$.

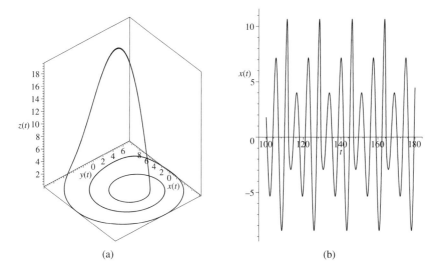

Figure 13.9: (a) A period-three limit cycle for system (13.6) when $c = 5.3$. (b) Period-three behavior for $x(t)$.

When $c = 5.3$, there is period-three behavior. Figure 13.9(a) shows the closed orbit in three-dimensional space, and the periodic behavior is shown in Figure 13.9(b). Note that there are three distinct amplitudes in Figure 13.9(b).

When $c = 6.3$, the system displays what appears to be *random behavior*. This type of behavior has been labeled *deterministic chaos*. A system is called *deterministic* if the behavior of the system is determined from the time evolution equations and the initial conditions alone, as in the case of the Rössler system. *Nondeterministic chaos* arises when there are no underlying equations, as in the United Kingdom national lottery, or there is noisy or random input. This text will be concerned with deterministic chaos only, and it will be referred to simply as chaos from now on.

13.3.II Chaos. Chaos is a multifaceted phenomenon that is not easily classified or identified. There is no universally accepted definition for chaos, but the following characteristics are nearly always displayed by the solutions of chaotic systems:

1. long-term *aperiodic (nonperiodic) behavior*;

2. *sensitivity to initial conditions*;

3. *fractal structure*.

Consider each of these items independently. Note, however, that a chaotic system generally displays all three types of behavior listed above.

Case 1. It is very difficult to distinguish between aperiodic behavior and periodic behavior with a very long period. For example, it is possible for a chaotic system to have a periodic solution of period 10^{100}.

Case 2. A simple method used to test whether a system is chaotic is to check for sensitivity to initial conditions. Figure 13.10(a) shows the trajectory in phase space and Figure 13.10(b) illustrates how the system is sensitive to the choice of initial conditions.

Definition 3. An *attractor* is a minimal closed invariant set that attracts nearby trajectories lying in the domain of stability (or basin of attraction) onto it.

Definition 4. A *strange attractor* (*chaotic attractor, fractal attractor*) is an attractor that exhibits sensitivity to initial conditions.

An example of a strange attractor is shown in Figure 13.10(a). Another method for establishing whether a system is chaotic is to use the *Lyapunov exponents* (see Chapter 2 for the discrete case). A system is chaotic if at least one of the Lyapunov exponents is positive. This implies that two trajectories that start close to each other on the strange attractor will diverge as time increases, as depicted in Figure 13.10(b). Note that an n-dimensional system will have n different Lyapunov exponents. Think of an infinitesimal sphere of perturbed initial conditions for a three-dimensional system. As time increases the sphere will evolve into an infinitesimal ellipsoid. If d_0 is the initial radius of the sphere, then $d_j = d_0 e^{\lambda_j}$ ($j = 1, \ldots, 3$) define the axes of the ellipsoid. For bounded chaotic attractors in

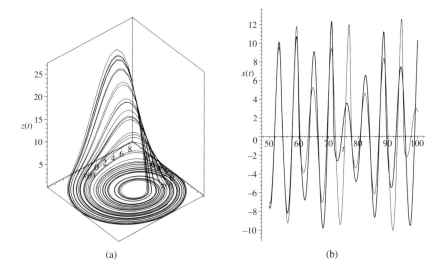

Figure 13.10: (a) The chaotic attractor for system (13.6) when $c = 6.3$. (b) Time series plot of $x(t)$ showing sensitivity to initial conditions; the initial conditions for one time series are $x(0) = y(0) = z(0) = 1$ and for the other are $x(0) = 1.01$, $y(0) = z(0) = 1$.

three dimensions $\lambda_1 > 0$, $\lambda_2 = 0$, and $\lambda_3 < 0$. A comparison of different methods for computing the Lyapunov exponents is given in [9]. One interesting feature of strange attractors is that it is sometimes possible to reconstruct the attractor from time series data alone; see [5], for example.

Gottwald and Melbourne [1] describe a new test for deterministic chaos that is applied to time series data alone, where the underlying equations are not required. Their diagnostic is the real valued function

$$p(t) = \int_0^t \phi(\mathbf{x}(s)) \cos(\omega_0 s) ds,$$

where ϕ is an observable on the dynamics $\mathbf{x}(t)$ and $\omega_0 \neq 0$ is a constant. They set

$$K = \lim_{t \to \infty} \frac{\log \mathbf{M}(t)}{\log(t)},$$

where $\mathbf{M}$ is the mean-square displacement for $p(t)$. Typically, $K = 0$ signifying regular dynamics, or $K = 1$ indicating chaotic dynamics. They state that the test works well for both continuous and discrete systems.

Case 3. The solution curves to chaotic systems generally display fractal structure (see Chapter 5). The structure of the strange attractors for general n-dimensional systems may be complicated and difficult to observe clearly. To overcome these problems, Poincaré maps, which exist in lower-dimensional spaces, can be used, as in Chapter 14.

13.4 The Lorenz Equations, Chua's Circuit, and the Belousov–Zhabotinski Reaction

Note that most nonlinear systems display steady-state behavior most of the time, so it is possible to predict, for example, the weather, motion of the planets, spread of an epidemic, motion of a driven pendulum, or beat of the human heart. However, nonlinear systems can also display chaotic behavior where prediction becomes impossible.

There are many examples of applications of three-dimensional autonomous systems to the real world. These systems obey the existence and uniqueness theorem from Chapter 7, but the dynamics can be much more complicated than in the two-dimensional case. The following examples taken from meteorology, electric circuit theory, and chemical kinetics have been widely investigated in recent years. There are more examples in the exercises at the end of the chapter.

13.4.I The Lorenz Equations. In 1963, the MIT meteorologist Edward Lorenz [14] constructed a highly simplified model of a convecting fluid. This simple model also displays a wide variety of behavior and for some parameter values is chaotic. The equations can be used to model convective flow up through the center and down on the sides of hexagonal columns. The system is given by

(13.7) $$\dot{x} = \sigma(y - x), \qquad \dot{y} = rx - y - xz, \qquad \dot{z} = xy - bz,$$

where x measures the rate of convective overturning, y measures the horizontal temperature variation, z measures the vertical temperature variation, σ is the Prandtl number, r is the Rayleigh number, and b is a scaling factor. The Prandtl number is related to the fluid viscosity, and the Rayleigh number is related to the temperature difference between the top and bottom of the column. Lorenz studied the system when $\sigma = 10$ and $b = \frac{8}{3}$.

The system can be considered to be a highly simplified model for the weather. Indeed, satellite photographs from space show hexagonal patterns on undisturbed desert floors. The astonishing conclusion derived by Lorenz is now widely labeled as the *butterfly effect*. Even this very simple model of the weather can display chaotic phenomena. Since the system is sensitive to initial conditions, small changes to wind speed (convective overturning), for example, generated by the flap of a butterfly's wings, can change the outcome of the results considerably. For example, a butterfly flapping its wings in Britain could cause or prevent a hurricane from occurring in the Bahamas in the not-so-distant future. Of course, there are many more variables that should be considered when trying to model weather systems, and this simplified model illustrates some of the problems meteorologists have to deal with.

13.4. The Lorenz Equations, Chua's Circuit, and the Belousov–Zhabotinski Reaction

Some simple properties of the Lorenz equations will now be listed, and all of these characteristics can be investigated with the aid of the MATLAB package:

1. System (13.7) has natural symmetry $(x, y, z) \to (-x, -y, z)$.

2. The z-axis is invariant.

3. The flow is volume contracting since div $\mathbf{X} = -(\sigma + b + 1) < 0$, where $\mathbf{X}$ is the vector field.

4. If $0 < r < 1$, the origin is the only critical point, and it is a global attractor.

5. At $r = 1$, there is a bifurcation, and there are two more critical points at $C_1 = (\sqrt{b(r-1)}, \sqrt{b(r-1)}, r-1)$ and $C_2 = (-\sqrt{b(r-1)}, -\sqrt{b(r-1)}, r-1)$.

6. At $r = r_H \approx 13.93$, there is a homoclinic bifurcation (see Chapter 15) and the system enters a state of transient chaos.

7. At $r \approx 24.06$, a strange attractor is formed.

8. If $1 < r < r_O$, where $r_O \approx 24.74$, the origin is unstable and C_1 and C_2 are both stable.

9. At $r > r_O$, C_1 and C_2 lose their stability by absorbing an unstable limit cycle in a subcritical Hopf bifurcation.

For more details, see the work of Sparrow [12] or most textbooks on nonlinear dynamics. Most of the results above can be observed by plotting phase portraits or time series using the MATLAB package. A strange attractor is shown in Figure 13.11.

The trajectories wind around the two critical points C_1 and C_2 in an apparently random unpredictable manner. The strange attractor has the following properties:

- The trajectory is aperiodic (or not periodic).
- The trajectory remains on the attractor forever (the attractor is invariant).
- The general form is independent of initial conditions.
- The sequence of windings is sensitive to initial conditions.
- The attractor has fractal structure.

A variation on the Lorenz model has recently been discovered by Guanrong Chen and Tetsushi Ueta (see Figure 13.12). The equations are

(13.8) $\quad \dot{x} = \sigma(y - x), \qquad \dot{y} = (r - \sigma)x + ry - xz, \qquad \dot{z} = xy - bz.$

13. Three-Dimensional Autonomous Systems and Chaos

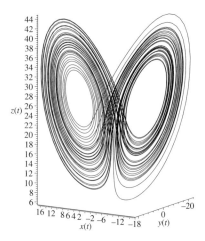

Figure 13.11: A strange attractor for the Lorenz system when $\sigma = 10$, $b = \frac{8}{3}$, and $r = 28$.

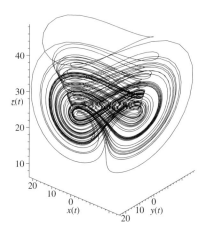

Figure 13.12: A strange attractor for system (13.8) when $\sigma = 35$, $b = 3$, and $r = 28$.

13.4.II Chua's Circuit. Elementary electric circuit theory was introduced in Chapter 7. In the mid-1980s Chua [8] modeled a circuit that was a simple oscillator exhibiting a variety of bifurcation and chaotic phenomena. The circuit diagram is given in Figure 13.13. The circuit equations are given by

(13.9)
$$\frac{dv_1}{dt} = \frac{(G(v_2 - v_1) - f(v_1))}{C_1}, \quad \frac{dv_2}{dt} = \frac{(G(v_1 - v_2) + i)}{C_2}, \quad \frac{di}{dt} = -\frac{v_2}{L},$$

13.4. The Lorenz Equations, Chua's Circuit, and the Belousov–Zhabotinski Reaction

Figure 13.13: Chua's electric circuit.

where v_1, v_2, and i are the voltages across C_1, C_2, and the current through L, respectively. The characteristic of the nonlinear resistor N_R is given by

$$f(v_1) = G_b v_1 + 0.5(G_a - G_b)(|v_1 + B_p| - |v_1 - B_p|),$$

where $G = 1/R$. Typical parameters used are $C_1 = 10.1 nF$, $C_2 = 101 nF$, $L = 20.8 mH$, $R = 1420\Omega$, $r = 63.8\Omega$, $G_a = -0.865 mS$, $G_b = -0.519 mS$, and $B_p = 1.85V$.

In the simple case, Chua's equations can be written in the following dimensionless form:

(13.10) $\qquad \dot{x} = a(y - x - g(x)), \qquad \dot{y} = x - y + z, \qquad \dot{z} = -by,$

where a and b are dimensionless parameters. The function $g(x)$ has the form

$$g(x) = cx + \frac{1}{2}(d - c)(|x + 1| - |x - 1|),$$

where c and d are constants.

Chua's circuit is investigated in some detail in [8] and exhibits many interesting phenomena including period-doubling cascades to chaos, intermittency routes to chaos, and *quasiperiodic* routes to chaos. For certain parameter values, the solutions lie on a *double-scroll attractor*, as shown in Figure 13.14.

The dynamics are more complicated than those appearing in either the Rössler or Lorenz attractors. Chua's circuit has proved to be a very suitable subject for study since laboratory experiments produce results that match very well with the results of the mathematical model. Recently, the author and Borresen [3] have shown the existence of a bistable cycle for Chua's electric circuit for the first time.

13.4.III   The Belousov–Zhabotinski Reaction. Periodic chemical reactions such as the Landolt clock and the Belousov–Zhabotinski reaction provide wonderful examples of relaxation oscillations in science (see [5], [6], [11]). They are often demonstrated in chemistry classes or used to astound the public at university open days. The first experiment was conducted by the Russian biochemist Boris Belousov in the 1950s, and the results were not confirmed until as late as 1968 by Zhabotinski.

Consider the following recipe for a Belousov periodic chemical reaction.

288   13. Three-Dimensional Autonomous Systems and Chaos

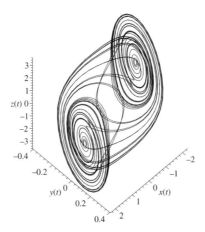

Figure 13.14: Chua's double-scroll attractor: Phase portrait for system (13.10) when $a = 15$, $b = 25.58$, $c = -\frac{5}{7}$, and $d = -\frac{8}{7}$. The initial conditions are $x(0) = -1.6$, $y(0) = 0$, and $z(0) = 1.6$.

Ingredients.

- Solution A: Malonic acid, 15.6 gm/l.
- Solution B: Potassium bromate, 41.75 gm/l, and potassium bromide, 0.006 gm/l.
- Solution C: Cerium IV sulphate, 3.23 gm/l in 6M sulphuric acid.
- Solution D: Ferroin indicator.

Procedure. Add 20 mls of solution A and 10 mls of solution B to a mixture of 10 mls of solution C and 1 ml of solution D. Stir continuously at room temperature. The mixture remains blue for about 10 minutes and then begins to oscillate blue-green-pink and back again with a period of approximately two minutes.

This reaction is used by the Chemistry Department at Manchester Metropolitan University during open days and is always a popular attraction.

Following the methods of Field and Noyes (see [7]) the chemical rate equations for an oscillating Belousov–Zhabotinski reaction are frequently written as

$$BrO_3^- + Br^- \to HBrO_2 + HOBr, \qquad \text{Rate} = k_1[BrO_3^-][Br^-],$$
$$HBrO_2 + Br^- \to 2HOBr, \qquad \text{Rate} = k_2[HBrO_2][Br^-],$$
$$BrO_3^- + HBrO_2 \to 2HBrO_2 + 2M_{OX}, \qquad \text{Rate} = k_3[BrO_3^-][HBrO_2],$$
$$2HBrO_2 \to BrO_3^- + HOBr, \qquad \text{Rate} = k_4[HBrO_2]^2,$$
$$OS + M_{OX} \to \tfrac{1}{2}CBr^-, \qquad \text{Rate} = k_5[OS][M_{OX}],$$

13.4. The Lorenz Equations, Chua's Circuit, and the Belousov–Zhabotinski Reaction

where OS represents all oxidizable organic species and C is a constant. Note that in the third equation, species $HBrO_2$ stimulates its own production, a process called *autocatalysis*. The reaction rate equations for the concentrations of intermediate species $x = [HBrO_2]$, $y = [Br^-]$, and $z = [M_{OX}]$ are

$$\dot{x} = k_1 ay - k_2 xy + k_3 ax - 2k_4 x^2,$$
$$\dot{y} = -k_1 ay - k_2 xy + \frac{1}{2} C k_5 bz,$$
(13.11) $$\dot{z} = 2k_3 ax - k_5 bz,$$

where $a = [BrO_3^-]$ and $b = [OS]$ are assumed to be constant. Taking the transformations

$$X = \frac{2k_4 x}{k_5 a}, \quad Y = \frac{k_2 y}{k_3 a}, \quad Z = \frac{k_5 k_4 bz}{(k_3 a)^2}, \quad \tau = k_5 bt,$$

system (13.11) becomes

$$\frac{dX}{d\tau} = \frac{qY - XY + X(1-X)}{\epsilon_1},$$
$$\frac{dY}{d\tau} = \frac{-qY - XY + CZ}{\epsilon_2},$$
(13.12) $$\frac{dZ}{d\tau} = X - Z,$$

where $\epsilon_1 = \frac{k_5 b}{k_3 a}$, $\epsilon_2 = \frac{2k_5 k_4 b}{k_2 k_3 a}$, and $q = \frac{2k_1 k_4}{k_2 k_3}$. Next, one assumes that $\epsilon_2 \ll 1$ so that $\frac{dY}{d\tau}$ is large unless the numerator $-qY - XY + CZ$ is also small. Assume that

$$Y = Y^* = \frac{CZ}{q+X}$$

at all times, so the bromide concentration $Y = [Br^-]$ is in a steady state compared to X. In this way, a three-dimensional system of differential equations is reduced to a two-dimensional system of autonomous ODEs

(13.13) $$\epsilon_1 \frac{dX}{d\tau} = X(1-X) - \frac{X-q}{X+q} CZ, \quad \frac{dZ}{d\tau} = X - Z.$$

For certain parameter values, system (13.13) has a limit cycle that represents an oscillating Belousov–Zhabotinski chemical reaction, as in Figure 13.15.

Example 9. Find and classify the critical points of system (13.13) when $\epsilon_1 = 0.05$, $q = 0.01$, and $C = 1$. Plot a phase portrait in the first quadrant.

13. Three-Dimensional Autonomous Systems and Chaos

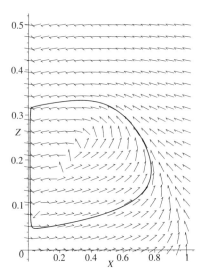

Figure 13.15: A limit cycle in the XZ plane for system (13.13) when $\epsilon_1 = 0.05$, $q = 0.01$, and $C = 1$.

Solution. There are two critical points, one at the origin and the other at $A \approx (0.1365, 0.1365)$. The Jacobian matrix is given by

$$J = \begin{pmatrix} \frac{1}{\epsilon_1}\left(1 - 2X - \frac{Z}{X+q} + \frac{(X-q)Z}{(X+q)^2}\right) & \frac{1}{\epsilon_1}\left(\frac{q-X}{X+q}\right) \\ 1 & -1 \end{pmatrix}.$$

It is not difficult to show that the origin is a saddle point and A is an unstable node. The direction vectors can be plotted using the M-file vectorfield.m given in Chapter 8. A phase portrait showing periodic oscillations is given in Figure 13.15.

The period of the limit cycle in Figure 13.15 is approximately 3.4. The trajectory moves quickly along the right and left branches of the limit cycle (up and down) and moves relatively slowly in the horizontal direction. This accounts for the rapid color changes and time spans between these changes.

It is important to note that chemical reactions are distinct from many other types of dynamical systems in that closed chemical reactions cannot oscillate about their chemical equilibrium state. This problem is easily surmounted by exchanging mass or introducing a flow with the chemical reaction and its surroundings. For example, the Belousov–Zhabotinski reaction used during university open days is stirred constantly and mass is exchanged. It is also possible for the Belousov–Zhabotinski reaction to display chaotic phenomena; see [6], for example. Multistable and bistable chemical reactions are also discussed in [7]. In these cases, there is an inflow and outflow of certain species and more than one steady state can coexist.

13.5 MATLAB Commands

```
% Programs_13a - 3D phase portrait (Figure 13.11).

Lorenz=inline(
'[10*(x(2)-x(1));28*x(1)-x(2)-x(1)*x(3);x(1)*x(2)-(8/3)
   *x(3)]','t','x');
options = odeset('RelTol',1e-4,'AbsTol',1e-4);
[t,xa]=ode45(Lorenz,[0 100],[15,20,30],options);
plot3(xa(:,1),xa(:,2),xa(:,3))
title('The Lorenz Attractor')

fsize=15;
xlabel('x(t)','Fontsize',fsize);
ylabel('y(t)','Fontsize',fsize);
zlabel('z(t)','FontSize',fsize);
% End of Programs_13a.
```

```
% Programs_13b - 3D phase portrait. Chua's chaotic attractor
% (Figure 13.14).

Chua=inline('[15*(x(2)-x(1)-(-(5/7)*x(1)+(1/2)*(-(8/7)
-(-5/7))*
(abs(x(1)+1)-abs(x(1)-1))));x(1)-x(2)+x(3);-25.58
*x(2)]','t','x');
options = odeset('RelTol',1e-4,'AbsTol',1e-4);
[t,xb]=ode45(Chua,[0 100],[-1.6,0,1.6],options);
plot3(xb(:,1),xb(:,2),xb(:,3))
title('Chua`s Double-Scroll Attractor')

fsize=15;
xlabel('x(t)','Fontsize',fsize);
ylabel('y(t)','Fontsize',fsize);
zlabel('z(t)','FontSize',fsize);
% End of Programs_13b.
```

```
% Programs_13c - The Chapman cycle. A stiff system.
% x=[singlet], y=[oxygen], z=[ozone]. Exercise 6 in Section
% 13.6.

% Simple model of ozone production.
clear
```

```
deq=inline('
[2*3e-12*x(2)+5.5e-4*x(3)-1.22e-33*x(1)*x(2)*9e17-6.86e-16
*x(1)*x(3);
5.5e-4*x(3)+2*6.86e-16*x(1)*x(3)-3e-12*x(2)-1.22e-33*x(1)
*x(2)*9e17;
1.22e-33*9e17*x(1)*x(2)-5.5e-4*x(3)-6.86e-16*x(1)
*x(3)]','t','x');
[t,xa]=ode23s(deq,[0 1e8],[4e16,2e16,2e16]);
last=size(xa,1)
O=xa(last,1)
O2=xa(last,2)
O3=xa(last,3)
% End of Programs_13c.
```

13.6 Exercises

1. Find the eigenvalues and eigenvectors of the matrix

$$A = \begin{pmatrix} 1 & 0 & -4 \\ 0 & 5 & 4 \\ -4 & 4 & 3 \end{pmatrix}.$$

   Hence show that the system $\dot{\mathbf{x}} = A\mathbf{x}$ can be transformed into $\dot{\mathbf{u}} = J\mathbf{u}$, where

$$J = \begin{pmatrix} 3 & 0 & 0 \\ 0 & -3 & 0 \\ 0 & 0 & 9 \end{pmatrix}.$$

   Sketch a phase portrait for the system $\dot{\mathbf{u}} = J\mathbf{u}$.

2. Classify the critical point at the origin for the system

$$\dot{x} = x + 2z, \qquad \dot{y} = y - 3z, \qquad \dot{z} = 2y + z.$$

3. Find and classify the critical points of the system

$$\dot{x} = x - y, \qquad \dot{y} = y + y^2, \qquad \dot{z} = x - z.$$

4. Consider the system

$$\dot{x} = -x + (\lambda - x)y, \qquad \dot{y} = x - (\lambda - x)y - y + 2z, \qquad \dot{z} = \frac{y}{2} - z,$$

   where $\lambda \geq 0$ is a constant. Show that the first quadrant is positively invariant and that the plane $x + y + 2z = \text{constant}$ is invariant. Find $\lambda^+(p)$ for p in the first quadrant given that there are no periodic orbits there.

5. (a) Prove that the origin of the system

$$\dot{x} = -x - y^2 + xz - x^3, \qquad \dot{y} = -y + z^2 + xy - y^3, \qquad \dot{z} = -z + x^2 + yz - z^3$$

is globally asymptotically stable.

(b) Determine the domain of stability for the system

$$\dot{x} = -ax + xyz, \qquad \dot{y} = -by + xyz, \qquad \dot{z} = -cz + xyz.$$

6. The chemical rate equations for the Chapman cycle modeling the production of ozone are

$$\begin{aligned} O_2 + hv &\to O + O, & \text{Rate} &= k_1, \\ O_2 + O + M &\to O_3 + M, & \text{Rate} &= k_2, \\ O_3 + hv &\to O_2 + O, & \text{Rate} &= k_3, \\ O + O_3 &\to O_2 + O_2, & \text{Rate} &= k_4, \end{aligned}$$

where O is a singlet, O_2 is oxygen, and O_3 is ozone. The reaction rate equations for species $x = [O]$, $y = [O_2]$, and $z = [O_3]$ are

$$\begin{aligned} \dot{x} &= 2k_1 y - k_3 z - k_2 xyM - k_4 xz, \\ \dot{y} &= k_3 z + 2k_4 xz - k_1 y - k_2 xyM, \\ \dot{z} &= k_2 xyM - k_3 z - k_4 xz. \end{aligned}$$

This is a stiff system of differential equations. Many differential equations applied in chemical kinetics are stiff. Given that $[M] = 9e17$, $k_1 = 3e{-}12$, $k_2 = 1.22e{-}33$, $k_3 = 5.5e{-}4$, $k_4 = 6.86e{-}16$, $x(0) = 4e16$, $y(0) = 2e16$, and $z(0) = 2e16$, show that the steady state reached is $[O] = 4.6806e7$, $[O_2] = 6.999e16$, and $[O_3] = 6.5396e12$.

7. A three-dimensional Lotka–Volterra model is given by

$$\dot{x} = x(1-2x+y-5z), \qquad \dot{y} = y(1-5x-2y-z), \qquad \dot{z} = z(1+x-3y-2z).$$

Prove that there is a critical point in the first quadrant at $P(\frac{1}{14}, \frac{3}{14}, \frac{3}{14})$. Plot possible trajectories and show that there is a solution plane $x + y + z = \frac{1}{2}$. Interpret the results in terms of species behavior.

8. Assume that a given population consists of susceptibles (S), exposed (E), infectives (I), and recovered/immune (R) individuals. Suppose that $S + E + I + R = 1$ for all time. A seasonally driven epidemic model is given by

$$\dot{S} = \mu(1-S) - \beta SI, \qquad \dot{E} = \beta SI - (\mu+\alpha)E, \qquad \dot{I} = \alpha E - (\mu+\gamma)I,$$

where β = contact rate, α^{-1} = mean latency period, γ^{-1} = mean infectivity period, and μ^{-1} = mean lifespan. The seasonality is introduced by assuming that $\beta = B(1 + A\cos(2\pi t))$, where $B \geq 0$ and $0 \leq A \leq 1$. Plot phase portraits when $A = 0.18$, $\alpha = 35.84$, $\gamma = 100$, $\mu = 0.02$, and $B = 1800$ for the initial conditions (i) $S(0) = 0.065$, $E(0) = 0.00075$, $I(0) = 0.00025$ and (ii) $S(0) = 0.038$, $E(0) = 3.27 \times 10^{-8}$, $I(0) = 1.35 \times 10^{-8}$. Interpret the results for the populations.

9. Plot some time series data for the Lorenz system (13.7) when $\sigma = 10$, $b = \frac{8}{3}$, and $166 \leq r \leq 167$. When $r = 166.2$, the solution shows intermittent behavior. There are occasional chaotic bursts in between what looks like periodic behavior.

10. Consider system (13.13) given in the text to model the periodic behavior of the Belousov–Zhabotinski reaction. By considering the isoclines and gradients of the vector fields, explain what happens to the solution curves for $\epsilon_1 \ll 1$ and appropriate values of q and C.

Recommended Reading

[1] G. A. Gottwald and I. Melbourne, A new test for chaos in deterministic systems, *Proc. Roy. Soc. London Ser.* A, **460**-2042 (2004), 603–611.

[2] S. Wiggins, *Introduction to Applied Nonlinear Dynamical Systems and Chaos*, 2nd ed., Texts in Applied Mathematics 2, Springer-Verlag, New York, 2003.

[3] J. Borresen and S. Lynch. Further investigation of hysteresis in Chua's circuit, *Internat. J. Bifurcation Chaos*, **12** (2002), 129–134.

[4] R. C. Hilborn, *Chaos and Nonlinear Dynamics: An Introduction for Scientists and Engineers*, 2nd ed., Oxford University Press, London, 2000.

[5] M. Zoltowski, An adaptive reconstruction of chaotic attractors out of their single trajectories, *Signal Process.*, **80**-6 (2000), 1099–1113.

[6] N. Arnold, *Chemical Chaos*, Hippo, London, 1997.

[7] S. K. Scott, *Oscillations, Waves, and Chaos in Chemical Kinetics*, Oxford Science Publications, Oxford, UK, 1994.

[8] R. N. Madan, *Chua's Circuit: A Paradigm for Chaos*, World Scientific, Singapore, 1993.

[9] K. Geist, U. Parlitz, and W. Lauterborn, Comparison of different methods for computing Lyapunov exponents, *Progr. Theoret. Phys.*, **83**-5 (1990), 875–893.

[10] J. P. Eckmann, S. O. Kamphorst, D. Ruelle, and S. Ciliberto, Lyapunov exponents from time series, *Phys. Rev.* A, **34**-6 (1986), 4971–4979.

[11] R. Field and M. Burger, eds., *Oscillations and Travelling Waves in Chemical Systems*, Wiley, New York, 1985.

[12] C. Sparrow, *The Lorenz Equations: Bifurcations, Chaos and Strange Attractors*, Springer-Verlag, New York, 1982.

[13] O. E. Rössler, An equation for continuous chaos, *Phys. Lett.*, **57**-A (1976), 397–398.

[14] E. N. Lorenz, Deterministic non-periodic flow, *J. Atmos. Sci.*, **20** (1963), 130–141.

14
Poincaré Maps and Nonautonomous Systems in the Plane

Aims and Objectives

- To introduce the theory of Poincaré maps.
- To compare periodic and quasiperiodic behavior.
- To introduce Hamiltonian systems with two degrees of freedom.
- To use Poincaré maps to investigate a nonautonomous system of differential equations.

On completion of this chapter the reader should be able to

- understand the basic theory of Poincaré maps;
- plot return maps for certain systems;
- use the Poincaré map as a tool for studying stability and bifurcations.

Poincaré maps are introduced via example using two-dimensional autonomous systems of differential equations. They are used extensively to transform complicated behavior in the phase space to discrete maps in a lower-dimensional space. Unfortunately, this nearly always results in numerical work since analytic solutions can rarely be found.

A periodically forced nonautonomous system of differential equations is introduced, and Poincaré maps are used to determine stability and plot bifurcation diagrams.

14.1 Poincaré Maps

When plotting the solutions to some nonlinear problems, the phase space can become overcrowded and the underlying structure may become obscured. To overcome these difficulties, a basic tool was proposed by Henri Poincaré [6] at the end of the 19th century. As a simple introduction to the theory of *Poincaré* (or *first return*) *maps* consider two-dimensional autonomous systems of the form

(14.1) $$\dot{x} = P(x, y), \qquad \dot{y} = Q(x, y).$$

Suppose that there is a curve or straight line segment, say, Σ, that is crossed *transversely* (no trajectories are tangential to Σ). Then Σ is called a *Poincaré section*. Consider a point r_0 lying on Σ. As shown in Figure 14.1, follow the flow of the trajectory until it next meets Σ at a point r_1. This point is known as the first return of the discrete Poincaré map $\mathbf{P} : \Sigma \to \Sigma$, defined by

$$r_{n+1} = \mathbf{P}(r_n),$$

where r_n maps to r_{n+1} and all points lie on Σ. Finding the function $\mathbf{P}$ is equivalent to solving the differential equations (14.1). Unfortunately, this is very seldom possible, and one must rely on numerical solvers to make any progress.

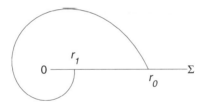

Figure 14.1: A first return on a Poincaré section, Σ.

Definition 1. A point r^* that satisfies the equation $\mathbf{P}(r^*) = r^*$ is called a *fixed point of period one*.

To illustrate the method for finding Poincaré maps, consider the following two simple examples (Examples 1 and 2), for which $\mathbf{P}$ may be determined explicitly.

Example 1. By considering the line segment $\Sigma = \{(x, y) \in \Re^2 : 0 \leq x \leq 1, y = 0\}$, find the Poincaré map for the system

(14.2) $$\dot{x} = -y - x\sqrt{x^2 + y^2}, \qquad \dot{y} = x - y\sqrt{x^2 + y^2}$$

and list the first eight returns on Σ given that $r_0 = 1$.

Solution. Convert to polar coordinates. System (14.2) then becomes

(14.3) $$\dot{r} = -r^2, \qquad \dot{\theta} = 1.$$

The origin is a stable focus and the flow is counterclockwise. A phase portrait showing the solution curve for this system is given in Figure 14.2.

Figure 14.2: A trajectory starting at $(1, 0)$, $(0 \leq t \leq 40)$ for system (14.3).

The set of equations (14.3) can be solved using the initial conditions $r(0) = 1$ and $\theta(0) = 0$. The solutions are given by

$$r(t) = \frac{1}{1+t}, \qquad \theta(t) = t.$$

Trajectories flow around the origin with a period of 2π. Substituting for t, the flow is defined by

$$r(t) = \frac{1}{1 + \theta(t)}.$$

The flow is counterclockwise, and the required successive returns occur when $\theta = 2\pi, 4\pi, \ldots$. A map defining these points is given by

$$r_n = \frac{1}{1 + 2n\pi}$$

on Σ, where $n = 1, 2, \ldots$. As $n \to \infty$, the sequence of points moves towards the fixed point at the origin as expected. Now

$$r_{n+1} = \frac{1}{1 + 2(n+1)\pi}.$$

Elementary algebra is used to determine the Poincaré return map **P**, which may be expressed as

$$r_{n+1} = \mathbf{P}(r_n) = \frac{r_n}{1 + 2\pi r_n}.$$

The first eight returns on the line segment Σ occur at the points $r_0 = 1$, $r_1 = 0.13730$, $r_2 = 0.07371$, $r_3 = 0.05038$, $r_4 = 0.03827$, $r_5 = 0.03085$, $r_6 = 0.02584$, $r_7 = 0.02223$, and $r_8 = 0.01951$, to five decimal places, respectively. Check these results for yourself using the MATLAB program at the end of the chapter.

Example 2. Use a one-dimensional map on the line segment $\Sigma = \{(x, y) \in \Re^2 : 0 \le x < \infty, y = 0\}$ to determine the stability of the limit cycle in the following system:

(14.4) $\quad \dot{x} = -y + x\left(1 - \sqrt{x^2 + y^2}\right), \qquad \dot{y} = x + y\left(1 - \sqrt{x^2 + y^2}\right).$

Solution. Convert to polar coordinates; then system (14.4) becomes

(14.5) $\qquad\qquad\qquad \dot{r} = r(1 - r), \qquad \dot{\theta} = 1.$

The origin is an unstable focus, and there is a limit cycle, say, Γ, of radius 1 centered at the origin. A phase portrait showing two trajectories is given in Figure 14.3.

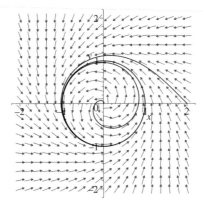

Figure 14.3: Two trajectories for system (14.5), one starting at (2, 0) and the other at (0.01, 0).

System (14.5) can be solved since both differential equations are separable. The solutions are given by

$$r(t) = \frac{1}{1 + Ce^{-t}}, \qquad \theta(t) = t + \theta_0,$$

where C and θ_0 are constants. Trajectories flow around the origin with a period of 2π.

Suppose that a trajectory starts outside Γ on Σ, say, at $r_0 = 2$. The solutions are then given by

$$r(t) = \frac{1}{1 - \frac{1}{2}e^{-t}}, \qquad \theta(t) = t.$$

Therefore, a return map can be expressed as

$$r_n = \frac{1}{1 - \frac{1}{2}e^{-2n\pi}},$$

where n is a natural number. If, however, a trajectory starts inside Γ at, say, $r_0 = \frac{1}{2}$, then

$$r(t) = \frac{1}{1 + e^{-t}}, \qquad \theta(t) = t,$$

and a return map is given by

$$r_n = \frac{1}{1 + e^{-2n\pi}}.$$

In both cases $r_n \to 1$ as $n \to \infty$. The limit cycle is stable on both sides, and the limit cycle Γ is hyperbolic stable since $r_n \to 1$ as $n \to \infty$ for any initial point apart from the origin. The next theorem gives a better method for determining the stability of a limit cycle.

Theorem 1. *Define the characteristic multiplier M to be*

$$M = \left.\frac{d\mathbf{P}}{dr}\right|_{r^*},$$

where r^* is a fixed point of the Poincaré map $\mathbf{P}$ corresponding to a limit cycle, say, Γ. Then if

1. *$|M| < 1$, Γ is a hyperbolic stable limit cycle;*
2. *$|M| > 1$, Γ is a hyperbolic unstable limit cycle;*
3. *$|M| = 1$ and $\frac{d^2\mathbf{P}}{dr^2} \neq 0$, then the limit cycle is stable on one side and unstable on the other; in this case Γ is called a semistable limit cycle.*

Definition 2. A fixed point of period one, say, r^*, of a Poincaré map $\mathbf{P}$ is called *hyperbolic* if $|M| \neq 1$.

Theorem 1 is sometimes referred to as the *derivative of the Poincaré map test*.

14. Poincaré Maps and Nonautonomous Systems in the Plane

Example 3. Use Theorem 1 to determine the stability of the limit cycle in Example 2.

Solution. Consider system (14.5). The return map along Σ is given by

$$(14.6) \qquad r_n = \frac{1}{1 + Ce^{-2n\pi}},$$

where C is a constant. Therefore,

$$(14.7) \qquad r_{n+1} = \frac{1}{1 + Ce^{-2(n+1)\pi}}.$$

Substituting $C = \frac{1-r_n}{r_n e^{2n\pi}}$ from equation (14.6) into (14.7) gives the Poincaré map

$$r_{n+1} = \mathbf{P}(r_n) = \frac{r_n}{r_n + (1 - r_n)e^{-2\pi}}.$$

The Poincaré map has two fixed points, one at zero (a trivial fixed point) and the other at $r^* = 1$, corresponding to the critical point at the origin and the limit cycle Γ, respectively. Now

$$\frac{d\mathbf{P}}{dr} = \frac{e^{-2\pi}}{(r + (1-r)e^{-2\pi})^2},$$

using elementary calculus, and

$$\left.\frac{d\mathbf{P}}{dr}\right|_{r^*=1} = e^{-2\pi} \approx 0.00187 < 1,$$

and so the limit cycle Γ is hyperbolic attracting.

Definition 3. A point r^* that satisfies the equation $\mathbf{P}^m(r^*) = r^*$ is called a *fixed point of period m*.

Example 4. Consider the circle map $\mathbf{P}$ defined by

$$r_{n+1} = \mathbf{P}(r_n) = e^{i2\pi \frac{q_1}{q_2}} r_n,$$

which maps points on the unit circle to itself. Assuming that $r_0 = 1$, plot iterates when

(a) $q_1 = 0, q_2 = 1$,

(b) $q_1 = 1, q_2 = 2$,

(c) $q_1 = 2, q_2 = 3$, and

(d) $q_1 = 1, q_2 = \sqrt{2}$.

14.1. Poincaré Maps   303

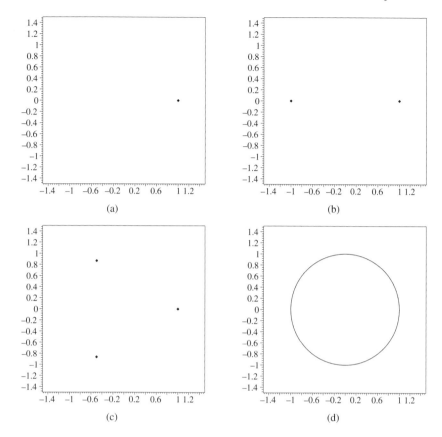

Figure 14.4: Fixed points of periods (a) one, (b) two, (c) three, and (d) quasiperiodic behavior for the circle map $r_{n+1} = \mathbf{P}(r_n) = e^{i2\pi \frac{q_1}{q_2}} r_n$.

Explain the results displayed in Figures 14.4(a)–(d).

Solution. In Figure 14.4(a), there is a fixed point of period one since $r_{n+1} = \mathbf{P} = r_n$. Similarly, in Figures 14.4(b)–(c), there are fixed points of periods two and three since $r_{n+2} = \mathbf{P}^2 = r_n$ and $r_{n+3} = \mathbf{P}^3 = r_n$.

For Figure 14.4(d), q_1 and q_2 are *rationally independent* since $c_1 q_1 + c_2 q_2 = 0$ with c_1 and c_2 integers is satisfied only by $c_1 = c_2 = 0$. This implies that the points on the circle map are never repeated and there is no periodic motion. (There is no integer c such that $r_{n+c} = \mathbf{P}^c = r_n$.) Figure 14.4(d) shows the first 1000 iterates of this mapping. If one were to complete the number of iterations to infinity then a closed circle would be formed as new points approach other points arbitrarily closely an infinite number of times. This new type of qualitative behavior is known as *quasiperiodicity*. Note that one has to be careful when distinguishing between quasiperiodic points and points that have very high periods. For example, Figure 14.4(d) could be depicting a very high-period trajectory. Systems displaying quasiperiodicity will be discussed in the next section.

14.2 Hamiltonian Systems with Two Degrees of Freedom

Hamiltonian systems with one degree of freedom were introduced in Chapter 11. These systems can always be integrated completely. Hamiltonian (or conservative) systems with two degrees of freedom will be discussed briefly in this section, but the reader should note that it is possible to consider Hamiltonian systems with N—or even an infinite number of—degrees of freedom.

In general, the set of Hamiltonian systems with two degrees of freedom are not completely integrable, and those that are form a very restricted but important subset. The trajectories of these systems lie in four-dimensional space, but the overall structure can be determined by plotting Poincaré maps. It is known that completely integrable systems display remarkable smooth regular behavior in all parts of the phase space, which is in stark contrast to what happens with nonintegrable systems, which can display a wide variety of phenomena including chaotic behavior. A brief definition of integrability is given below, and Hamiltonian systems with two degrees of freedom will now be defined.

Definition 4. A Hamiltonian system with two degrees of freedom is defined by

$$(14.8) \quad \dot{p}_1 = -\frac{\partial H}{\partial q_1}, \quad \dot{q}_1 = \frac{\partial H}{\partial p_1}, \quad \dot{p}_2 = -\frac{\partial H}{\partial q_2}, \quad \dot{q}_2 = \frac{\partial H}{\partial p_2},$$

where H is the Hamiltonian of the system. In physical applications, q_1 and q_2 are generalized coordinates and p_1 and p_2 represent a generalized momentum. The Hamiltonian may be expressed as

$$H(\mathbf{p}, \mathbf{q}) = K_E(\mathbf{p}, \mathbf{q}) + P_E(\mathbf{q}),$$

where K_E and P_E are the kinetic and potential energies, respectively.

Definition 5. The Hamiltonian system with two degrees of freedom given by (14.8) is *integrable* if the system has two integrals, say, F_1 and F_2, such that

$$\{F_1, H\} = 0, \{F_2, H\} = 0, \{F_1, F_2\} = 0,$$

where F_1 and F_2 are functionally independent and $\{,\}$ are the so-called *Poisson brackets* defined by

$$\{F_1, F_2\} = \frac{\partial F_1}{\partial \mathbf{q}} \frac{\partial F_2}{\partial \mathbf{p}} - \frac{\partial F_1}{\partial \mathbf{p}} \frac{\partial F_2}{\partial \mathbf{q}}.$$

14.2. Hamiltonian Systems with Two Degrees of Freedom

Some of the dynamics involved in these type of systems will now be described using some simple examples.

Example 5. Consider the Hamiltonian system with two degrees of freedom given by

(14.9) $$H(\mathbf{p},\mathbf{q}) = \frac{\omega_1}{2}(p_1^2 + q_1^2) + \frac{\omega_2}{2}(p_2^2 + q_2^2),$$

which is integrable with integrals given by $F_1 = p_1^2 + q_1^2$ and $F_2 = p_2^2 + q_2^2$. This system can be used to model a linear harmonic oscillator with two degrees of freedom.

Plot three-dimensional and two-dimensional projections of the Poincaré surface-of-section for system (14.9) given the following set of initial conditions for p_1, p_2 and q_1, q_2:

(i) $\omega_1 = \omega_2 = 2$ with the initial conditions $t = 0$, $p_1 = 0.5$, $p_2 = 1.5$, $q_1 = 0.5$, $q_2 = 0$;

(ii) $\omega_1 = 8$, $\omega_2 = 3$ with the initial conditions $t = 0$, $p_1 = 0.5$, $p_2 = 1.5$, $q_1 = 0.3$, $q_2 = 0$;

(iii) $\omega_1 = \sqrt{2}$, $\omega_2 = 1$ with the initial conditions $t = 0$, $p_1 = 0.5$, $p_2 = 1.5$, $q_1 = 0.3$, $q_2 = 0$.

Solutions. A MATLAB program is listed in Section 14.4 (see Figure 14.5).

The results may be interpreted as follows: in cases (i) and (ii) the solutions are periodic, and in case (iii) the solution is quasiperiodic. For the quasiperiodic solution, a closed curve will be formed in the p_1q_1 plane as the number of iterations goes to infinity. The quasiperiodic cycle never closes on itself; however, the motion is not chaotic. Once more the trajectories are confined to invariant tori.

Example 6. Consider the Hénon–Heiles Hamiltonian system (which may be used as a simple model of the motion of a star inside a galaxy) given by

$$H(\mathbf{p},\mathbf{q}) = \frac{1}{2}(p_1^2 + q_1^2 + p_2^2 + q_2^2) + q_1^2 q_2 - \frac{q_2^3}{3}.$$

This Hamiltonian represents two simple harmonic oscillators (see Example 5(i)) coupled with a cubic term. The Hamiltonian in this case is nonintegrable. Plot three-dimensional and two-dimensional projections of the Poincaré surface-of-section of the Hamiltonian system for the set of initial conditions given by $t = 0$, $p_1 = 0.06$, $p_2 = 0.1$, $q_1 = -0.2$, $q_2 = -0.2$.

Solution. See Figure 14.6.

A rich variety of behavior is observed in the Poincaré section for the Hénon–Heiles system as the energy levels increase. For example, Figure 14.7 shows

14. Poincaré Maps and Nonautonomous Systems in the Plane

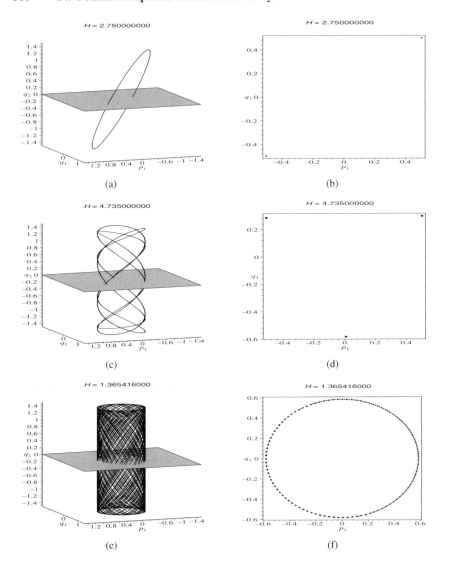

Figure 14.5: Projections of the Poincaré surface-of-section for system (14.9) when (a)–(b) $\omega_1 = \omega_2 = 2$, (c)–(d) $\omega_1 = 8$ and $\omega_2 = 3$, and (e)–(f) $\omega_1 = \sqrt{2}$ and $\omega_2 = 1$. The initial conditions are listed in (i)–(iii) of Example 5.

how the Poincaré section changes as the energy level increases from 0.041666 to 0.166666.

As the energy levels increase the closed orbits, representing quasiperiodic behavior, are replaced by irregular patterns, and eventually the Poincaré plane

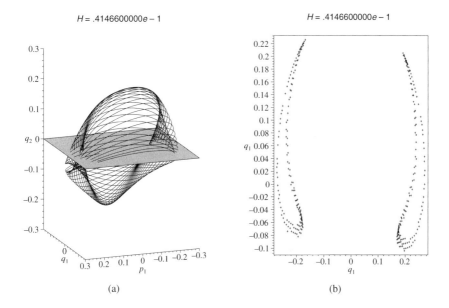

Figure 14.6: A three-dimensional and two-dimensional projection of the Poincaré surface-of-section for the Hénon–Heiles system with the initial conditions $t = 0$, $p_1 = 0.06$, $p_2 = 0.1$, $q_1 = -0.2$, $q_2 = -0.2$. Note that the energy level is equal to 0.041466 in this case.

seems to be swamped by chaos. In fact, there is a famous theorem due to Kolmogorov, Arnold, and Moser, now known as the KAM theorem. Interested readers are referred to the book of Guckenheimer and Holmes [4].

Theorem 2. *Suppose that a Hamiltonian system with two degrees of freedom is given by $H = H_0 + \epsilon H_1$, where ϵ is a small parameter, H_0 is integrable, and H_1 makes H nonintegrable. The quasiperiodic cycles (also known as KAM tori), which exist for $\epsilon = 0$, will also exist for $0 < \epsilon \ll 1$ but will be deformed by the perturbation. The KAM tori dissolve one by one as ϵ increases and points begin to scatter around the Poincaré plane. A similar pattern of behavior can be seen in Figure* 14.7.

14.3 Nonautonomous Systems in the Plane

The existence and uniqueness theorems introduced in Chapter 7 hold for autonomous systems of differential equations. This means that trajectories cannot cross, and the Poincaré–Bendixson theorem implies that there is no chaos in two dimensions. However, chaos can be displayed in three-dimensional autonomous systems as shown in Chapter 13, where various strange attractors were plotted

14. Poincaré Maps and Nonautonomous Systems in the Plane

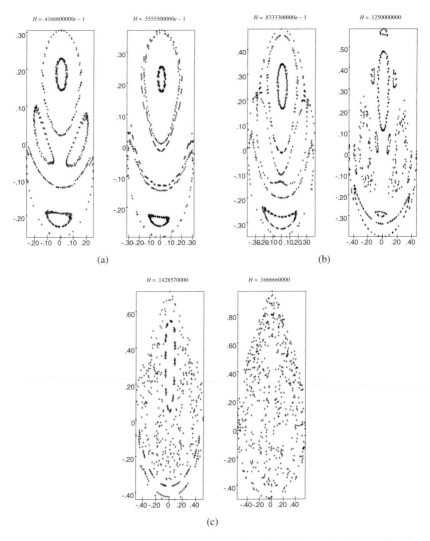

Figure 14.7: The Poincaré transversal plane for the Hénon–Heiles Hamiltonian system with different energy levels. The smoothness of the curves in both (p, q) planes is related to the integrability of the system.

using the MATLAB package. This section is concerned with *nonautonomous* (or *forced*) systems of differential equations of the form

$$\ddot{x} = f(x, \dot{x}, t),$$

where the function f depends explicitly on t. There is no longer uniqueness of the solutions, and trajectories can cross in the phase plane. For certain parameter

14.3. Nonautonomous Systems in the Plane

values, the phase portrait can become entangled with trajectories criss-crossing one another. By introducing a Poincaré map, it becomes possible to observe the underlying structure of the complicated flow.

As a particular example, consider the *Duffing equation* given by

$$\ddot{x} + k\dot{x} + (x^3 - x) = \Gamma \cos(\omega t),$$

where, in physical models, k is a damping coefficient, Γ represents a driving amplitude, and ω is the frequency of the driving force. Let $\dot{x} = y$; then the Duffing equation can be written as a system of the form

(14.10) $\qquad \dot{x} = y, \qquad \dot{y} = x - ky - x^3 + \Gamma \cos(\omega t).$

The equations may be used to model a periodically forced pendulum that has a cubic restoring force, where $x(t)$ represents displacement and $\dot{x}$ the speed of a simple mass. The equation can also be used to model periodically forced resistor–inductor–capacitor circuits with nonlinear circuit elements, where $x(t)$ would represent the charge oscillating in the circuit at time t.

Systems of the form (14.10) have been studied extensively in terms of, for example, stability, harmonic solutions, subharmonic solutions, transients, chaotic output, chaotic control, and Poincaré maps. The work here will be restricted to considering the Poincaré maps and bifurcation diagrams for system (14.10) as the driving amplitude Γ varies when $k = 0.3$ and $\omega = 1.25$ are fixed.

Any periodically forced nonautonomous differential equation can be represented in terms of an autonomous flow in a torus. To achieve this transformation, simply introduce a third variable $\theta = \omega t$. System (14.10) then becomes a three-dimensional autonomous system given by

(14.11) $\qquad \dot{x} = y, \qquad \dot{y} = x - ky - x^3 + \Gamma \cos(\theta), \qquad \dot{\theta} = \omega.$

A flow in this state space corresponds to a trajectory flowing around a torus with period $\frac{2\pi}{\omega}$. This naturally leads to a Poincaré mapping of a $\theta = \theta_0$ plane to itself as depicted in Figure 14.8.

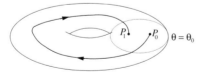

Figure 14.8: The first return of a point P_0 to P_1 in the plane $\theta = \theta_0$. The trajectories flow inside a torus in three-dimensional space.

When $\Gamma = 0$, system (14.10) has three critical points at $M = (-1, 0)$, $N = (1, 0)$, and $O = (0, 0)$. The points M and N are stable foci when $0 < k < 2\sqrt{2}$

and O is a saddle point. As Γ is increased from zero, stable periodic cycles appear from M and N and there are bifurcations of subharmonic oscillations. The system can also display chaotic behavior for certain values of Γ.

Only periodic cycles initially appearing from the critical point N will be considered here. A gallery of phase portraits along with their respective Poincaré return maps are presented in Figures 14.9 and 14.10.

When $\Gamma = 0.2$, there is a period-one harmonic solution of period $\frac{2\pi}{\omega}$, which is depicted as a closed curve in the phase plane and as a single point in the $\theta = 0$ plane (see Figure 14.9(a)). When $\Gamma = 0.3$, a period-two cycle of period $\frac{4\pi}{\omega}$ appears; this is a subharmonic of order $\frac{1}{2}$. A period-two cycle is represented by two points in the Poincaré section (see Figure 14.9(b)); note that the trajectory crosses itself in this case. A period-four cycle of period $\frac{8\pi}{\omega}$ is present when $\Gamma = 0.31$ (see Figure 14.9(c)). When $\Gamma = 0.37$, there is a period-five cycle that is centered at O and also surrounds both M and N (see Figure. 14.10(a)). When $\Gamma = 0.5$, the system becomes chaotic. A single trajectory plotted in the phase plane intersects itself many times, and the portrait soon becomes very messy. However, if one plots the first returns on the Poincaré section, then a strange attractor is formed that demonstrates some underlying structure (see Figure 14.10(b)). It must be noted that the chaotic attractor will have different forms on different Poincaré sections. This strange (or chaotic) attractor has fractal structure. At $\Gamma = 0.8$, there is once more a stable period-one solution. However, it is now centered at O (see Figure 14.10(c)).

Figures 14.9 and 14.10 display some of the behavior possible for the Duffing equation for specific values of the parameter Γ. Of course, it would be far better to summarize all of the possible behaviors as the parameter Γ varies on one diagram. To achieve this goal, one must plot bifurcation diagrams. There are basically two ways in which bifurcation diagrams may be produced; one involves a feedback mechanism, the other does not. The first and second iterative methods are described in Section 4.5.

Figure 14.11 shows a possible bifurcation diagram for system (14.10) for forcing amplitudes in the range $0 < \Gamma < 0.4$ near the critical point at N. The vertical axis labeled r represents the distance of the point in the Poincaré map from the origin ($r = \sqrt{x^2 + y^2}$). The first iterative method was employed in this case. For each value of Γ, the last 10 of 50 iterates were plotted, and the step length used in this case was 0.01. The initial values were chosen close to one of the existing periodic solutions. The diagram shows period-one behavior for $0 < \Gamma < 0.28$, approximately. For values of $\Gamma > 0.28$, there is period-two behavior, and then the results become a little obscure.

Figure 14.12 shows a possible bifurcation diagram produced using the second iterative method. The parameter Γ is increased from zero to 0.4 and then decreased from $\Gamma = 0.4$ back to zero. A similar study was carried out in Chapter 4. There were 2000 iterates used as Γ was increased and then decreased. The solid curve lying approximately between $0 \leq \Gamma < 0.32$ represents steady-state behavior. As

14.3. Nonautonomous Systems in the Plane

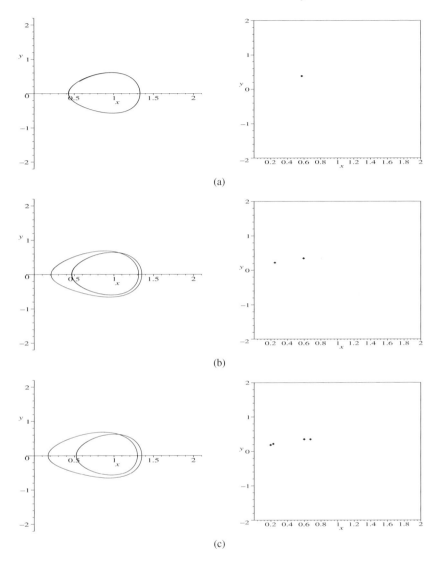

Figure 14.9: A gallery of phase portraits and Poincaré maps for system (14.10) when $k = 0.3$ and $\omega = 1.25$: (a) $\Gamma = 0.2$ (forced period one); (b) $\Gamma = 0.3$ (a period-two subharmonic); (c) $\Gamma = 0.31$ (a period-four subharmonic).

Γ increases beyond 0.32, the system goes through a chaotic regime and returns to periodic behavior before $\Gamma = 0.4$. As the parameter Γ is decreased, the system returns through the periodic paths, enters a chaotic region, and period undoubles back to the steady-state solution at $\Gamma \approx 0.28$. Note that on the ramp-up part of the

312   14. Poincaré Maps and Nonautonomous Systems in the Plane

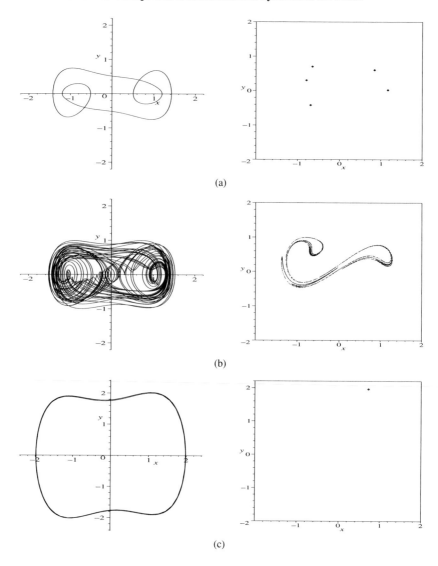

Figure 14.10: A gallery of phase portraits and Poincaré maps for system (14.10) when $k = 0.3$ and $\omega = 1.25$: (a) $\Gamma = 0.37$ (a period-five subharmonic); (b) $\Gamma = 0.5$ (chaos), 4,000 points are plotted; (c) $\Gamma = 0.8$ (forced period one).

iterative scheme, the steady state overshoots into the region where the system is of period two, roughly where $0.28 < \Gamma < 0.32$.

Figure 14.13 shows a bifurcation diagram produced as Γ is increased from zero to 0.45 and then decreased back to zero. Once more as Γ is increased, there is

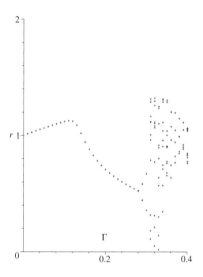

Figure 14.11: A bifurcation diagram for system (14.10) produced using the first iterative method.

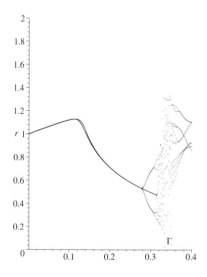

Figure 14.12: A bifurcation diagram for system (14.10) produced using the second iterative method.

steady-state behavior for Γ lying between zero and approximately 0.32. However, as the parameter is decreased a different steady state is produced and a large bistable region is present.

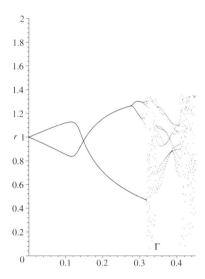

Figure 14.13: A bifurcation diagram for system (14.10) produced using the second iterative method. There is a large bistable region.

Note that there will also be steady-state behavior and bifurcations associated with the critical point at M. The flow near to saddle fixed points will now be considered.

Homoclinic and Heteroclinic Bifurcations. Some of the theory involved in the bifurcations to chaos for flows and maps is a result of the behavior of the stable and unstable manifolds of saddle points. Discrete maps have been discussed in some detail in earlier chapters. The stable and unstable manifolds can form homoclinic and heteroclinic orbits as a parameter is varied. Homoclinic and heteroclinic orbits were introduced in Chapter 11, and the stable and unstable branches of saddle points were discussed in Chapter 6. It is also possible for the stable and unstable manifolds to approach one another and eventually intersect as a parameter varies. When this occurs, there is said to be a homoclinic (or heteroclinic) intersection. The intersection is homoclinic if a stable/unstable branch of a saddle point crosses the unstable/stable branch of the same saddle point, and it is heteroclinic if the stable/unstable branches of one saddle point cross the unstable/stable branches of a different saddle point. If the stable and unstable branches of saddle points intersect once, then it is known that there must be an infinite number of intersections, and a so-called *homoclinic* (or *heteroclinic*) *tangle* is formed. In 1967, Smale [5] provided an elegant geometric construction to describe this phenomenon. The mapping function used is now known as the *Smale horseshoe map*. Consider a small square, say, S, of initial points surrounding a saddle point in the Poincaré

14.3. Nonautonomous Systems in the Plane

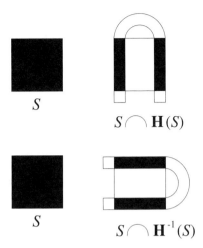

Figure 14.14: The mappings **H** and **H**$^{-1}$.

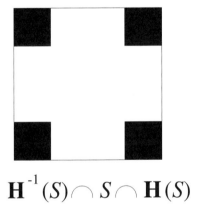

Figure 14.15: The first stage of the Smale horseshoe map.

section. Under the iterative scheme, this square of points will be stretched out in the direction of the unstable manifold and compressed along the stable branch of the saddle point. In Smale's construction, a square of initial points is stretched in one direction and then compressed in an orthogonal direction. Suppose that the map is given by $\mathbf{H} : S \to \Re^2$ and that **H** contracts S in the horizontal direction, expands S in the vertical direction, and then folds the rectangle back onto itself to form a horseshoe, as in Figure 14.14. Similarly, the action of $\mathbf{H}^{-1}$ on S is also given in Figure 14.14. The result of the intersection of these first two sets is given in Figure 14.15.

As this process is iterated to infinity, points fall into the area contained by the original square in smaller and smaller subareas. The result is an *invariant Cantor set* that contains a countable set of periodic orbits and an uncountable set of bounded nonperiodic orbits.

The *Smale–Birkhoff theorem* states that homoclinic tangles guarantee that a dynamical system will display *horseshoe dynamics*. For more details, the reader is directed once again to the excellent textbook of Guckenheimer and Holmes [4].

14.4 MATLAB Commands

```
% Programs_14a - Solving an initial value problem.
% Solve a differential equation (Example 1).
% Use the Symbolic Math Toolbox.

r=dsolve('Dr=-r^2','r(0)=1');

% List the first eight returns on the segment {y=0, 0<x<1}.
% There may be small inaccuracies due to the numerical
% solution.
deq=inline('[-(r(1))^2]','t','r');
options=odeset('RelTol',1e-6,'AbsTol',1e-6);
[t,returns]=ode45(deq,[0:2*pi:16*pi],[1]);
returns
% End of Programs_14a.
```

```
% Programs_14b - Phase portraits for Hamiltonian systems
% with two degrees of freedom.
% Poincare surfaces of section (Figures 14.5(e)-(f)).
% Quasiperiodic behavior.
% The Hamiltonian equations.
% Here p1=p(1),p2=p(2),q1=p(3),q2=p(4).

deq=inline('[-sqrt(2)*p(3);-p(4);sqrt(2)*p(1);p(2)]','t','p');
options=odeset('RelTol',1e-4,'AbsTol',1e-4);
[t,pp]=ode45(deq,[0 200],[.5,1.5,.5,0],options);

% A three-dimensional projection (Figure 14.5(e)).
subplot(2,1,1)
fsize=15;
plot3(pp(:,1),pp(:,2),pp(:,4))

deq=inline('[-sqrt(2)*p(3);-p(4);sqrt(2)*p(1);p(2)]','t','p');
options=odeset('RelTol',1e-4,'AbsTol',1e-4);
[t,pq]=ode45(deq,[0 600],[.5,1.5,.5,0],options);
```

14.4. MATLAB Commands

```
% A two-dimensional projection (Figure 14.5(f)).
% Determine where trajectory crosses q2=0 plane.
k=0;
for i=1:size(pq)
    if abs(pq(i,4))<0.1
        k=k+1;
        p1_0(k)=pq(i,1);
        q1_0(k)=pq(i,3);
    end
end

subplot(2,1,2)
hold on
axis([-1 1 -1 1])
set(gca,'xtick',[-1:.5:1],'FontSize',fsize)
set(gca,'ytick',[-1:.5:1],'FontSize',fsize)
xlabel('p_1','FontSize',fsize)
ylabel('q_1','FontSize',fsize)
plot(p1_0(1:k),q1_0(1:k),'+','MarkerSize',3)
hold off
% End of Programs_14b.
```

```
% Programs_14c - Phase portraits for nonautonomous systems
% (Figure 14.10(b)).
% Phase portrait for the Duffing system.

deq=inline('[x(2);x(1)-0.3*x(2)-(x(1))^3+0.5
*cos(1.25*t)]','t','x');
options=odeset('RelTol',1e-4,'AbsTol',1e-4);
[t,xx]=ode45(deq,[0 200],[1,0],options);
plot(xx(:,1),xx(:,2))

fsize=15;
axis([-2 2 -2 2])
xlabel('x','FontSize',fsize)
ylabel('y','FontSize',fsize)
% End of Programs_14c.
```

```
% Programs_14d - Poincare section for the Duffing system
% (Figure 14.10(b)).

clear
deq=inline('[x(2);x(1)-0.3*x(2)-(x(1))^3+0.5
*cos(1.25*t)]','t','x');
options=odeset('RelTol',1e-4,'AbsTol',1e-4);
[t,xx]=ode45(deq,[0:(2/1.25)*pi:(4000/1.25)*pi],[1,0]);
```

318   14. Poincaré Maps and Nonautonomous Systems in the Plane

```
plot(xx(:,1),xx(:,2),'.','MarkerSize',1)
fsize=15;
axis([-2 2 -2 2])
xlabel('x','FontSize',fsize)
ylabel('y','FontSize',fsize)
title('Poincare Section of the Duffing System')
% End of Programs_14d.
```

```
% Programs_14e.

% Periodic behavior (Figures 14.5(a) and (b)).
deq=inline('[-2*p(3);-2*p(4);2*p(1);2*p(2)]','t','p');
options=odeset('RelTol',1e-4,'AbsTol',1e-4);
[t,pp]=ode45(deq,[0 100],[.5,1.5,.5,0],options);

% A three-dimensional projection.
subplot(2,1,1)
fsize=15;
plot3(pp(:,1),pp(:,2),pp(:,4))

% A two-dimensional projection.
k=0;
for n=1:1000
    if abs(pp(n,4))<0.01
        k=k+1;
        p1_0(k)=pp(n,1);
        q1_0(k)=pp(n,3);
    end
end

subplot(2,1,2)
hold on
axis([-1 1 -1 1])
set(gca,'xtick',[-1:.5:1],'FontSize',fsize)
set(gca,'ytick',[-1:.5:1],'FontSize',fsize)
xlabel('p_1','FontSize',fsize)
ylabel('q_1','FontSize',fsize)
plot(p1_0(1:k),q1_0(1:k),'+','MarkerSize',3)
hold off
% End of Programs_14e.
```

```
% Programs_14f.

function xdot=Programs_14f(t,x)
% The Duffing System.
global Gamma;
```

```
xdot(1)=x(2);
xdot(2)=x(1)-0.1*x(2)-(x(1))^3+Gamma*cos(1.25*t);
xdot=[xdot(1);xdot(2)];
% End of Programs_14f.
```

```
% Programs_14g - Bifurcation diagram for the Duffing
% equation.

% Make sure Programs_14f is in your directory.
clear
global Gamma;
Max=120;step=0.001;interval=Max*step;a=1;b=0;
% Ramp the amplitude up.
for n=1:Max
    Gamma=step*n;
    [t,x]=ode45('Programs_14f',[0:(2*pi/1.25):(4*pi/1.25)],
                [a,b]);
    a=x(2,1);
    b=x(2,2);
    rup(n)=sqrt((x(2,1))^2+(x(2,2))^2);
end
% Ramp the amplitude down.
for n=1:Max
    Gamma=interval-step*n;
    [t,x]=ode45('Programs_14f',[0:(2*pi/1.25):(4*pi/1.25)],
                [a,b]);
    a=x(2,1);
    b=x(2,2);
    rdown(n)=sqrt((x(2,1))^2+(x(2,2))^2);
end
hold on
rr=step:step:interval;
plot(rr,rup)
plot(interval-rr,rdown)
hold off
fsize=15;
axis([0 .12 0 2])
xlabel('\Gamma','FontSize',fsize)
ylabel('r','FontSize',fsize)
title('Bifurcation Diagram of the Duffing System')

% End of Programs_14g.
```

14.5 Exercises

1. Consider the system

$$\dot{x} = -y - 0.1x\sqrt{x^2 + y^2}, \qquad \dot{y} = x - 0.1y\sqrt{x^2 + y^2}.$$

   By considering the line segment $\Sigma = \{(x, y) \in \Re^2 : 0 \le x \le 4, y = 0\}$, list the first 10 returns on Σ given that a trajectory starts at the point $(4, 0)$.

2. Obtain a Poincaré map for the system

$$\dot{x} = \mu x + y - x\sqrt{x^2 + y^2}, \qquad \dot{y} = -x + \mu y - y\sqrt{x^2 + y^2}$$

   on the Poincaré section $\Sigma = \{(x, y) \in \Re^2 : 0 \le x < \infty, y = 0\}$.

3. Use the characteristic multiplier to determine the stability of the limit cycle in Example 2.

4. Solve the following differential equations:

$$\dot{r} = r(1 - r^2), \qquad \dot{\theta} = 1.$$

   By considering the line segment $\Sigma = \{(x, y) \in \Re^2 : 0 \le x \le \infty\}$, find the Poincaré map for this system.

5. Use the characteristic multiplier to determine the stability of the limit cycle in Example 4.

6. Consider the two degrees of freedom Hamiltonian given by

$$H(\mathbf{p}, \mathbf{q}) = \frac{\omega_1}{2}(p_1^2 + q_1^2) + \frac{\omega_2}{2}(p_2^2 + q_2^2).$$

   Plot three-dimensional and two-dimensional Poincaré sections when $\omega_1 = 3$ and $\omega_2 = 7$ for the set of initial conditions $t = 0$, $p_1 = 0.5$, $p_2 = 1.5$, $q_1 = 0.5$, $q_2 = 0$.

7. Consider the two degrees of freedom Hamiltonian given by

$$H(\mathbf{p}, \mathbf{q}) = \frac{\omega_1}{2}(p_1^2 + q_1^2) + \frac{\omega_2}{2}(p_2^2 + q_2^2).$$

   Plot three-dimensional and two-dimensional Poincaré sections when $\omega_1 = \sqrt{2}$ and $\omega_2 = 3$ for the set of initial conditions $t = 0$, $p_1 = 0.5$, $p_2 = 1.5$, $q_1 = 0.5$, $q_2 = 0$.

8. Plot three-dimensional and two-dimensional Poincaré sections of the Toda Hamiltonian given by

$$H = \frac{p_1^2}{2} + \frac{p_2^2}{2} + \frac{e^{2q_2+2\sqrt{3}q_1}}{24} + \frac{e^{2q_2-2\sqrt{3}q_1}}{24} + \frac{e^{-4q_2}}{24} - \frac{1}{8},$$

for several different energy levels of your choice.

9. A damped driven pendulum may be modeled using the nonautonomous system of differential equations defined by

(14.12) $\qquad \frac{d^2\theta}{dt^2} + k\frac{d\theta}{dt} + \frac{g}{l}\sin(\theta) = \Gamma\cos(\omega t),$

where k is a measure of the frictional force, Γ and ω are the amplitude and frequency of the driving force, g is the acceleration due to gravity, and l is the length of the pendulum. Plot a Poincaré map for this system when $k = 0.3$, $\Gamma = 4.5$, $\omega = 0.6$, and $\frac{g}{l} = 4$.

10. (a) Consider system (14.10) with $k = 0.1$ and $\omega = 1.25$. Plot a bifurcation diagram for $0 \leq \Gamma \leq 0.1$ and show that there is a *clockwise hysteresis loop* at approximately $0.04 < \Gamma < 0.08$. Note that there is *ringing* (oscillation) at the ramp-up and ramp-down parts of the bistable region.

(b) Plot the two stable limit cycles in the bistable region for Exercise 10(a) on one phase portrait. This shows that the system is multistable. For example, take $\Gamma = 0.07$. These limit cycles correspond to steady states on the upper and lower branches of the bistable cycle.

Recommended Reading

[1] S. Lynch, *Dynamical Systems with Applications Using Maple*, Birkhäuser, Boston, 2001.

[2] C-ODE-E (Consortium for ODE Experiments), *ODE Architect: The Ultimate ODE Power Tool*, Wiley, New York, 1999.

[3] E. S. Cheb-Terrab and H. P. de Oliveira, Poincaré sections of Hamiltonian systems, *Comput. Phys. Comm.*, **95** (1996), 171.

[4] J. Guckenheimer and P. Holmes, *Nonlinear Oscillations, Dynamical Systems, and Bifurcations of Vector Fields*, 3rd ed., Springer-Verlag, New York, 1990.

[5] S. Smale, Differentiable dynamical systems, *Bull. Amer. Math. Soc.*, **73** (1967), 747–817.

[6] H. Poincaré, Mémoire sur les courbes définies par une equation différentielle, *J. Math.*, **7** (1881), 375–422; oeuvre, Gauthier-Villars, Paris, 1890.

15
Local and Global Bifurcations

Aims and Objectives

- To introduce some local and global bifurcation theory in the plane.
- To bifurcate limit cycles in the plane.

On completion of this chapter the reader should be able to

- bifurcate small-amplitude limit cycles from fine foci;
- bifurcate limit cycles from a center;
- investigate limit cycle bifurcation from homoclinic loops, numerically.

The problem of determining the maximum number of limit cycles for planar differential systems dates back more than 100 years and will be discussed in more detail in Chapter 16. Local limit cycles can be analyzed in terms of local behavior of the system near to a relevant critical point or limit cycle. The theory involved in global bifurcations is not so well developed and involves larger-scale behavior in the plane.

An algorithm is presented for bifurcating small-amplitude limit cycles out of a critical point. The Melnikov function is used to determine the approximate location and number of limit cycles when a parameter is small. The limit cycles are bifurcated from a center. A bifurcation from a homoclinic loop is investigated numerically using MATLAB.

15.1 Small-Amplitude Limit Cycle Bifurcations

The general problem of determining the maximum number and relative configurations of limit cycles in the plane has remained unresolved for over a century. The problem will be stated in Chapter 16. Both local and global bifurcations have been studied to create vector fields with as many limit cycles as possible. All of these techniques rely heavily on symbolic manipulation packages such as MATLAB and its Symbolic Math Toolbox. Unfortunately, the results in the global case number relatively few. Only in recent years have many more results been found by restricting the analysis to *small-amplitude limit cycle* bifurcations; see, for example, Chapter 16.

Consider systems of the form

$$(15.1) \qquad \dot{x} = P(x, y), \qquad \dot{y} = Q(x, y),$$

where P and Q are polynomials in x and y. It is well known that a nondegenerate critical point, say, $\mathbf{x_0}$, of center or focus type can be moved to the origin by a linear change of coordinates to give

$$(15.2) \qquad \dot{x} = \lambda x - y + p(x, y), \qquad \dot{y} = x + \lambda y + q(x, y),$$

where p and q are at least quadratic in x and y. If $\lambda \neq 0$, then the origin is structurally stable for all perturbations.

Definition 1. A critical point, say, $\mathbf{x_0}$, is called a *fine focus* of system (15.1) if it is a center for the linearized system at $\mathbf{x_0}$. Equivalently, if $\lambda = 0$ in system (15.2), then the origin is a fine focus.

In the work to follow, assume that the unperturbed system does not have a center at the origin. The technique used here is entirely local; limit cycles bifurcate out of a fine focus when its stability is reversed by perturbing λ and the coefficients arising in p and q. These are said to be local or small-amplitude limit cycles. How close the origin is to being a center of the nonlinear system determines the number of limit cycles that may be obtained from bifurcation. The method for bifurcating limit cycles will now be summarized and is given in detail in [3].

By a classical result, there exists a Lyapunov function, say, $V(x, y) = V_2(x, y) + V_4(x, y) + \cdots + V_k(x, y) + \cdots$, where V_k is a homogeneous polynomial of degree k, such that

$$(15.3) \qquad \frac{dV}{dt} = \eta_2 r^2 + \eta_4 r^4 + \cdots + \eta_{2i} r^{2i} + \cdots,$$

where $r^2 = x^2 + y^2$. The η_{2i} are polynomials in the coefficients of p and q and are called the *focal values*. The origin is said to be a fine focus of order k if $\eta_2 = \eta_4 = \cdots = \eta_{2k} = 0$ but $\eta_{2k+2} \neq 0$. Take an analytic transversal through the origin parameterized by some variable, say, c. It is well known that the return map of (15.2), $c \mapsto h(c)$, is analytic if the critical point is nondegenerate. Limit cycles of system (15.2) then correspond to zeros of the *displacement function*, $d(c) = h(c) - c$; see Chapter 14. Hence at most k limit cycles can bifurcate from the fine focus. The stability of the origin is clearly dependent on the sign of the first nonzero focal value, and the origin is a nonlinear center if and only if all of the focal values are zero. Consequently, it is the reduced values, or *Lyapunov quantities*, say, $L(j)$, that are significant. One needs only consider the value η_{2k} reduced modulo the ideal $(\eta_2, \eta_4, \ldots, \eta_{2k-2})$ to obtain the Lyapunov quantity $L(k-1)$. To bifurcate limit cycles from the origin, select the coefficients in the Lyapunov quantities such that

$$|L(m)| \ll |L(m+1)| \quad \text{and} \quad L(m)L(m+1) < 0,$$

for $m = 0, 1, \ldots, k-1$. At each stage, the origin reverses stability and a limit cycle bifurcates in a small region of the critical point. If all of these conditions are satisfied, then there are exactly k small-amplitude limit cycles. Conversely, if $L(k) \neq 0$, then at most k limit cycles can bifurcate. Sometimes it is not possible to bifurcate the full complement of limit cycles; an example is given in [3].

The algorithm for bifurcating small-amplitude limit cycles may be split into the following four steps:

1. computation of the focal values using a mathematical package;

2. reduction of the focal values to Lyapunov quantities using substitutions;

3. checking that the origin is a center when all of the relevant Lyapunov quantities are zero;

4. bifurcation of the limit cycles by suitable perturbations.

For some systems, the following theorems can be used to prove that the origin is a center.

The Divergence Test. Suppose that the origin of system (15.1) is a critical point of focus type. If

$$\operatorname{div}(\psi \mathbf{X}) = \frac{\partial(\psi P)}{\partial x} + \frac{\partial(\psi Q)}{\partial y} = 0,$$

where $\psi : \Re^2 \to \Re^2$, then the origin is a center.

The Classical Symmetry Argument. Suppose that $\lambda = 0$ in system (15.2) and that either

(i) $p(x, y) = -p(x, -y)$ and $q(x, y) = q(x, -y)$ or

(ii) $p(x, y) = p(-x, y)$ and $q(x, y) = -q(-x, y)$.

15. Local and Global Bifurcations

Then the origin is a center.

Adapting the classical symmetry argument, it is also possible to prove the following theorem.

Theorem 1. *The origin of the system*

$$\dot{x} = y - F(G(x)), \qquad \dot{y} = -\frac{G'(x)}{2} H(G(x)),$$

where F and H are polynomials, $G(x) = \int_0^x g(s)ds$ with $g(0) = 0$ and $g(x)\,\mathrm{sgn}(x) > 0$ for $x \neq 0$, is a center. (See the exercises at the end of the chapter.)

To demonstrate the method for bifurcating small-amplitude limit cycles, consider Liénard equations of the form

(15.4) $$\dot{x} = y - F(x), \qquad \dot{y} = -g(x),$$

where $F(x) = a_1 x + a_2 x^2 + \cdots + a_u x^u$ and $g(x) = x + b_2 x^2 + b_3 x^3 + \cdots + b_v x^v$. This system has proved very useful in the investigation of limit cycles when showing existence, uniqueness, and hyperbolicity of a limit cycle. There have also been many local results; see, for example, [2] and Chapter 16. Therefore, it seems sensible to use this class of system to illustrate the method.

The computation of the first three focal values will be given. Write $V_k(x, y) = \sum_{i+j=k} V_{i,j} x^i y^j$ and denote $V_{i,j}$ as being odd or even according to whether i is odd or even and that $V_{i,j}$ is 2-odd or 2-even according to whether j is odd or even, respectively. Solving equation (15.3), it is easily seen that $V_2 = \frac{1}{2}(x^2 + y^2)$ and $\eta_2 = -a_1$. Therefore, set $a_1 = 0$. The odd and even coefficients of V_3 are then given by the two pairs of equations

$$3V_{3,0} - 2V_{1,2} = b_2,$$
$$V_{1,2} = 0$$

and

$$-V_{2,1} = a_2,$$
$$2V_{2,1} - 3V_{0,3} = 0,$$

respectively. Solve the equations to give

$$V_3 = \frac{1}{3}b_2 x^3 - a_2 x^2 y - \frac{2}{3} a_2 y^3.$$

Both η_4 and the odd coefficients of V_4 are determined by the equations

$$-\eta_4 - V_{3,1} = a_3,$$
$$-2\eta_4 + 3V_{3,1} - 3V_{1,3} = -2a_2 b_2,$$
$$-\eta_4 + V_{1,3} = 0.$$

15.1. Small-Amplitude Limit Cycle Bifurcations

The even coefficients are determined by the equations

$$4V_{4,0} - 2V_{2,2} = b_3 - 2a_2^2,$$
$$2V_{2,2} - 4V_{0,4} = 0$$

and the supplementary condition $V_{2,2} = 0$. In fact, when computing subsequent coefficients for V_{4m}, it is convenient to require that $V_{2m,2m} = 0$. This ensures that there will always be a solution. Solving these equations gives

$$V_4 = \frac{1}{4}(b_3 - 2a_2^2)x^4 - (\eta_4 + a_3)x^3 y + \eta_4 xy^3$$

and

$$\eta_4 = \frac{1}{8}(2a_2 b_2 - 3a_3).$$

Suppose that $\eta_4 = 0$ so that $a_3 = \frac{2}{3} a_2 b_2$. It can be checked that the two sets of equations for the coefficients of V_5 give

$$V_5 = \left(\frac{b_4}{5} - \frac{2a_2^2 b_2}{3} \right) x^5 + (2a_2^3 - a_4)x^4 y + \left(\frac{8a_2^3}{3} - \frac{4a_4}{3} + \frac{2a_2 b_3}{3} \right) x^2 y^3$$
$$+ \left(\frac{16 a_2^3}{15} - \frac{8 a_4}{15} - \frac{4 a_2 b_3}{15} \right) y^5.$$

The coefficients of V_6 may be determined by inserting the extra condition $V_{4,2} + V_{2,4} = 0$. In fact, when computing subsequent even coefficients for V_{4m+2}, the extra condition $V_{2m-2,2m+2} + V_{2m+2,2m-2} = 0$ is applied, which guarantees a solution. The polynomial V_6 contains 27 terms and will not be listed here. However, η_6 leads to the Lyapunov quantity

$$L(2) = 6a_2 b_4 - 10 a_2 b_2 b_3 + 20 a_4 b_2 - 15 a_5.$$

Lemma 1. *The first three Lyapunov quantities for system* (15.4) *are* $L(0) = -a_1$, $L(1) = 2a_2 b_2 - 3a_3$, *and* $L(2) = 6a_2 b_4 - 10 a_2 b_2 b_3 + 20 a_4 b_2 - 15 a_5$.

Let $\hat{H}(u, v)$ denote the maximum number of small-amplitude limit cycles that can be bifurcated from the origin for system (15.4).

Example 1. Prove that

(i) $\hat{H}(3, 2) = 1$ and

(ii) $\hat{H}(3, 3) = 2$

for system (15.4).

Solutions. (i) Consider the case where $u = 3$ and $v = 2$. Now L(0)=0 if $a_1 = 0$ and $L(1) = 0$ if $a_3 = \frac{2}{3}a_2b_2$. Thus system (15.4) becomes

$$\dot{x} = y - a_2x^2 - \frac{2}{3}a_2b_2x^3, \qquad \dot{y} = -x - b_2x^2,$$

and the origin is a center by Theorem 1. Therefore, the origin is a fine focus of order one if and only if $a_1 = 0$ and $2a_2b_2 - 3a_3 \neq 0$. The conditions are consistent. Select a_3 and a_1 such that

$$|L(0)| \ll |L(1)| \quad \text{and} \quad L(0)L(1) < 0.$$

The origin reverses stability once and a limit cycle bifurcates. The perturbations are chosen such that the origin reverses stability once and the limit cycles that bifurcate persist. Thus $\hat{H}(3, 2) = 1$.

(ii) Consider system (15.4) with $u = 3$ and $v = 3$. Now $L(0) = 0$ if $a_1 = 0$, $L(1) = 0$ if $a_3 = \frac{2}{3}a_2b_2$, and $L(2) = 0$ if $a_2b_2b_3 = 0$. Thus $L(2) = 0$ if

(a) $a_2 = 0$,

(b) $b_3 = 0$, or

(c) $b_2 = 0$.

If condition (a) holds, then $a_3 = 0$ and the origin is a center by the divergence test (div $\mathbf{X} = 0$). If condition (b) holds, then the origin is a center since $\hat{H}(3, 2) = 1$. If condition (c) holds, then $a_3 = 0$ and system (15.3) becomes

$$\dot{x} = y - a_2x^2, \qquad \dot{y} = -x - b_3x^3,$$

and the origin is a center by the classical symmetry argument. The origin is thus a fine focus of order two if and only if $a_1 = 0$ and $2a_2b_2 - 3a_3 = 0$ but $a_2b_2b_3 \neq 0$. The conditions are consistent. Select b_3, a_3, and a_1 such that

$$|L(1)| \ll |L(2)|, \quad L(1)L(2) < 0, \quad \text{and} \quad |L(0)| \ll |L(1)|, \quad L(0)L(1) < 0.$$

The origin has changed stability twice, and there are two small-amplitude limit cycles. The perturbations are chosen such that the origin reverses stability twice and the limit cycles that bifurcate persist. Thus $\hat{H}(3, 3) = 2$.

15.2 Melnikov Integrals and Bifurcating Limit Cycles from a Center

Consider perturbed two-dimensional differential systems of the form

(15.5) $$\dot{\mathbf{x}} = \mathbf{f}(\mathbf{x}) + \epsilon \mathbf{g}(\mathbf{x}, \epsilon, \mu).$$

15.2. Melnikov Integrals and Bifurcating Limit Cycles from a Center

Assume that the unperturbed system

(15.6) $$\dot{\mathbf{x}} = \mathbf{f}(\mathbf{x})$$

has a one-parameter family of periodic orbits given by

$$\Gamma_r : \mathbf{x} = \gamma_r(t),$$

where the functions $\gamma_r(t)$ have minimum periods T_r and r belongs to an indexing set, say, I, that is either a finite or semiinfinite open interval of $\Re$.

Definition 2. The *Melnikov function* for system (15.5) along the cycle $\Gamma_r : \mathbf{x} = \gamma_r(t), 0 \leq t \leq T_r$, of (15.6) is given by

$$M(r, \mu) = \int_0^{T_r} \exp\left(-\int_0^t \nabla \cdot \mathbf{f}(\gamma_r(s))ds\right) \mathbf{f} \wedge \mathbf{g}(\gamma_r(t), 0, \mu) dt.$$

Theorem 2. *Suppose that*

$$M(r_0, \mu_0) = 0 \quad \text{and} \quad \left.\frac{\partial M}{\partial r}\right|_{(r_0, \mu_0)} \neq 0,$$

where $r_0 \in I$. Then for $0 < \epsilon \ll 1$, system (15.5) has a unique hyperbolic limit cycle close to Γ_{r_0}. System (15.5) has no limit cycle close to Γ_{r_0} if $M(r_0, \mu_0) \neq 0$ and ϵ is small.

Theorem 3. *Suppose that $M(r, \mu_0) = 0$ has exactly k solutions $r_1, r_2, \ldots, r_k \in I$ with*

$$\left.\frac{\partial M}{\partial r}\right|_{(r_i, \mu_0)} \neq 0,$$

for some i from 1 to k. Then for $0 < \epsilon \ll 1$, exactly k one-parameter families of hyperbolic limit cycles bifurcate from the period annulus of (15.6) at the points $r_1, r_2, \ldots, r_k$. If $M(r, \mu_0) \neq 0$, then there are no limit cycles.

Melnikov-type integrals have been widely used since Poincaré's investigations at the end of the 19th century. It is well known that the Melnikov function for system (15.5) is proportional to the derivative of the Poincaré map for (15.5) with respect to ϵ. The interested reader may consult [1] for more details; the paper also deals with limit cycles of multiplicity greater than one and the bifurcation of limit cycles from separatrix cycles. To avoid elliptic integrals, only systems with $\gamma_r(t) = (x(t), y(t)) = (r \cos t, r \sin t)$ will be considered in this book.

Example 2. Consider the van der Pol system

$$\dot{x} = y, \qquad \dot{y} = -x - \epsilon(1 - x^2)y.$$

Prove that there is a limit cycle asymptotic to the circle of radius two when ϵ is small.

Solution. In this case, $\mathbf{f}(\mathbf{x}) = (y, -x)^T$, $T_r = 2\pi$, $\mathbf{g}(\mathbf{x}, \epsilon) = (0, -\epsilon y(1 - x^2))^T$, $x = r\cos(t)$, $y = r\sin(t)$, and $\nabla \cdot \mathbf{f}(\mathbf{x}) = 0$. Therefore,

$$M(r, \mu) = \int_0^{T_r} \mathbf{f} \wedge \mathbf{g}(\gamma_r(t), 0, \mu) dt.$$

Thus

$$M(r, \mu) = \int_0^{2\pi} -r^2(\sin^2 t(1 - r^2 \cos^2 t))dt$$

and

$$M(r, \mu) = \frac{\pi}{4} r^2(r^2 - 4).$$

Hence $M(r_0, \mu) = 0$ when $r_0 = 2$ and $\frac{\partial M}{\partial r}|_{(r_0, 0)} = \pi r_0(r_0^2 - 2) \neq 0$. Therefore, there exists a unique hyperbolic limit cycle asymptotic to a circle of radius two for the van der Pol system when ϵ is sufficiently small.

Example 3. Consider the Liénard system

(15.7) $\qquad \dot{x} = -y + \epsilon(a_1 x + a_3 x^3 + a_5 x^5), \qquad \dot{y} = x.$

Determine the maximum number and approximate location of the limit cycles when ϵ is sufficiently small.

Solution. Again, $\mathbf{f}(\mathbf{x}) = (-y, x)^T$, $T_r = 2\pi$, $\mathbf{g}(\mathbf{x}, \epsilon) = (\epsilon(a_1 x + a_3 x^3 + a_5 x^5), 0)^T$, and $\nabla \cdot \mathbf{f}(\mathbf{x}) = 0$. Therefore,

$$M(r, \mu) = \int_0^{2\pi} -a_1 r^2 \cos^2 t - a_3 r^4 \cos^4 t - a_5 r^6 \cos^6 t \, dt$$

and

$$M(r, \mu) = -\pi r^2 \left(a_1 + \frac{3a_3}{4} r^2 + \frac{5a_5}{8} r^4 \right).$$

The polynomial $m(r) = a_1 + \frac{3a_3}{4} r^2 + \frac{5a_5}{8} r^4$ has at most two positive roots. Therefore, when ϵ is sufficiently small, system (15.7) has at most two hyperbolic limit cycles asymptotic to circles of radii r_j ($j = 1, 2$), where r_j are the positive roots of $m(r)$.

15.3 Homoclinic Bifurcations

Global bifurcations of limit cycles from centers were investigated in Section 15.2. The global bifurcation of limit cycles from a homoclinic loop will now be discussed via example. The analysis involved in bifurcating limit cycles from separatrix

15.3. Homoclinic Bifurcations

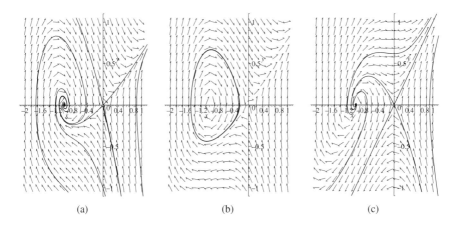

Figure 15.1: Phase portraits for Example 4 when (a) $\lambda = -1.5$, (b) $\lambda = -0.9$, and (c) $\lambda = 0$.

cycles is beyond the scope of this book; however, interested readers are referred to [1]. Both homoclinic and heteroclinic bifurcations are used to obtain polynomial systems with a number of limit cycles; see Chapter 16. The MATLAB package can be used to investigate some of these systems numerically.

Example 4. Investigate the system

$$\dot{x} = y, \qquad \dot{y} = x + x^2 - xy + \lambda y$$

as the parameter λ varies and plot possible phase portraits.

Solution. There are two critical points at $O = (0, 0)$ and $P = (-1, 0)$. The Jacobian is given by

$$J = \begin{pmatrix} 0 & 1 \\ 1 + 2x - y & -x + \lambda \end{pmatrix}.$$

The origin is a saddle point, and it can be shown that the point P is a node or focus. Since trace $J_P = 1 + \lambda$, it follows that P is stable if $\lambda < -1$ and unstable if $\lambda > -1$. The point P is also stable if $\lambda = -1$.

It can be shown that a limit cycle exists for $-1 < \lambda < \lambda_0$, where $\lambda_0 \approx -0.85$. Since the limit cycle appears from a homoclinic loop, which exists at a value, say, λ_0, this is known as a homoclinic bifurcation. More details can be found in [1]. Phase portraits for three values of λ are shown in Figure 15.1.

Another example is given in the exercises in Section 15.5.

15.4 MATLAB Commands

```
% Programs_15a - Computing the coefficients of V3 and V4
% Lyapunov quantities for the Lienard system.
% V3=[V30;V21;V12;V03] and V4=[V40;V31;V22;V13;V04;eta4]

syms a2 b2 a3 b3 a4 b4 a5 b5;
A=[3 0 -2 0;0 0 1 0;0 -1 0 0;0 2 0 -3];
B=[b2; 0; a2; 0];

V3=A\B

A=[0 -1 0 0  0 -1;0 3 0 -3 0 -2;0 0 0 1 0 -1;4 0 -2 0 0 0;
   0 0 2 0 -4 0;0 0 1 0 0 0];
B=[a3; -2*a2*b2; 0; b3-2*a2^2;0;0];

V4=A\B

% End of Programs_15a.
```

15.5 Exercises

1. Prove that the origin of the system

$$\dot{x} = y - F(G(x)), \qquad \dot{y} = -\frac{G'(x)}{2}H(G(x))$$

   is a center using the transformation $u^2 = G(x)$ and the classical symmetry argument.

2. Write a program to compute the first seven Lyapunov quantities of the Liénard system

   (15.8) $\qquad \dot{x} = y - (a_1 x + a_2 x^2 + \cdots + a_{13} x^{13}), \qquad \dot{y} = -x.$

3. Using the results from Exercise 2, prove that at most six small-amplitude limit cycles can be bifurcated from the origin of system (15.8).

4. Consider the system

   $$\dot{x} = y - (a_1 x + a_3 x^3 + \cdots + a_{2n+1} x^{2n+1}), \qquad \dot{y} = -x.$$

   Prove by induction that at most n small-amplitude limit cycles can be bifurcated from the origin.

5. Write a program to compute the first five Lyapunov quantities for the Liénard system

$$\dot{x} = y-(a_1x+a_2x^2+\cdots+a_7x^7), \qquad \dot{y} = -(x+b_2x^2+b_3x^3+\cdots+b_6x^6).$$

6. Using the results from Exercise 5, prove that $\hat{H}(4, 2) = 2$, $\hat{H}(7, 2) = 4$, and $\hat{H}(3, 6) = 4$. Note that in $\hat{H}(u, v)$, u is the degree of F and v is the degree of g.

7. Consider the generalized mixed Rayleigh–Liénard oscillator equations given by

$$\dot{x} = y, \qquad \dot{y} = -x - a_1y - b_{30}x^3 - b_{21}x^2y - b_{41}x^4y - b_{03}y^3.$$

Prove that at most three small-amplitude limit cycles can be bifurcated from the origin.

8. Plot a phase portrait for the system

$$\dot{x} = y, \qquad \dot{y} = x + x^2.$$

Determine an equation for the curve on which the homoclinic loop lies.

9. Consider the Liénard system given by

$$\dot{x} = y - \epsilon(a_1x + a_2x^2 + a_3x^3), \qquad \dot{y} = -x.$$

Prove that for sufficiently small ϵ, there is at most one limit cycle that is asymptotic to a circle of radius

$$r = \sqrt{\frac{4|a_1|}{3|a_3|}}.$$

10. Using the MATLAB package, investigate the system

$$\dot{x} = y, \qquad \dot{y} = x - x^3 + \epsilon(\lambda y + x^2 y)$$

when $\epsilon = 0.1$ for values of λ from -1 to -0.5. How many limit cycles are there at most?

Recommended Reading

[1] T. R. Blows and L. M. Perko, Bifurcation of limit cycles from centers and separatrix cycles of planar analytic systems, *SIAM Rev.*, **36** (1994), 341–376.

[2] N. G. Lloyd and S. Lynch, Small-amplitude limit cycles of certain Liénard systems, *Proc. Roy. Soc. London Ser.* A, **418** (1988), 199–208.

[3] N. G. Lloyd, Limit cycles of polynomial systems, in T. Bedford and J. Swift, eds., *New Directions in Dynamical Systems*, L.M.S. Lecture Notes Series 127, Cambridge University Press, Cambridge, UK, 1988, 192–234.

16
The Second Part of Hilbert's Sixteenth Problem

Aims and Objectives

- To describe the second part of Hilbert's sixteenth problem.
- To review the main results on the number of limit cycles of planar polynomial systems.
- To consider the flow at infinity after Poincaré compactification.
- To review the main results on the number of limit cycles of Liénard systems.
- To prove two theorems concerning limit cycles of certain Liénard systems.

On completion of this chapter the reader should be able to

- state the second part of Hilbert's sixteenth problem;
- describe the main results for this problem;
- compactify the plane and construct a global phase portrait that shows the behavior at infinity for some simple systems;
- compare local and global results;
- prove that certain systems have a unique limit cycle;
- prove that a limit cycle has a certain shape for a large parameter value.

16. The Second Part of Hilbert's Sixteenth Problem

The second part of Hilbert's sixteenth problem is stated and the main results are listed. To understand these results, it is necessary to introduce Poincaré compactification, where the plane is mapped onto a sphere and the behavior on the equator of the sphere represents the behavior at infinity for planar systems.

Many autonomous systems of two-dimensional differential equations can be transformed to systems of Liénard type. In recent years, there have been many results published associated with Liénard systems. The major results for both global and local bifurcations of limit cycles for these systems are listed in the following sections.

A method for proving the existence, uniqueness, and hyperbolicity of a limit cycle is illustrated in this chapter, and the Poincaré–Bendixson theorem is applied to determine the shape of a limit cycle when a parameter is large.

16.1 Statement of Problem and Main Results

Poincaré began investigating isolated periodic cycles of planar polynomial vector fields in the 1880s. However, the general problem of determining the maximum number and relative configurations of limit cycles in the plane has remained unresolved for over a century. Recall that limit cycles in the plane can correspond to steady-state behavior for a physical system (see Chapter 12), so it is important to know how many possible steady states there are.

In 1900, David Hilbert presented a list of 23 problems to the International Congress of Mathematicians in Paris. Most of the problems have been solved, either completely or partially. However, the second part of the sixteenth problem remains unsolved. Il'yashenko [1] presents a centennial history of Hilbert's sixteenth problem.

The Second Part of Hilbert's Sixteenth Problem. Consider planar polynomial systems of the form

(16.1) $$\dot{x} = P(x, y), \qquad \dot{y} = Q(x, y),$$

where P and Q are polynomials in x and y. The question is to estimate the maximal number and relative positions of the limit cycles of system (16.1). Let H_n denote the maximum possible number of limit cycles that system (16.1) can have when P and Q are of degree n. More formally, the Hilbert numbers H_n are given by

$$H_n = sup\{\pi(P, Q) : \partial P, \partial Q \leq n\},$$

where ∂ denotes "the degree of" and $\pi(P, Q)$ is the number of limit cycles of system (16.1).

Dulac's theorem states that a given polynomial system cannot have infinitely many limit cycles. This theorem has only recently been proved independently by Ecalle et al. [13] and Il'yashenko [11], respectively. Unfortunately, this does not imply that the Hilbert numbers are finite.

16.1. Statement of Problem and Main Results

Of the many attempts to make progress in this question, one of the more fruitful approaches has been to create vector fields with as many isolated periodic orbits as possible using both local and global bifurcations. There are relatively few results in the case of general polynomial systems even when considering local bifurcations. Bautin [21] proved that no more than three small-amplitude limit cycles could bifurcate from a critical point for a quadratic system. For a homogeneous cubic system (no quadratic terms), Sibirskii [20] proved that no more than five small-amplitude limit cycles could be bifurcated from one critical point. Zoladek [7] recently found an example where 11 limit cycles could be bifurcated from the origin of a cubic system, but he was unable to prove that this was the maximum possible number.

Although easily stated, Hilbert's sixteenth problem remains almost completely unsolved. For quadratic systems, Shi Songling [16] has obtained a lower bound for the Hilbert number $H_2 \geq 4$. A possible global phase portrait is given in Figure 16.1. The line at infinity is included and the properties on this line are determined using Poincaré compactification, which is described in Section 16.2. There are three small-amplitude limit cycles around the origin and at least one other surrounding another critical point. Some of the parameters used in this example are very small.

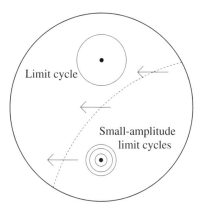

Figure 16.1: A possible configuration for a quadratic system with four limit cycles: one of large amplitude and three of small amplitude.

Blows and Rousseau [8] consider the bifurcation at infinity for polynomial vector fields and give examples of cubic systems having the following configurations:

$$\{(4), 1\}, \ \{(3), 2\}, \ \{(2), 5\}, \ \{(4), 2\}, \ \{(1), 5\}, \text{ and } \{(2), 4\},$$

where $\{(l), L\}$ denotes the configuration of a vector field with l small-amplitude limit cycles bifurcated from a point in the plane and L large-amplitude limit cycles simultaneously bifurcated from infinity. There are many other configurations possible, some involving other critical points in the finite part of the plane as shown in Figure 16.2. Recall that a limit cycle must contain at least one critical point.

By considering cubic polynomial vector fields, in 1985 Li Jibin and Li Chunfu [14] produced an example showing that $H_3 \geq 11$ by bifurcating limit cycles out of homoclinic and heteroclinic orbits; see Figure 16.2.

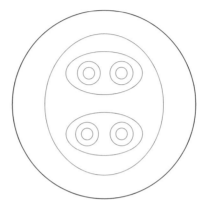

Figure 16.2: A possible configuration for a cubic system with 11 limit cycles.

Returning to the general problem, in 1995 Christopher and Lloyd [6] considered the rate of growth of H_n as n increases. They showed that H_n grows at least as rapidly as $n^2 \log n$.

In recent years, the focus of research in this area has been directed at a small number of classes of systems. Perhaps the most fruitful has been the Liénard system.

16.2 Poincaré Compactification

The method of compactification was introduced by Henri Poincaré at the end of the 19th century. By making a simple transformation, it is possible to map the phase plane onto a sphere. Note that the plane can be mapped to both the upper and lower hemispheres. In this way, the points at infinity are transformed to the points on the equator of the sphere. Suppose that a point (x, y) in the plane is mapped to a point (X, Y, Z) on the upper hemisphere of a sphere, say, $S^2 = \{(X, Y, Z) \in \Re^3 : X^2 + Y^2 + Z^2 = 1\}$. (Note that it is also possible to map onto the lower hemisphere.) The equations defining (X, Y, Z) in terms of (x, y) are given by

$$X = \frac{x}{\sqrt{1+r^2}}, \quad Y = \frac{y}{\sqrt{1+r^2}}, \quad Z = \frac{1}{\sqrt{1+r^2}},$$

where $r^2 = x^2 + y^2$. A central projection is illustrated in Figure 16.3.

16.2. Poincaré Compactification

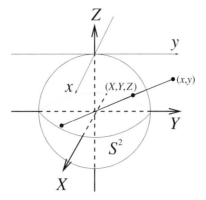

Figure 16.3: A mapping of (x, y) in the plane onto (X, Y, Z) on the upper part of the sphere.

Consider the autonomous system (16.1). Convert to polar coordinates. Thus system (16.1) transforms to

(16.2)
$$\dot{r} = r^n f_{n+1}(\theta) + r^{n-1} f_{n-1}(\theta) + \cdots + f_1(\theta)$$
$$\dot{\theta} = r^{n-1} g_{n+1}(\theta) + r^{n-2} g_{n-1}\theta + \cdots + r^{-1} g_1(\theta),$$

where f_m and g_m are polynomials of degree m in $\cos\theta$ and $\sin\theta$.

Let $\rho = \frac{1}{r}$. Hence $\dot{\rho} = -\frac{\dot{r}}{r^2}$, and system (16.2) becomes

$$\dot{\rho} = -\rho f_{n+1}(\theta) + O(\rho^2), \qquad \dot{\theta} = g_{n+1}(\theta) + O(\rho).$$

Theorem 1. *The critical points at infinity are found by solving the equations $\dot{\rho} = \dot{\theta} = 0$ on $\rho = 0$, which is equivalent to solving*

$$g_{n+1}(\theta) = \cos\theta\, Q_n(\cos\theta, \sin\theta) - \sin\theta\, P_n(\cos\theta, \sin\theta) = 0,$$

where P_n and Q_n are homogeneous polynomials of degree n. Note that the solutions are given by the pairs θ_i and $\theta_i + \pi$. As long as $g_{n+1}(\theta)$ is nonzero, there are $n+1$ pairs of roots and the flow is clockwise when $g_{n+1}(\theta) < 0$ and it is counterclockwise when $g_{n+1}(\theta) > 0$.

To determine the flow near the critical points at infinity, one must project the hemisphere with $X > 0$ onto the plane $X = 1$ with axes y and z or project the hemisphere with $Y > 0$ onto the plane $Y = 1$ with axes x and z. The projection of the sphere S^2 onto these planes is depicted in Figure 16.4.

If n is odd, the antinodal points on S^2 are qualitatively equivalent. If n is even, the antinodal points are qualitatively equivalent but the direction of the flow is reversed.

16. The Second Part of Hilbert's Sixteenth Problem

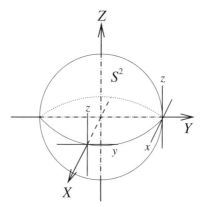

Figure 16.4: The projections used to determine the behavior at infinity.

The flow near a critical point at infinity can be determined using the following theorem.

Theorem 2. *The flow defined on the yz plane* ($X = \pm 1$), *except the points* $(0, \pm 1, 0)$, *is qualitatively equivalent to the flow defined by*

$$\pm \dot{y} = yz^n P\left(\frac{1}{z}, \frac{y}{z}\right) - z^n Q\left(\frac{1}{z}, \frac{y}{z}\right), \qquad \pm \dot{z} = z^{n+1} P\left(\frac{1}{z}, \frac{y}{z}\right),$$

where the direction of the flow is determined from $g_{n+1}(\theta)$.

In a similar way, the flow defined on the xz plane ($Y = \pm 1$), *except the points* $(\pm 1, 0, 0)$, *is qualitatively equivalent to the flow defined by*

$$\pm \dot{x} = xz^n Q\left(\frac{x}{z}, \frac{1}{z}\right) - z^n P\left(\frac{x}{z}, \frac{1}{z}\right), \qquad \pm \dot{z} = z^{n+1} Q\left(\frac{x}{z}, \frac{1}{z}\right),$$

where the direction of the flow is determined from $g_{n+1}(\theta)$.

Example 1. Construct global phase portraits, including the flow at infinity, for the following linear systems:

(a) $\dot{x} = -x + 2y, \dot{y} = 2x + 2y$;

(b) $\dot{x} = x + y, \dot{y} = -x + y$.

Solutions. (a) The origin is a saddle point or col with eigenvalues and corresponding eigenvectors given by $\lambda_1 = 3, (1, 2)^T$ and $\lambda_2 = -2, (2, -1)^T$. The critical points at infinity satisfy the equation $g_2(\theta) = 0$, where

$$g_2(\theta) = \cos\theta \, Q_1(\cos\theta, \sin\theta) - \sin\theta \, P_1(\cos\theta, \sin\theta).$$

16.2. Poincaré Compactification

Now

$$g_2(\theta) = 2\cos^2\theta + 3\cos\theta\sin\theta - 2\sin^2\theta.$$

The roots are given by $\theta_1 = \tan^{-1}(2)$ radians, $\theta_2 = \tan^{-1}(2) + \pi$ radians, $\theta_3 = \tan^{-1}(-\frac{1}{2})$ radians, and $\theta_4 = \tan^{-1}(-\frac{1}{2}) + \pi$ radians.
A plot of $g_2(\theta)$ is given in Figure 16.5.

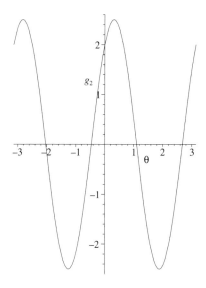

Figure 16.5: The function $g_2(\theta)$.

The flow near a critical point at infinity is qualitatively equivalent to the flow of the system

$$\pm\dot{y} = yz\left(-\frac{1}{2} + \frac{2y}{z}\right) - z\left(\frac{2}{z} - \frac{2y}{z}\right), \qquad \pm\dot{z} = z^2\left(-\frac{1}{z} + \frac{2y}{z}\right).$$

From Figure 16.5, the flow is counterclockwise if $\tan^{-1}(-\frac{1}{2}) < \theta < \tan^{-1}(2)$. Therefore, the flow at infinity is determined by the system

$$-\dot{y} = -3y + 2y^2 - 2, \qquad -\dot{z} = -z + 2yz.$$

There are critical points at $A = (2, 0)$ and $B = (-\frac{1}{2}, 0)$ in the yz plane. Point A is a stable node and point B is an unstable node. A phase portrait is given in Figure 16.6.

Since n is odd, the antinodal points are qualitatively equivalent. A global phase portrait is shown in Figure 16.7.

16. The Second Part of Hilbert's Sixteenth Problem

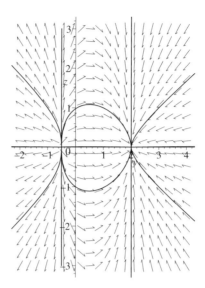

Figure 16.6: Some trajectories in the yz plane (X=1) that define the flow at infinity.

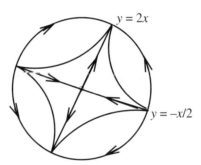

Figure 16.7: A global phase portrait for Example 1(a).

(b) The origin is an unstable focus and the flow is clockwise. The critical points at infinity satisfy the equation $g_2(\theta) = 0$, where

$$g_2(\theta) = \cos\theta\, Q_1(\cos\theta, \sin\theta) - \sin\theta\, P_1(\cos\theta, \sin\theta) = -(\cos^2\theta + \sin^2\theta).$$

There are no roots for $g_2(\theta)$, so there are no critical points at infinity. A global phase portrait is given in Figure 16.8.

Example 2. Show that the system given by

$$\dot{x} = -\frac{x}{2} - y - x^2 + xy + y^2, \qquad \dot{y} = x(1 + x - 3y)$$

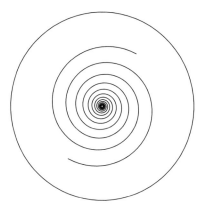

Figure 16.8: A global phase portrait for Example 1(b). There are no critical points at infinity and the flow is clockwise.

has at least two limit cycles.

Solution. There are two critical points at $O = (0, 0)$ and $A = (0, 1)$. The Jacobian matrix is given by

$$J = \begin{pmatrix} -\frac{1}{2} - 2x + y & -1 + x + 2y \\ 1 + 2x - 3y & -3x \end{pmatrix}.$$

Now

$$J_O = \begin{pmatrix} -\frac{1}{2} & -1 \\ 1 & 0 \end{pmatrix} \quad \text{and} \quad J_A = \begin{pmatrix} \frac{1}{2} & 1 \\ -2 & 0 \end{pmatrix}.$$

Therefore, O is a stable focus and A is an unstable focus. On the line $L_1 : 1 + x - 3y = 0$, $\dot{y} = 0$ and $\dot{x} < 0$, so the flow is transverse to L_1.

The critical points at infinity satisfy the equation $g_3(\theta) = 0$, where

$$g_3(\theta) = \cos\theta \, Q_2(\cos\theta, \sin\theta) - \sin\theta \, P_2(\cos\theta, \sin\theta).$$

Now

$$g_3(\theta) = \cos^3\theta - 2\cos^2\theta \sin\theta - \cos\theta \sin^2\theta - \sin^3\theta.$$

A plot for $g_3(\theta)$ is given in Figure 16.9.

There are two roots for $g_3(\theta)$: $\theta_1 = 0.37415$ radians and $\theta_2 = 3.51574$ radians. The flow near a critical point at infinity is qualitatively equivalent to the flow of the system

$$\pm\dot{y} = -\frac{yz}{2} - y^2 z + 2y + y^2 + y^3 - z - 1,$$

$$\pm\dot{z} = -\frac{z^2}{2} - yz^2 - z + yz + y^2 z.$$

344   **16. The Second Part of Hilbert's Sixteenth Problem**

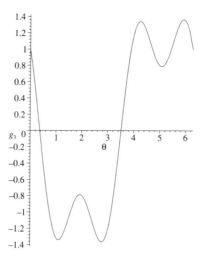

Figure 16.9: The function $g_3(\theta)$.

There is one critical point at $(y, z) = (0.39265, 0)$, which is a col. Since n is even, the antinodal point is also a col, but the direction of the flow is reversed. The direction of the flow may be established by inspecting $g_3(\theta)$ in Figure 16.9.

Part of the global phase portrait is shown in Figure 16.10, and from the corollary to the Poincaré–Bendixson theorem, there are at least two limit cycles

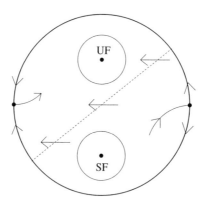

Figure 16.10: A global phase portrait showing at least two limit cycles.

If the system is nonlinear and there are no critical points at infinity, it is also possible to bifurcate limit cycles from infinity; see, for example, the work of Blows and Rousseau [8].

16.3 Global Results for Liénard Systems

Consider polynomial Liénard equations of the form

(16.3) $$\ddot{x} + f(x)\dot{x} + g(x) = 0,$$

where $f(x)$ is known as the damping coefficient and $g(x)$ is called the restoring coefficient. Equation (16.3) corresponds to the class of systems

(16.4) $$\dot{x} = y, \qquad \dot{y} = -g(x) - f(x)y,$$

in the phase plane. Liénard applied the change of variable $Y = y + F(x)$, where $F(x) = \int_0^x f(s)ds$, to obtain an equivalent system in the so-called Liénard plane:

(16.5) $$\dot{x} = Y - F(x), \qquad \dot{Y} = -g(x).$$

For the critical point at the origin to be a nondegenerate focus or center, the conditions $g(0) = 0$ and $g'(0) > 0$ are imposed. Periodic solutions of (16.5) correspond to limit cycles of (16.2) and (16.5). There are many examples in both the natural sciences and technology where these and related systems are applied. The differential equation is often used to model either mechanical systems or electric circuits, and in the literature, many systems are transformed to Liénard type to aid in the investigations. For a list of applications to the real world, see, for example, Moreira [10]. In recent years, the number of results for this class of system has been phenomenal, and the allocation of this topic to a whole chapter of the book is well justified.

These systems have proved very useful in the investigation of multiple limit cycles and also when proving existence, uniqueness, and hyperbolicity of a limit cycle. Let ∂ denote the degree of a polynomial, and let $H(i, j)$ denote the maximum number of global limit cycles, where i is the degree of f and j is the degree of g. The main global results for systems (16.2) and (16.5) to date are listed below:

- In 1928, Liénard (see [5, Chapter 10]) proved that when $\partial g = 1$ and F is a continuous odd function, which has a unique root at $x = a$ and is monotone increasing for $x \geq a$, then (16.5) has a unique limit cycle.

- In 1973, Rychkov [19] proved that if $\partial g = 1$ and F is an odd polynomial of degree five, then (16.5) has at most two limit cycles.

- In 1976, Cherkas [18] gave conditions in order for a Liénard equation to have a center.

- In 1977, Lins, de Melo, and Pugh [17] proved that $H(2, 1) = 1$. They also conjectured that $H(2m, 1) = H(2m + 1, 1) = m$, where m is a natural number.

- In 1988, Coppel [12] proved that $H(1, 2) = 1$.

16. The Second Part of Hilbert's Sixteenth Problem

- In 1992, Zhang Zhifen et al. [9] proved that a certain generalized Liénard system has a unique limit cycle.
- In 1996, Dumortier and Chengzhi Li [5] proved that $H(1, 3) = 1$.
- In 1997, Dumortier and Chengzhi Li [4] proved that $H(2, 2) = 1$.

More recently, Giacomini and Neukirch [3] introduced a new method to investigate the limit cycles of Liénard systems when $\partial g = 1$ and $F(x)$ is an odd polynomial. They are able to give algebraic approximations to the limit cycles and obtain information on the number and bifurcation sets of the periodic solutions even when the parameters are not small. Other work has been carried out on the algebraicity of limit cycles, but it is beyond the scope of this book.

Limit cycles were discussed in some detail in Chapter 10, and a method for proving the existence and uniqueness of a limit cycle was introduced. Another method for proving the existence, uniqueness, and hyperbolicity of a limit cycle will be illustrated in Theorem 4.

Consider the general polynomial system

$$\dot{x} = P(x, y), \qquad \dot{y} = Q(x, y),$$

where P and Q are polynomials in x and y, and define $\mathbf{X} = (P, Q)$ to be the vector field. Let a limit cycle, say, $\Gamma(t) = (x(t), y(t))$, have period T.

Definition 1. The quantity $\int_\Gamma \mathrm{div}(\mathbf{X}) dt$ is known as the *characteristic exponent*.

Theorem 3. *Suppose that*

$$\int_\Gamma \mathrm{div}(\mathbf{X}) dt = \int_0^T \left(\frac{\partial P}{\partial x} + \frac{\partial Q}{\partial y} \right) (x(t), y(t)) dt.$$

Then

(i) Γ *is hyperbolic attracting if* $\int_\Gamma \mathrm{div}(\mathbf{X}) dt < 0$;

(ii) Γ *is hyperbolic repelling if* $\int_\Gamma \mathrm{div}(\mathbf{X}) dt > 0$.

Theorem 4. *Consider the Liénard system*

(16.6) $$\dot{x} = y - (a_1 x + a_2 x^2 + a_3 x^3), \qquad \dot{y} = -x.$$

There exists a unique hyperbolic limit cycle if $a_1 a_3 < 0$.

Proof. The method is taken from the paper of Lins, de Melo, and Pugh [17]. Note that the origin is the only critical point. The flow is horizontal on the line $x = 0$ and vertical on the curve $y = a_1 x + a_2 x^2 + a_3 x^3$. It is not difficult to prove that a trajectory starting on the positive (or negative) y-axis will meet the negative (or positive) y-axis. The solution may be divided into three stages:

I. Every limit cycle of system (16.6) must cross both of the lines given by

$$L_1 : x_0 = -\sqrt{-\frac{a_1}{a_3}} \quad \text{and} \quad L_2 : x_1 = \sqrt{-\frac{a_1}{a_3}}.$$

II. System (16.6) has at least one and at most two limit cycles; one of them is hyperbolic.

III. System (16.6) has a unique hyperbolic limit cycle.

Stage I. Consider the Lyapunov function given by

$$V(x, y) = e^{-2a_2 y}\left(y - a_2 x^2 + \frac{1}{2a_2}\right).$$

Now

$$\frac{dV}{dt} = 2a_2 e^{-2a_2 y} x^2 (a_1 + a_3 x^2).$$

The Lyapunov function is symmetric with respect to the y-axis since $V(x, y) = V(-x, y)$, and there is a closed level curve $V(x, y) = C$ that is tangent to both L_1 and L_2. Since $\frac{dV}{dt}$ does not change sign inside the disc $V(x, y) = C$, no limit cycle can intersect the disk, which proves Stage I.

Stage II. Suppose that there are two limit cycles $\gamma_1 \subset \gamma_2$ surrounding the origin as in Figure 16.11.

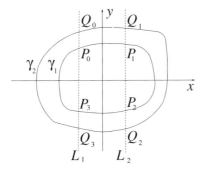

Figure 16.11: Two limit cycles crossing the lines L_1 and L_2.

Suppose that $a_1 < 0$ and $a_3 > 0$. Then the origin is unstable. Let γ_1 be the innermost periodic orbit, which must be attracting on the inside. Therefore,

$$\int_{\gamma_1} \text{div}(\mathbf{X}) dt = \int_{\gamma_1} -(a_1 + 2a_2 x + 3a_3 x^2) \leq 0.$$

16. The Second Part of Hilbert's Sixteenth Problem

Let P_i and Q_i, $i = 0, 1, 2, 3$, be the points of intersection of γ_1 and γ_2, respectively, with the lines L_1 and L_2. Now $\int_{\gamma_1} x \, dt = \int_{\gamma_1} -\frac{dy}{dt} dt = 0$, and similarly for the periodic orbit γ_2.

Consider the branches $P_0 P_1$ and $Q_0 Q_1$ on γ_1 and γ_2, respectively. The flow is never vertical on these branches. Hence one may parameterize the integrals by the variable x. Thus

$$\int_{P_0 P_1} -(a_1 + 3a_3 x^2) dt = \int_{x_0}^{x_1} \frac{-(a_1 + 3a_3 x^2)}{y_{\gamma_1}(x) - F(x)} dx$$

and

$$\int_{Q_0 Q_1} -(a_1 + 3a_3 x^2) dt = \int_{x_0}^{x_1} \frac{-(a_1 + 3a_3 x^2)}{y_{\gamma_2}(x) - F(x)} dx.$$

In the region $x_0 < x < x_1$, the quantity $-(a_1 + 3a_3 x^2) > 0$ and $y_{\gamma_2}(x) - F(x) > y_{\gamma_1}(x) - F(x) > 0$. It follows that

$$\int_{Q_0 Q_1} -(a_1 + 3a_3 x^2) dt < \int_{P_0 P_1} -(a_1 + 3a_3 x^2) dt.$$

Using similar arguments, it is not difficult to show that

$$\int_{Q_2 Q_3} -(a_1 + 3a_3 x^2) dt < \int_{P_2 P_3} -(a_1 + 3a_3 x^2) dt.$$

Consider the branches $P_1 P_2$ and $Q_1 Q_2$ on γ_1 and γ_2, respectively. The flow is never horizontal on these branches. Hence one may parameterize the integrals by the variable y. Thus

$$\int_{P_1 P_2} -(a_1 + 3a_3 x^2) dt = \int_{y_1}^{y_2} \frac{(a_1 + 3a_3 (x_{\gamma_1}(y))^2)}{x_{\gamma_1}} dy$$

and

$$\int_{Q_1 Q_2} -(a_1 + 3a_3 x^2) dt = \int_{y_1}^{y_2} \frac{-(a_1 + 3a_3 (x_{\gamma_2}(y))^2)}{x_{\gamma_2}} dy.$$

In the region $y_1 < y < y_2$, $x_{\gamma_2}(y) > x_{\gamma_1}(y)$. It follows that

$$\int_{Q_1 Q_2} -(a_1 + 3a_3 x^2) dt < \int_{P_1 P_2} -(a_1 + 3a_3 x^2) dt.$$

Using similar arguments, it is not difficult to show that

$$\int_{Q_3 Q_0} -(a_1 + 3a_3 x^2) dt < \int_{P_3 P_0} -(a_1 + 3a_3 x^2) dt.$$

16.3. Global Results for Liénard Systems

Thus adding all of the branches together,

$$\int_{\gamma_2} \text{div}(\mathbf{X})dt < \int_{\gamma_1} \text{div}(\mathbf{X})dt \le 0,$$

which proves Stage II.

Stage III. Since the origin is unstable and $\int_{\gamma_2} \text{div}(\mathbf{X})dt < \int_{\gamma_1} \text{div}(\mathbf{X})dt \le 0$, the limit cycle γ_2 is hyperbolic stable and the limit cycle γ_1 is semistable. By introducing a small perturbation such as $\dot{x} = y - F(x) - \epsilon x$, it is possible to bifurcate a limit cycle from γ_1 that lies between γ_2 and γ_1. Therefore, system (16.6) has at least three limit cycles, which contradicts the result at Stage II. Hence system (16.6) has a unique hyperbolic limit cycle. □

A Liénard System with a Large Parameter. Consider the parameterized cubic Liénard equation given by

$$\ddot{x} + \mu f(x)\dot{x} + g(x) = 0,$$

where $f(x) = -1 + 3x^2$ and $g(x) = x$, which becomes

(16.7) $\qquad\qquad \dot{x} = \mu y - \mu F(x), \qquad \mu\dot{y} = -g(x),$

where $F(x) = \int_0^x f(s)ds = -x + x^3$, in the Liénard plane. Liénard (see [5, Chapter 10]) proved that system (16.7) has a unique limit cycle. Systems containing small parameters were considered in Chapter 15 using Melnikov integrals.

The obvious question then is, what happens when μ is large? Figure 16.12 shows the limit cycle behavior in the Liénard and tx planes, when $\mu = 20$ for system (16.7).

Let $\mu = \frac{1}{\epsilon}$. Then system (16.7) can be written as an equivalent system in the form

(16.8) $\qquad\qquad \epsilon\dot{x} = y - F(x), \qquad \dot{y} = -\epsilon g(x).$

Theorem 5. *Consider system (16.8) and the Jordan curve J shown in Figure 16.13. As $\mu \to \infty$ or, alternatively, $\epsilon \to 0$, the limit cycle tends towards the piecewise analytic Jordan curve J.*

Proof. The method of proof involves the Poincaré–Bendixson theorem from Chapter 10. Thus everything is reduced to the construction of an annular region A that is positively invariant and that contains no critical points. The construction is shown in Figure 16.14.

Note that system (16.8) is symmetric about the y-axis, so we need only consider one half of the plane.

First, consider the outer boundary. The arc 1–2 is a horizontal line, and 2–3 is a vertical line from the graph $y = F(x)$ to the graph $y = F(x) - h$, where h is

16. The Second Part of Hilbert's Sixteenth Problem

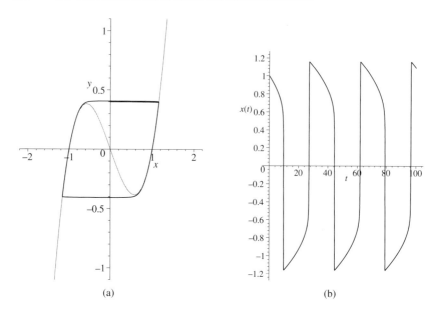

Figure 16.12: (a) A limit cycle for the cubic system when $F(x) = -x + x^3$; the function $y = F(x)$ is also shown. (b) Periodic behavior in the tx plane.

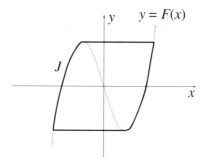

Figure 16.13: The Jordan curve and the function $y = F(x)$.

a small constant. The arc 3–4 follows the $y = F(x) - h$ curve, and the line 4–5 is a tangent.

Now consider the inner boundary. The line 6–7 is sloped below the horizontal, and the line 7–8 is vertical and meets the curve $y = F(x)$. The arc 8–9 follows the curve $y = F(x)$, and the line 9–10 is horizontal.

To prove that the region is positively invariant, one must show that the marked arrows point in the directions indicated in Figure 16.14. Consider each arc separately.

16.3. Global Results for Liénard Systems

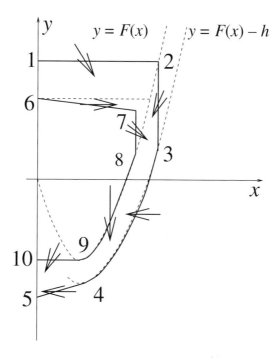

Figure 16.14: Construction of the inner and outer boundaries of the annular region that forms a positively invariant set in one half of the plane. A similar construction is used in the other half of the plane using symmetry arguments.

Notation. For any point n in Figure 16.14, let $F'(n)$ and $g(n)$ be the values of these functions at the abscissa of n.

Arc 1–2. On this line, $\dot{y} < 0$ since $\dot{y} = -x$ and $x > 0$.

Arc 2–3. On this line, $y \leq F(x)$, so $\epsilon \dot{x} = y - F(x) \leq 0$. Note that $\dot{y} < 0$ at point 2.

Arc 3–4. Suppose that p is a point on this arc. The slope of a trajectory crossing this arc is given by

$$\left.\frac{dy}{dx}\right|_p = \frac{-\epsilon^2 g(p)}{-h} < \frac{\epsilon^2 g(3)}{h},$$

and

$$\left.\frac{dy}{dx}\right|_p \to 0$$

16. The Second Part of Hilbert's Sixteenth Problem

as $\epsilon \to 0$. Therefore, for ϵ small enough,

$$\left.\frac{dy}{dx}\right|_p < F'(4) < F'(p)$$

on the arc. Since $\dot{x} < 0$ along the arc, trajectories cross the boundary inwards.

Arc 4–5. Since $|y - F(x)| > h$, the slope of the curve 4–5 is

$$\left.\frac{dy}{dx}\right|_4 < \frac{\epsilon^2 g(4)}{h},$$

which tends to zero as $\epsilon \to 0$. Once more $\dot{x} < 0$ on this arc, for ϵ small enough, and the pointing is inward.

Arc 6–7. Let d_1 be the vertical distance of the line 7–8. For d_1 small enough, along the line 6–7, $|y - F(x)| > d_1$. Thus the slope of the curve at a point say, q, on the line 6–7 is given by

$$\left.\frac{dy}{dx}\right|_q < \frac{\epsilon^2 g(q)}{d_1} < \frac{\epsilon^2 g(7)}{d_1},$$

which tends to zero as $\epsilon \to 0$. Since $\dot{x} > 0$ on this arc, for ϵ small enough, the pointing will be as indicated in Figure 16.14.

Arc 7–8. On this line, $y - F(x) > 0$, so $\dot{x} > 0$.

Arc 8–9. On the curve, $y = F(x)$ with $x > 0$, $\dot{y} < 0$, and $\dot{x} = 0$.

Arc 9–10. On this line, $y - F(x) < 0$ and $\dot{y} < 0$.

Using similar arguments on the left-hand side of the y-axis, a positively invariant annulus can be constructed. Since system (16.8) has a unique critical point at the origin, the Poincaré–Bendixson theorem can be applied to prove that there is a limit cycle in the annular region A. For suitably small values of h and d_1, the annular region will be arbitrarily near the Jordan curve J. Therefore, if $\Gamma(\epsilon)$ is the limit cycle, then $\Gamma(\epsilon) \to J$ as $\epsilon \to 0$. □

16.4 Local Results for Liénard Systems

Although the Liénard equation (16.5) appears simple enough, the known global results on the maximum number of limit cycles are scant. By contrast, if the analysis is restricted to local bifurcations, then many more results may be obtained. The method for bifurcating small-amplitude limit cycles is given in Chapter 15. Consider the Liénard system

(16.9) $$\dot{x} = y, \qquad \dot{y} = -g(x) - f(x)y,$$

16.4. Local Results for Liénard Systems

where $f(x) = a_0 + a_1 x + a_2 x^2 + \cdots + a_m x^m$ and $g(x) = x + b_2 x^2 + b_3 x^3 + \cdots + b_n x^n$; m and n are natural numbers. Let $\hat{H}(m, n)$ denote the maximum number of small-amplitude limit cycles that can be bifurcated from the origin for system (16.9), where m is the degree of f and n is the degree of g.

In 1984, Blows and Lloyd [15] proved the following results for system (16.9):

- If $\partial f = m = 2i$ or $2i + 1$, then $\hat{H}(m, 1) = i$.

- If g is odd and $\partial f = m = 2i$ or $2i + 1$, then $\hat{H}(m, n) = i$.

Note that the first result seems to support the conjecture of Lins, de Melo, and Pugh [17] for global limit cycles. In addition to the above the author has proved the following results by induction.

- If $\partial g = n = 2j$ or $2j + 1$, then $\hat{H}(1, n) = j$.

- If f is even, $\partial f = 2i$, then $\hat{H}(2i, n) = i$.

- If f is odd, $\partial f = 2i + 1$ and $\partial g = n = 2j + 2$ or $2j + 3$; then $\hat{H}(2i+1, n) = i + j$.

- If $\partial f = 2$, $g(x) = x + g_e(x)$, where g_e is even and $\partial g = 2j$; then $\hat{H}(2, 2j) = j$.

Christopher and the author [2] have more recently developed a new algebraic method for determining the Lyapunov quantities, and this has allowed further computations. Let $\lfloor . \rfloor$ denote the integer part. Then the new results are listed below:

- $\hat{H}(2, n) = \lfloor \frac{2n+1}{3} \rfloor$.

- $\hat{H}(m, 2) = \lfloor \frac{2m+1}{3} \rfloor$.

- $\hat{H}(3, n) = 2 \lfloor \frac{3n+6}{8} \rfloor$ for all $1 < n \leq 50$.

- $\hat{H}(m, 3) = 2 \lfloor \frac{3m+6}{8} \rfloor$ for all $1 < m \leq 50$.

Complementing these results is the calculation of $\hat{H}(m, n)$ for specific values of m and n. The results are presented in Table 16.1.

The ultimate aim is to establish a general formula for $\hat{H}(m, n)$ as a function of the degrees of f and g. Future work will concentrate on simultaneous bifurcations, when there is more than one fine focus and determining a relationship, if any, between global and local results.

16. The Second Part of Hilbert's Sixteenth Problem

Table 16.1: The values of $\hat{H}(m, n)$ for varying values of m and n.

degree of f																		
50	↑	↑	38															
49	24	33	38															
48	24	32	36															
⋮	⋮	⋮	⋮															
13	6	9	10															
12	6	8	10															
11	5	7	8															
10	5	7	8															
9	4	6	8	9														
8	4	5	6	9														
7	3	5	6	8														
6	3	4	6	7														
5	2	3	4	6	6													
4	2	3	4	4	6	7	8	9	9									
3	1	2	2	4	4	6	6	6	8	8	8	10	10	⋯	36	38	38	
2	1	1	2	3	3	4	5	5	6	7	7	8	9	⋯	32	33	→	
1	0	1	1	2	2	3	3	4	4	5	5	6	6	⋯	24	24	→	
	1	2	3	4	5	6	7	8	9	10	11	12	13	⋯	48	49	50	

degree of g

16.5 Exercises

1. Draw a global phase portrait for the linear system
$$\dot{x} = y, \qquad \dot{x} = -4x - 5y$$
including the flow at infinity.

2. Draw a global phase portrait for the system
$$\dot{x} = -3x + 4y, \qquad \dot{y} = -2x + 3y$$
and give the equations defining the flow near critical points at infinity.

3. Determine a global phase portrait for the quadratic system given by
$$\dot{x} = x^2 + y^2 - 1, \qquad \dot{y} = 5xy - 5.$$

4. Draw a global phase portrait for the Liénard system
$$\dot{x} = y - x^3 - x, \qquad \dot{y} = -y.$$

5. Draw a global phase portrait for the Liénard system
$$\dot{x} = y - x^3 + x, \qquad \dot{y} = -y.$$

6. Use the M-files in Chapters 8 and 10 to compare the limit cycles for Liénard systems in the phase plane and in the Liénard plane. Plot the periodic orbits in xt as well.

7. Use the M-files in Chapter 8 to investigate the system

$$\dot{x} = y - (a_1 x + a_2 x^2 + a_3 x^3), \qquad \dot{y} = -x$$

for varying values of the parameters a_1, a_2, and a_3.

8. Use MATLAB to investigate the limit cycles, if they exist, of the system

$$\dot{x} = y - \epsilon(a_1 x + a_2 x^2 + \cdots + a_M x^M), \qquad \dot{y} = -x,$$

as the parameter ϵ varies from zero to infinity. When ϵ is large the stepsize has to be small, in fact, it would be better to use a variable step length in the numerical solver.

9. Prove Liénard's theorem, that when $\partial g = 1$ and $F(x)$ is a continuous odd function that has a unique root at $x = a$ and is monotone increasing for $x \geq a$, (16.5) has a unique limit cycle.

10. This is a quite difficult question. Consider the Liénard system

(16.10) $\qquad \dot{x} = y - F(x), \qquad \dot{y} = -x,$

where $F(x) = (a_1 x + a_3 x^3 + x^5)$ is odd. Prove that system (16.10) has at most two limit cycles.

Recommended Reading

[1] Y. Il'yashenko, Centennial history of Hilbert's 16th problem, *Bull. Amer. Math. Soc.*, **39**-3 (2002), 301–354.

[2] C. J. Christopher and S. Lynch, Small-amplitude limit cycle bifurcations for Liénard systems with quadratic or cubic damping or restoring forces, *Nonlinearity*, **12** (1999), 1099–1112.

[3] H. Giacomini and S. Neukirch, Improving a method for the study of limit cycles of the Liénard equation, *Phys. Rev. E*, **57** (1998), 6573–6576.

[4] F. Dumortier and L. Chengzhi, Quadratic Liénard equations with quadratic damping, *J. Differential Equations*, **139** (1997), 41–59.

[5] F. Dumortier and L. Chengzhi, On the uniqueness of limit cycles surrounding one or more singularities for Liénard equations, *Nonlinearity*, **9** (1996), 1489–1500.

[6] C. J. Christopher and N. G. Lloyd, Polynomial systems: A lower bound for the Hilbert numbers, *Proc. Roy. Soc. London Ser. A*, **450** (1995), 219–224.

[7] H. Zoladek, Eleven small limit cycles in a cubic vector field, *Nonlinearity*, **8** (1995), 843–860.

[8] T. R. Blows and C. Rousseau, Bifurcation at infinity in polynomial vector fields, *J. Differential Equations*, **104** (1993), 215–242.

[9] Zhang Zhifen, Ding Tongren, Huang Wenzao, and Dong Zhenxi, *Qualitative Theory of Differential Equations*, Translation of Mathematical Monographs 102, American Mathematical Society, Providence, 1992.

[10] H. N. Moreira, Liénard-type equations and the epidemiology of malaria, *Ecol. Model.*, **60** (1992), 139–150.

[11] Yu. Il'yashenko, *Finiteness Theorems for Limit Cycles*, Translations of Mathematical Monographs 94, American Mathematical Society, Providence, 1991.

[12] W. A. Coppel, Some quadratic systems with at most one limit cycle, in U. Kirchgraber and H. O. Walther, eds., *Dynamics Reported*, Vol. 2, Wiley/Teubner, New York/Stuttgart, 1988, 61–68.

[13] J. Ecalle, J. Martinet, J. Moussu, and J. P. Ramis, Non-accumulation des cycles-limites I, *C. R. Acad. Sci. Paris Sér. I Math.*, **304** (1987), 375–377.

[14] Li Jibin and Li Chunfu, Global bifurcation of planar disturbed Hamiltonian systems and distributions of limit cycles of cubic systems, *Acta. Math. Sinica*, **28** (1985), 509–521.

[15] T. R. Blows and N. G. Lloyd, The number of small-amplitude limit cycles of Liénard equations, *Math. Proc. Cambridge Philos. Soc.*, **95** (1984), 359–366.

[16] Shi Songling, A concrete example of the existence of four limit cycles for plane quadratic systems, *Sci. Sinica* A, **23** (1980), 153–158.

[17] A. Lins, W. de Melo, and C. Pugh, On Liénards equation with linear damping, in J. Palis and M. do Carno, eds., *Geometry and Topology* III, Lecture Notes in Mathematics 597, Springer-Verlag, Berlin, 1977, 335–357.

[18] L. A. Cherkas, Conditions for a Liénard equation to have a center, *Differentsial'nye Uravneniya*, **12** (1976), 201–206.

[19] G. S. Rychkov, The maximum number of limit cycles of the system $\dot{x} = y - a_0 x - a_1 x^3 - a_2 x^5$, $\dot{y} = -x$ is two, *Differentsial'nye Uravneniya*, **11** (1973), 380–391.

[20] K. S. Sibirskii, The number of limit cycles in the neighbourhood of a critical point, *Differential Equations*, **1** (1965), 36–47.

[21] N. Bautin, On the number of limit cycles which appear with the variation of the coefficients from an equilibrium point of focus or center type, *Amer. Math. Soc. Trans.*, **5** (1962), 396–414.

17
Neural Networks

Aims and Objectives

- To provide a brief historical background to neural networks.
- To investigate simple neural network architectures.
- To consider applications in the real world.
- To present working MATLAB program files for some neural networks.
- To introduce neurodynamics.

On completion of this chapter the reader should be able to

- use the generalized delta learning rule with backpropagation of errors to train a network;
- determine the stability of Hopfield networks using a suitable Lyapunov function;
- use the Hopfield network as an associative memory;
- study the dynamics of a neuromodule in terms of bistability, chaos, periodicity, and chaos control.

17. Neural Networks

Neural networks are being used to solve all kinds of problems from a wide range of disciplines. Some neural networks work better than others on specific problems and the models are run using continuous, discrete, and stochastic methods. For more information on stochastic methods, the reader is directed to the textbooks at the end of this chapter. The topic is highly interdisciplinary in nature, and so it is extremely difficult to develop an introductory and comprehensive treatise on the subject in one short chapter of a textbook. A brief historical introduction is given in Section 17.1 and the fundamentals are reviewed. Real-world applications are then discussed. The author has decided to concentrate on three types of neural network—the feedforward multilayer network and backpropagation of errors using the generalized delta learning rule, the recurrent Hopfield neural network, and the minimal chaotic neuromodule. The first network is probably the most widely used in applications in the real world; the second is a much studied network in terms of stability and Lyapunov functions; and the third provides a useful introduction to neurodynamics.

For a more detailed historical introduction and review of the theory of neural networks, the reader is once more directed to the textbooks in the reference section of this chapter.

Some of the MATLAB program files listed in Section 17.5 are quite long. Remember that you can download the M-files from the MATLAB Central File Exchange at

$$\text{http://www.mathworks.com/matlabcentral/.}$$

17.1 Introduction

This textbook has thus far been concerned with deterministic dynamical systems where the underlying equations are known. This chapter provides a means of tackling nondeterministic systems, where the equations used to model the system are not known. Unfortunately, many real-world problems do not come prepackaged with mathematical equations, and often the equations derived might not be accurate or suitable. Throughout history, scientists have attempted to model physical systems using mathematical equations. This has been quite successful in some scientific fields, but not in all. For example, what equations would a doctor use to diagnose an illness and then prescribe a treatment? How does a bank manager determine whether to issue a mortgage? How can we tell whether somebody is telling the truth? These questions have been successfully dealt with by the adoption of *neural networks*, or *artificial neural networks*, as they are sometimes referred to, using machine learning or data mining. Applications of this theory will be dealt with in more detail at the end of this section.

Definition 1. A neural network is a parallel information-processing system that has certain characteristics in common with certain brain functions. It is composed of *neurons* and *synaptic weights* and performs complex computations through a *learning process*.

The brain is a highly complex nonlinear information-processing system. It is a parallel computer, infinitely more powerful than traditional, electronic, sequential, logic-based digital computers and powerful parallel and vector computers on the market today. The average human brain consists of some 10^{11} neurons, each about 100μm in size, and approximately 10^{14} synapses. The synapses, or dendrites, are mainly chemical, converting electrical signals into chemical signals and back to electrical again. The synapses connecting neurons store acquired knowledge and can be excitory or inhibitory. It should be pointed out that the numbers of neurons and synaptic weights do not remain constant in the human brain. Scientists are attempting to incorporate some features of the way the brain works into modern computing.

Network Architecture. The neuronal model is made up of four basic components: an input vector, a set of synaptic weights, a summing junction with an *activation*, or *(transfer), function*, and an output. The *bias* increases or decreases the net input of the activation function. Synapses receive input signals that they send to the neural cell body; the soma (summing junction) sums these signals; and the axon transmits the signal to synapses of connecting neurons. A schematic illustrating a simple mathematical model of a neuron is shown in Figure 17.1.

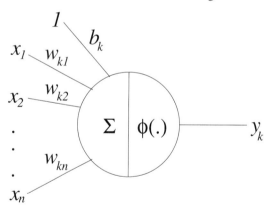

Figure 17.1: A simple nonlinear model of a single neuron k. The vector $\mathbf{x} = (x_1, x_2, \ldots, x_n)^T$ represents the input; the synaptic weights are denoted by $\mathbf{w}_k = w_{kj}$, $j = 1, 2, \ldots, n$; b_k is the bias; $\phi(.)$ is the activation function applied after a summation of the products of weights with inputs; and y_k is the output of neuron k.

The neuron has bias b_k, which is added to the summation of the products of weights with inputs to give

$$v_k = \mathbf{w}_k \mathbf{x} + b_k,$$

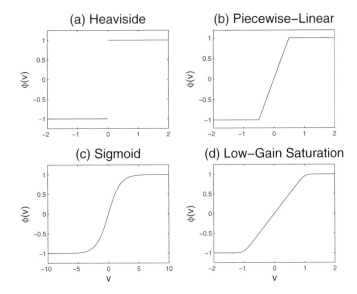

Figure 17.2: Some activation functions: (a) a Heaviside function; (b) a piecewise linear function; (c) a sigmoid function; (d) a low-gain saturation function.

where v_k is the *activation potential*. The neuron output is written as

$$y_k = \phi(v_k).$$

Note in this case that $\mathbf{w}_k$ is a vector. The activation function $\phi(.)$ typically ranges from -1 to $+1$ (is *bipolar*) in applications, and has an antisymmetric form with respect to the origin. This textbook will be concerned mainly with bipolar activation functions. There are *unipolar* activation functions, where the function ranges from 0 to $+1$, but bipolar functions are predominantly used in applications. Some bipolar activation functions are shown in Figure 17.2. They are defined by the following equations:

(a) $$\phi(v) = \begin{cases} 1, & v \geq 0, \\ -1, & v < 0; \end{cases}$$

(b) $$\phi(v) = \begin{cases} 1, & v \geq 0.5, \\ v, & -0.5 < v < 0.5, \\ -1, & v \leq -0.5; \end{cases}$$

(c) $$\phi(v) = \tanh(av);$$

(d) $$\phi(v) = \frac{1}{2a} \log \frac{\cosh(a(v+1))}{\cosh(a(v-1))}.$$

17.1. Introduction

The all-or-none law model of a neuron devised by McCulloch and Pitts [23] in the early 1940s is widely acknowledged as the origin of the modern theory of neural networks. They showed, in principle, that the neuron could compute any arithmetic or logical function. Indeed, even today, the McCulloch–Pitts neuron is the one most widely used as a logic circuit. In 1949 Hebb [22] proposed the first learning law for neural networks used to modify synaptic weights. He suggested that the strength of the synapse connecting two simultaneously active neurons should be increased. There are many variations of Hebb's learning law, and they are being applied to a variety of neural network architectures; see Section 17.3, for example. In 1958 Rosenblatt [21] introduced a class of neural network called the *perceptron*. A typical architecture is shown in Figure 17.3. It was found that the perceptron learning rule was more powerful than the Hebb rule. Unfortunately, shortly afterwards it was shown that the basic perceptron could only solve problems that were linearly separable. One simple example of a problem that is not linearly separable is the exclusive or (XOR) gate. An XOR gate is a circuit in a computer that fires only if one of its inputs fire.

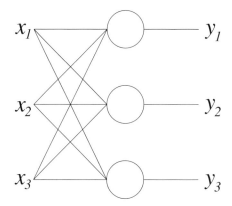

Figure 17.3: A feedforward single layer network.

Training. In 1960, Widrow and Hoff [20] introduced the ADALINE (ADAptive LInear NEuron) network, and a learning rule labeled as the *delta learning rule* or the *least mean squared* (LMS) algorithm. The perceptron learning rule adjusts synaptic weights whenever the response is incorrect, whereas the delta learning rule adjusts synaptic weights to reduce the error between the output vector and the target vector. This led to an improved ability of the network to generalize. Neither the ADALINE nor the perceptron were able to solve problems that were not linearly separable, as reported in the widely publicized book of Minsky and Papert [19]. Rumelhart and McClelland [14] edited a book that brought together the work of several researchers on backpropagation of errors using multilayer feedforward networks with hidden layers (see Figure 17.4). This algorithm partially addressed

17. Neural Networks

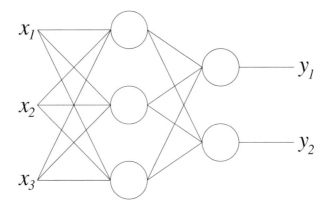

Figure 17.4: A feedforward neural network with one hidden layer; there are three neurons in the hidden layer and two in the output layer.

the problems raised by Minsky and Papert in the 1960s. Nowadays, over 90% of the applications to real-world problems use the backpropagation algorithm with *supervised learning*. Supervised learning is achieved by presenting a sequence of training vectors to the network, each with a corresponding known target vector. A complete set of input vectors with known targets is known as an *epoch*; it is usually loaded as a data file. A backpropagation algorithm using a supervised generalized delta learning rule is discussed in more detail in Section 17.2. Throughout the 1980s, Kohonen [18] developed self-organizing feature maps to form clusters for *unsupervised learning*. No target vectors are required for this algorithm—similar input vectors are assigned the same output cluster.

The seminal paper of Hopfield [17] published in 1982 used statistical mechanics to explain the operation of a recurrent neural network used as an associative memory. The architecture of a recurrent Hopfield neural network comprising three neurons is shown in Figure 17.5. The main difference between a feedforward network and a recurrent network is that there is feedback in the latter case. Figure 17.5 illustrates the multiple-loop feedback for a three-neuron module. Note that the output of each neuron is fed back to each of the other neurons in the network.

The network operation can be analyzed using Lyapunov functions. Both continuous and discrete recurrent Hopfield networks are discussed in more detail in Section 17.3.

Applications. The field of neural networks has generated a phenomenal amount of interest from a broad range of scientific disciplines. One of the reasons for this is adaptability. Innovative architectures and new training rules have been tested on powerful computers, and it is difficult to predict where this research will take us in the future. As mentioned earlier, the vast majority of real-world applications have relied on the backpropagation algorithm for training multilayer networks, and

17.1. Introduction

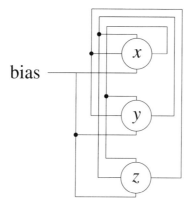

Figure 17.5: A recurrent Hopfield neural network with feedback. Note that there is no self-feedback in this case.

recently *kernel machines* have proved to be useful for a wide range of applications, including document classification and gene analysis, for example. In general, more than one network is required and each network is designed to perform a specific task. Some well-known applications are listed and a more in-depth account is given for the research carried out on psychological profiling in the Department of Computing and Mathematics at Manchester Metropolitan University. See the Web pages at

http://www.doc.mmu.ac.uk/RESEARCH/Intelgrp/

for more information. The list is by no means exhaustive and it will not be difficult for the reader to find examples applied in their own research area.

Neural networks are being used extensively in the fields of aeronautics, banking, defense, engineering, finance, insurance, marketing, manufacturing, medicine, robotics, psychology, security, and telecommunications. One of the early applications was in signal processing; the ADALINE was used to suppress noise on a telephone line. Many neural networks are being used as associative memories for pattern and speech production and recognition, for example. Simple networks can be set up as instant physicians. The expertise of many general practitioners can be used to train a network using symptoms to diagnose an illness and even suggest a possible treatment. In engineering, neural networks are being used extensively as controllers, and in banking they are being used in mortgage assessment. Scientists find them very useful as function approximators. They can test whether the mathematical equations (which could have been used for many years) used to model a system are correct.

The Artificial Intelligence Group at Manchester Metropolitan University has developed a machine for automatic psychological profiling. The work has generated a huge amount of interest and recently was reported on national television in many countries around the world. Bandar et al. [3] have patented the machine, and the expectations are high for future applications. The machine could be used in police questionning, at airport customs, and by doctors diagnosing schizophrenia, depression, and stress. A short article on using the machine as a lie detector has recently appeared in *New Scientist* [2]. The group claims that the lie detector is accurate in 80% of test cases. Their machine uses about 20 independent neural networks; each one using the generalized delta learning rule and backpropagation of errors. Some of the channels used in the machine include eye gaze, blinking, head movement forward, hand movement, and blushing.

The same group has also carried out extensive work on conversational agents. It will not be long before we are all able to have conversations with our computers.

This introductory section has given a brief overview of neural networks. For more detailed information the reader is directed to the textbooks [5]–[8], [10], and [12] listed in the reference section of this chapter.

17.2 The Delta Learning Rule and Backpropagation

Widrow and Hoff [20] generalized the perceptron training algorithm to continuous inputs and outputs and presented the delta rule (or LMS rule). Consider a single neuron as in Figure 17.1. If the activation function is linear, then

$$y_k = \sum_j w_{kj} x_j + b_k.$$

Define an *error function* by the mean squared error, so

$$E = \frac{1}{2N} \sum_{\mathbf{x}} (E_k^{\mathbf{x}})^2 = \frac{1}{2N} \sum_{\mathbf{x}} (t_k - y_k)^2,$$

where the index $\mathbf{x}$ ranges over all input vectors, N is the number of neurons, $E^{\mathbf{x}}$ is the error on vector $\mathbf{x}$, and t_k is the target (or desired) output when vector $\mathbf{x}$ is presented. The aim is to minimize the error function E with respect to the weights w_{kj}. It is an unconstrained optimization problem; parameters w_{kj} are sought to minimize the error. The famous *method of steepest descent* is applied to the error function. Theorem 1 gives the delta rule when the activation function is linear. There are two ways to update the synaptic weights using the generalized delta rule. One is instantaneously (a weight is updated on each iteration) and the other is batch (where the weights are updated based on the average error for one epoch).

Theorem 1. *The iterative method of steepest descent for adjusting the weights in a neural network with a linear activation function is given by*

$$w_{kj}(n+1) = w_{kj}(n) - \eta g_{kj},$$

17.2. The Delta Learning Rule and Backpropagation

where n is the number of iterates, $g = -(t_k - y_k)x_j$ *is the gradient vector, and* η *is a small positive constant called the* learning rate.

Proof. Partially differentiating the error with respect to the weight vector gives

$$\frac{\partial E(w_{kj})}{\partial w_{kj}} = \frac{\partial E}{\partial E_k^x} \frac{\partial E_k^x}{\partial y_k} \frac{\partial y_k}{\partial w_{kj}}.$$

Now

$$\frac{\partial E}{\partial E_k^x} = E_k^x = (t_k - y_k),$$

and

$$\frac{\partial E_k^x}{\partial y_k} = -1,$$

and

$$\frac{\partial y_k}{\partial w_{kj}} = x_j.$$

An estimate for the gradient vector is

$$g_{kj} = (y_k - t_k)x_j.$$

The delta rule for a linear activation function is thus formulated as

$$w_{kj}(n+1) = w_{kj}(n) - \eta g_{kj},$$

where η is the learning rate parameter. The choice of η is important in applications. If it is too large the algorithm can become unstable. One normally experiments with η; it is not desirable for the algorithm to converge too slowly. $\square$

Note that there are other optimization methods available, such as Newton's method and the Gauss–Newton method, which converge quicker and are less sensitive to the choice of η.

Theorem 2. *When the activation function is nonlinear, say,* $y_k = \phi(v_k)$, *the generalized delta rule can be formulated as*

(17.1) $$w_{kj}(n+1) = w_{kj}(n) - \eta g_{kj},$$

where

(17.2) $$g_{kj} = (y_k - t_k)\frac{\partial \phi}{\partial v_k}x_j.$$

Proof. The proof will be left as an exercise for the reader in Section 17.6. $\square$

17. Neural Networks

Figure 17.6: An output neuron k connected to a hidden neuron j.

The Backpropagation Algorithm. If neuron k is an output neuron, then Theorem 2 can be applied to adjust the weights of the synapses. However, if neuron j is a hidden neuron in a layer below neuron k, as depicted in Figure 17.6, then a new algorithm is required.

Theorem 3. *When neuron j is in a hidden layer, the error backpropagation rule is formulated as*

(17.3) $$w_{ji}(n+1) = w_{ji}(n) - \eta g_{ji},$$

where

(17.4) $$g_{ji} = \sum_{k}\left((y_k - t_k)\frac{\partial \phi}{\partial v_k}w_{kj}\right)\frac{\partial \phi}{\partial v_j}u_i.$$

Proof. The proof is left as an exercise for the reader. The error is backpropagated through the network, layer by layer—back to the input layer, using gradient descent. □

The generalized delta rule and backpropagation will now be applied to examples for estimating the value of owner-occupied homes in Boston, Massachusetts, in the 1970s.

The Boston housing data was downloaded from the UCI Machine Learning Repository on the Web at

http://www.ics.uci.edu/~mlearn/MLRepository.html.

Other databases at the site include arrhythmia data, automobile miles per gallon data, breast cancer data, and credit screening data.

The Boston housing data was created by D. Harrison and D. L. Rubinfeld (Hedonic prices and the demand for clean air, *J. Environmental Econ. Management*, **5** (1978), 81–102). They reported on housing values in the suburbs of Boston. There are 506 input vectors and 14 attributes including per capita crime rate by town, average number of rooms per dwelling, and pupil–teacher ratio by town. The data is saved as housing.txt in the files that can be downloaded at the Mathworks Web site.

Example 1. Write a MATLAB program to apply the generalized delta learning rule to the Boston housing data for three attributes: columns six (average number of

17.2. The Delta Learning Rule and Backpropagation

rooms), nine (index of accessibility to radial highways), and 13 (percentage lower status of population), using the target data presented in column 14 (median value of owner-occupied homes in thousands of dollars). Use the activation function $\phi(v) = \tanh(v)$ and show how the weights are adjusted as the number of iterations increases. This is a simple three-neuron feedforward network; there are no hidden layers; and there is only one output (see Figure 17.1).

Solution. The MATLAB program file is listed as Programs 17a in Section 17.5. A summary of the algorithm is listed below to aid in understanding the program:

1. Scale the data to zero mean, unit variance, and introduce a bias on the input.

2. Set small random weights.

3. Set the learning rate, say, η, and the number of epochs.

4. Calculate model outputs y_k, the error $t_k - y_k$, the gradients g, and perform the gradient descent to evaluate $w_{kj}(n+1) = w_{kj}(n) - \eta g_{kj}$ for each weight (see (17.1)).

5. Plot a graph of weight values versus number of iterations.

Note that $\phi'(v) = 1 - (\phi(v))^2$, since $\phi(v) = \tanh(v)$. The reader will be asked to verify this in the exercises. The synaptic weights converge to the following approximate values: $b_1 \approx -0.27$, $w_{11} \approx 0.2$, $w_{12} \approx -0.04$, and $w_{13} \approx -0.24$, as shown in Figure 17.7.

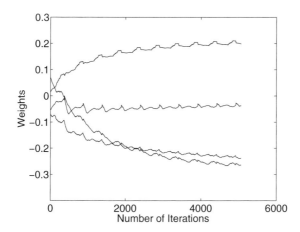

Figure 17.7: Updates of the four weights (including the bias) against the number of iterations.

Example 2. Use the generalized delta rule with batch backpropagation of errors on the full data set listed in housing.txt for the Boston house data. Use the same activation function as in Example 1 and introduce one hidden layer in the neural network. Compare performance for one and two neurons in the hidden layer, when $\eta = 0.05$. One epoch consists of 506 input vectors, each with one target, and there are 13 input vectors.

Solution. The MATLAB program file is listed as Programs 17b in Section 17.5. A summary of the algorithm is listed below to aid in understanding the program:

1. Scale the data to zero mean, unit variance, and introduce a bias on the input.
2. Iterate over the number of neurons in the hidden layer.
3. Set random weights for the hidden and output layers.
4. Iterate over a number of epochs using batch error backpropagation.
    (a) Compute model outputs and the error.
    (b) Compute output and hidden gradients and perform gradient descent.
    (c) Determine the mean squared error for each epoch.
5. Plot a graph of mean squared error versus the number of epochs for each number of neurons in the hidden layer.

Note that it is possible to work with any number of hidden layers, but in general one hidden layer suffices. Indeed, it has been shown that one hidden layer is sufficient to approximate any continuous function. Often the functionality that comes from extra hidden layers causes the network to overfit. The results on the full data set are shown in Figure 17.8.

17.3 The Hopfield Network and Lyapunov Stability

This section is concerned with recurrent neural networks that have fixed synaptic weights but where the activation values undergo relaxation processes through feedback. A primary application of the Hopfield network is as an associative memory, where the network is used to store patterns for future retrieval. The synaptic weights are set such that the stable points of the system correspond with the input patterns to be stored. One can think of these states as local minima in energy space. When a noisy or incomplete test pattern is input, the system should settle onto a stable state that corresponds to a stored pattern. A discrete Hopfield network is discussed in some detail later in this section, where it is used as an associative memory on some patterns. It should be noted that another famous problem addressed by Hopfield and Tank [15] was in optimisation and is known as the traveling salesman problem. Simple continuous Hopfield networks are considered before the applications in order to highlight stability properties using Lyapunov functions.

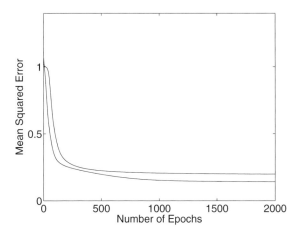

Figure 17.8: Number of epochs versus mean squared error for the Boston housing data. The upper curve is the error with one hidden neuron (settles to approximately 0.2); the lower curve is the error with two hidden neurons (stabilizes to approximately 0.14). The learning rate used in this case was $\eta = 0.05$.

The Continuous Hopfield Model. A Hopfield network does not require training data with targets. A network consisting of three neurons is shown in Figure 17.5, and a two-neuron module is shown in Figure 17.6. In 1984, Hopfield [16] showed how an analog electrical circuit could behave as a small network of neurons with graded response. He derived a Lyapunov function for the network to check for stability and used it as a content-addressable memory. The differential equations derived by Hopfield for the electrical circuit using Kirchhoff's laws could be reduced to the following system of differential equations

$$(17.5) \qquad \frac{d}{dt}\mathbf{x}(t) = -\mathbf{x}(t) + \mathbf{W}\mathbf{a}(t) + \mathbf{b},$$

where $\mathbf{x}(t)$ is a vector of neuron activation levels, $\mathbf{W}$ is the weight matrix representing synaptic connections, $\mathbf{b}$ are the biases, and $\mathbf{a}(t) = \phi(\mathbf{x}(t))$ are the nonlinear input/output activation levels. Hopfield derived the following theorem for stability properties.

Theorem 4. *A Lyapunov function for the n-neuron Hopfield network defined by equation (17.5) is given by*

$$(17.6) \qquad \mathbf{V}(\mathbf{a}) = -\frac{1}{2}\mathbf{a}^T\mathbf{W}\mathbf{a} + \sum_{i=1}^{n}\left(\int_0^{a_i}\phi^{-1}(u)du\right) - \mathbf{b}^T\mathbf{a}$$

as long as

17. Neural Networks

1. $\phi^{-1}(a_i)$ is an increasing function, that is,

$$\frac{d}{da_i}\phi^{-1}(a_i) > 0, \quad \text{and}$$

2. the weight matrix **W** is symmetric.

Proof. The proof is left as an exercise for the reader (see Section 17.6). □

Consider the following two-neuron module taken from Hopfield's original paper [16].

Example 3. A schematic of the two-neuron module is shown in Figure 17.9. The differential equations used in Hopfield's model are given by

$$\dot{x} = -x + \frac{2}{\pi}\tan^{-1}\left(\frac{\gamma\pi y}{2}\right), \quad \dot{y} = -y + \frac{2}{\pi}\tan^{-1}\left(\frac{\gamma\pi x}{2}\right),$$

where the activation functions are arctan. Determine the stable critical points and derive a Lyapunov function.

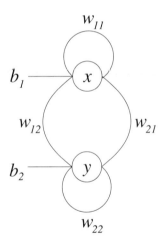

Figure 17.9: A simple recurrent Hopfield neural network, a two-neuron module.

Solution. In this case,

$$\mathbf{W} = \begin{pmatrix} 0 & 1 \\ 1 & 0 \end{pmatrix}, \quad \mathbf{b} = \begin{pmatrix} 0 \\ 0 \end{pmatrix},$$

$$a_1 = \frac{2}{\pi}\tan^{-1}\left(\frac{\gamma\pi x}{2}\right), \quad a_2 = \frac{2}{\pi}\tan^{-1}\left(\frac{\gamma\pi y}{2}\right).$$

17.3. The Hopfield Network and Lyapunov Stability

A Lyapunov function, derived using equation (17.6), is given by

$$V(\mathbf{a}) = -\frac{1}{2}(a_1 \; a_2)\begin{pmatrix} 0 & 1 \\ 1 & 0 \end{pmatrix}\begin{pmatrix} a_1 \\ a_2 \end{pmatrix} + \int_0^{a_1} \phi^{-1}(u)du +$$

$$\int_0^{a_2} \phi^{-1}(u)du - (0 \; 0)\begin{pmatrix} a_1 \\ a_2 \end{pmatrix}.$$

Therefore,

$$V(\mathbf{a}) = -a_1 a_2 - \frac{4}{\gamma \pi^2}(\log(\cos(\pi a_1/2)) + \log(\cos(\pi a_2/2))).$$

Vector field plots for the differential equations are shown in Figure 17.10. The corresponding Lyapunov functions can be plotted using MATLAB when γ is given. Plot the surface for $|a_i| \leq 1, i = 1, 2$.

When $0 < \gamma \leq 1$, there is one stable critical point at the origin (see Figure 17.10(a)). As γ passes through one, two stable critical points bifurcate from the origin and the critical point at the origin becomes unstable (see Figure 17.10(b)). As $\gamma \to \infty$, the stable critical points approach corners of the unit square as depicted in Figure 17.10(c).

Example 4. Consider the recurrent Hopfield network modeled using the differential equations

$$\dot{x} = -x + 2\left(\frac{2}{\pi}\tan^{-1}\left(\frac{\gamma \pi x}{2}\right)\right), \quad \dot{y} = -y + 2\left(\frac{2}{\pi}\tan^{-1}\left(\frac{\gamma \pi y}{2}\right)\right).$$

Plot a vector field portrait and derive a suitable Lyapunov function.

Solution. In this case,

$$\mathbf{W} = \begin{pmatrix} 2 & 0 \\ 0 & 2 \end{pmatrix} \quad \text{and} \quad \mathbf{b} = \begin{pmatrix} 0 \\ 0 \end{pmatrix}.$$

A vector field plot is shown in Figure 17.11. There are four stable critical points and five unstable critical points.

A Lyapunov function is given by

$$V(\mathbf{a}) = -(a_1^2 + a_2^2) - \frac{4}{\gamma \pi^2}(\log(\cos(\pi a_1/2)) + \log(\cos(\pi a_2/2))).$$

You can plot the Lyapunov function using MATLAB and rotate it in the figure window.

Continuous Hopfield networks with self-feedback loops can be Lyapunov stable. However, discrete systems must have no self-feedback to guarantee Lyapunov stability.

374   **17. Neural Networks**

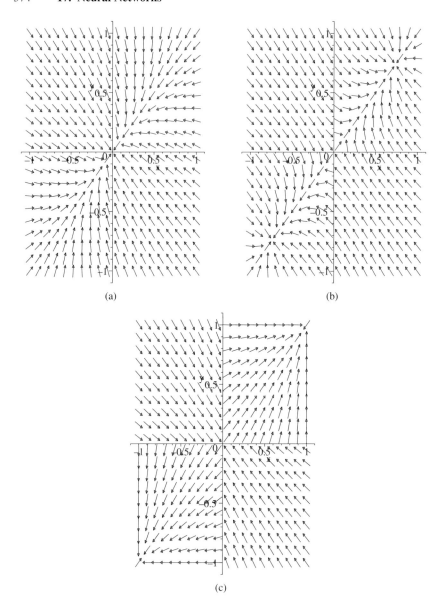

Figure 17.10: Vector field plots when (a) $0 < \gamma \leq 1$, (b) $\gamma > 1$, and (c) $\gamma \to \infty$.

The Discrete Hopfield Model. Hopfield [15], [16], [17] used his network as a content-addressable memory using fixed points as attractors for certain fundamental memories. The Hopfield model can be summarized using the following four-step algorithm. There is no self-feedback in this case.

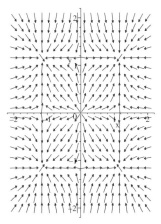

Figure 17.11: A vector field plot for Example 4 when $\gamma = 0.7$. There are nine critical points.

1. **Hebb's postulate of learning.** Let $\mathbf{x}_1, \mathbf{x}_2, \ldots, \mathbf{x}_M$ denote a set of N-dimensional fundamental memories. The synaptic weights of the network are determined using the formula

$$\mathbf{W} = \frac{1}{N} \sum_{r=1}^{M} \mathbf{x}_r \mathbf{x}_r^T - \frac{M}{N} \mathbf{I}_n$$

where $\mathbf{I}_n$ is the $N \times N$ identity matrix. Once computed, the synaptic weights remain fixed.

2. **Initialization.** Let $\mathbf{x}_p$ denote the unknown probe vector to be tested. The algorithm is initialized by setting

$$x_i(0) = x_{ip}, \quad i = 1, 2, \ldots, N,$$

where $x_i(0)$ is the state of neuron i at time $n = 0$, x_{ip} is the ith element of vector $\mathbf{x}_p$, and N is the number of neurons.

3. **Iteration.** The elements are updated asynchronously (i.e., one at a time in a random order) according to the rule

$$x_i(n+1) = \text{hsgn}\left(\sum_{j=1}^{N} w_{ij} x_j(n)\right), \quad i = 1, 2, \ldots, N,$$

where

$$\text{hsgn}(v_i(n+1)) = \begin{cases} 1, & v_i(n+1) > 0, \\ v_i(n), & v_i(n+1) = 0, \\ -1, & v_i(n+1) < 0 \end{cases}$$

and $v_i(n+1) = \sum_{j=1}^{N} w_{ij} x_j(n)$. The iterations are repeated until the vector converges to a stable value. Note that at least N iterations are carried out to guarantee convergence.

4. **Result.** The stable vector, say, x_{fixed}, is the result.

The algorithm above uses asynchronous updating of synaptic weights. *Synchronous updating* is the procedure by which weights are updated simultaneously. The fundamental memories should first be presented to the Hopfield network. This tests the network's ability to recover the stored vectors using the computed synaptic weight matrix. The desired patterns should be recovered after one iteration; if not, then an error has been made. Distorted patterns or patterns missing information can then be tested using the above algorithm. There are two possible outcomes:

1. The network converges to one of the fundamental memories.

2. The network converges to a *spurious steady state*. Spurious steady states include the following:

    (a) *Reversed fundamental memories*—e.g., if x_f is a fundamental memory then so is $-x_f$.

    (b) *Mixed fundamental memories*—a linear combination of fundamental memories.

    (c) *Spin-glass states*—local minima not correlated with any fundamental memories.

Before looking at an application of a Hopfield network as a content-addressable memory, a simple example is shown below to illustrate the algorithm.

Example 5. A five-neuron discrete Hopfield network is required to store the following fundamental memories:

$$x_1 = (1, 1, 1, 1, 1)^T, \quad x_2 = (1, -1, -1, 1, -1)^T, \quad x_3 = (-1, 1, -1, 1, 1)^T.$$

(a) Compute the synaptic weight matrix W.

(b) Use asynchronous updating to show that the three fundamental memories are stable.

(c) Test the following vectors on the Hopfield network (the random orders affect the outcome):

$$\mathbf{x}_4 = (1, -1, 1, 1, 1)^T, \qquad \mathbf{x}_5 = (0, 1, -1, 1, 1)^T, \qquad \mathbf{x}_6 = (-1, 1, 1, 1, -1)^T.$$

Solution.

(a) The synaptic weight matrix is given by

$$\mathbf{W} = \frac{1}{5}(\mathbf{x}_1\mathbf{x}_1^T + \mathbf{x}_2\mathbf{x}_2^T + \mathbf{x}_3\mathbf{x}_3^T) - \frac{3}{5}\mathbf{I}_5,$$

so

$$\mathbf{W} = \frac{1}{5}\begin{pmatrix} 0 & -1 & 1 & 1 & -1 \\ -1 & 0 & 1 & 1 & 3 \\ 1 & 1 & 0 & -1 & 1 \\ 1 & 1 & -1 & 0 & 1 \\ -1 & 3 & 1 & 1 & 0 \end{pmatrix}.$$

(b) *Step 1.* First input vector, $\mathbf{x}_1 = \mathbf{x}(0) = (1, 1, 1, 1, 1)^T$.

Step 2. Initialize $x_1(0) = 1$, $x_2(0) = 1$, $x_3(0) = 1$, $x_4(0) = 1$, $x_5(0) = 1$.

Step 3. Update in random order $x_3(1), x_4(1), x_2(1), x_1(1), x_5(1)$, one at a time:

$$x_3(1) = \text{hsgn}(0.4) = 1, \qquad x_4(1) = \text{hsgn}(0.4) = 1,$$
$$x_1(1) = \text{hsgn}(0) = x_1(0) = 1,$$
$$x_5(1) = \text{hsgn}(0.8) = 1, \qquad x_2(1) = \text{hsgn}(0.8) = 1.$$

Thus $\mathbf{x}(1) = \mathbf{x}(0)$ and the net has converged.

Step 4. The net has converged to the steady state $\mathbf{x}_1$.

Step 1. Second input vector, $\mathbf{x}_2 = \mathbf{x}(0) = (1, -1, -1, 1, -1)^T$.

Step 2. Initialize $x_1(0) = 1, x_2(0) = -1, x_3(0) = -1, x_4(0) = 1, x_5(0) = -1$.

Step 3. Update in random order $x_5(1), x_3(1), x_4(1), x_1(1), x_2(1)$, one at a time:

$$x_5(1) = \text{hsgn}(-0.8) = -1, \qquad x_3(1) = \text{hsgn}(-0.4) = -1,$$
$$x_4(1) = \text{hsgn}(0) = x_4(0) = 1, \qquad x_1(1) = \text{hsgn}(0.4) = 1,$$
$$x_2(1) = \text{hsgn}(-0.8) = -1.$$

Thus $\mathbf{x}(1) = \mathbf{x}(0)$ and the net has converged.

17. Neural Networks

Step 4. The net has converged to the steady state $\mathbf{x}_2$.

Step 1. Third input vector, $\mathbf{x}_3 = \mathbf{x}(0) = (-1, 1, -1, 1, 1)^T$.

Step 2. Initialize $x_1(0) = -1, x_2(0) = 1, x_3(0) = -1, x_4(0) = 1, x_5(0) = 1$.

Step 3. Update in random order $x_5(1), x_1(1), x_4(1), x_2(1), x_3(1)$, one at a time:

$$x_5(1) = \text{hsgn}(0.8) = 1, \qquad x_1(1) = \text{hsgn}(-0.4) = -1,$$
$$x_4(1) = \text{hsgn}(0.4) = 1,$$
$$x_2(1) = \text{hsgn}(0.8) = 1, \qquad x_3(1) = \text{hsgn}(0) = x_3(0) = -1.$$

Thus $\mathbf{x}(1) = \mathbf{x}(0)$ and the net has converged.

Step 4. The net has converged to the steady state $\mathbf{x}_3$.

(c) *Step 1.* Fourth input vector, $\mathbf{x}_4 = \mathbf{x}(0) = (1, -1, 1, 1, 1)^T$.

Step 2. Initialize $x_1(0) = 1, x_2(0) = -1, x_3(0) = 1, x_4(0) = 1, x_5(0) = 1$.

Step 3. Update in random order $x_2(1), x_4(1), x_3(1), x_5(1), x_1(1)$, one at a time:

$$x_2(1) = \text{hsgn}(0.8) = 1, \qquad x_4(1) = \text{hsgn}(0.4) = 1,$$
$$x_3(1) = \text{hsgn}(0.4) = 1,$$
$$x_5(1) = \text{hsgn}(0.8) = 1, \qquad x_1(1) = \text{hsgn}(0) = x_1(0) = 1.$$

Thus $\mathbf{x}(1) = \mathbf{x}_1$ and the net has converged.

Step 4. The net has converged to the steady state $\mathbf{x}_1$.

Step 1. Fifth input vector, $\mathbf{x}_5 = \mathbf{x}(0) = (0, 1, -1, 1, 1)^T$, information is missing in the first row.

Step 2. Initialize $x_1(0) = 0, x_2(0) = 1, x_3(0) = -1, x_4(0) = 1, x_5(0) = 1$.

Step 3. Update in random order $x_2(1), x_4(1), x_3(1), x_5(1), x_1(1)$, one at a time:

$$x_4(1) = \text{hsgn}(0.6) = 1, \qquad x_5(1) = \text{hsgn}(0.6) = 1,$$
$$x_1(1) = \text{hsgn}(-0.4) = -1,$$
$$x_2(1) = \text{hsgn}(0.8) = 1, \qquad x_3(1) = \text{hsgn}(0) = x_3(0) = -1.$$

Thus $\mathbf{x}(1) = \mathbf{x}_3$ and the net has converged.

Step 4. The net has converged to the steady state $\mathbf{x}_3$.

Step 1. Sixth input vector, $\mathbf{x}_6 = \mathbf{x}(0) = (-1, 1, 1, 1, -1)^T$.

Step 2. Initialize $x_1(0) = -1$, $x_2(0) = 1$, $x_3(0) = 1$, $x_4(0) = 1$, $x_5(0) = -1$.

Step 3. Update in random order $x_3(1), x_2(1), x_5(1), x_4(1), x_1(1)$, one at a time:

$$x_3(1) = \mathrm{hsgn}(-0.4) = -1, \quad x_2(1) = \mathrm{hsgn}(-0.4) = -1,$$
$$x_5(1) = \mathrm{hsgn}(-0.4) = -1, \quad x_4(1) = \mathrm{hsgn}(-0.4) = -1,$$
$$x_1(1) = \mathrm{hsgn}(0) = x_1(0) = -1.$$

Step 3 (again). Update in random order $x_2(1), x_1(1), x_5(1), x_4(1), x_3(1)$:

$$x_2(2) = \mathrm{hsgn}(-0.8) = -1, \quad x_1(2) = \mathrm{hsgn}(0) = x_1(1) = -1,$$
$$x_5(2) = \mathrm{hsgn}(-0.8) = -1, \quad x_4(2) = \mathrm{hsgn}(-0.4) = -1,$$
$$x_3(2) = \mathrm{hsgn}(-0.4) = -1.$$

Thus $\mathbf{x}(2) = \mathbf{x}(1)$ and the net has converged.

Step 4. The net has converged to the spurious steady state $-\mathbf{x}_1$.

Example 6. Write a MATLAB program that illustrates the behavior of the discrete Hopfield network as a content-addressable memory using $N = 81$ neurons and the set of handcrafted patterns displayed in Figure 17.12.

Solution. See Programs 17c listed in Section 17.6. Note that in this case, the noise level is set at $\frac{1}{3}$ and on average the network will converge after $\frac{1}{3} \times 81 = 27$ iterations. In order for this algorithm to work, the vectors defining the patterns have to be as orthogonal as possible. If some patterns are similar, the network will not perform very well.

17.4 Neurodynamics

It is now understood that chaos, oscillations, synchronization effects, wave patterns, and feedback are present in higher-level brain functions and on different levels of signal processing. In recent years, the disciplines of neuroscience and nonlinear dynamics have increasingly coalesced, leading to a new branch of science called *neurodynamics*. This section will concentrate on a minimal chaotic *neouromodule*, studied in some detail by Pasemann and his group [9], [11] (see also [2] in Chapter 2). They have considered chaos control and synchronization effects for this simple model, and the author has recently demonstrated bistability for this class of system [1].

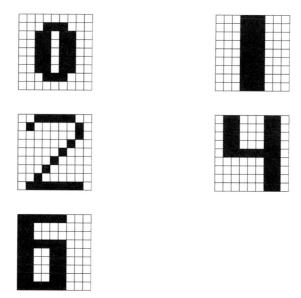

Figure 17.12: The patterns to be used as fundamental memories for the discrete Hopfield model.

A Minimal Chaotic Neuromodule. The discrete two-dimensional system investigated by Pasemann is defined by the map

(17.7)
$$x_{n+1} = b_1 + w_{11}\phi_1(x_n) + w_{12}\phi_2(y_n), \qquad y_{n+1} = b_2 + w_{21}\phi_1(x_n) + w_{22}\phi_2(y_n),$$

where its activity at time n is given by (x_n, y_n), b_1, b_2 are biases, w_{ij} are the synaptic weights connecting neurons, and ϕ represents the transfer function defined by

(17.8)
$$\phi_1(x) = \phi_2(x) = \frac{1}{1+e^{-x}}.$$

The simple network architecture of this recurrent module with an excitory neuron and an inhibitory neuron with self-connection is shown in Figure 17.9. Pasemann and Stollenwerk (see [2] in Chapter 2) considered the model with the following parameter values

(17.9)
$$b_1 = -2, \quad b_2 = 3, \quad w_{11} = -20, \quad w_{21} = -6, \quad w_{12} = 6, \quad \text{and} \quad w_{22} = 0.$$

Figure 17.13 shows the chaotic attractor for system (17.7) using the transfer function in equation (17.8) and the parameters listed in (17.9).

Figure 17.13: The chaotic attractor for a minimal chaotic neuromodule.

The fixed points of periods one and two may be found in the usual way. Fixed points of period one satisfy the simultaneous equations $x_{n+1} = x_n = x$ and $y_{n+1} = y_n = y$. There is one fixed point of period one at $P_{11} = (-1.2804, 1.6951)$, working to four decimal places. The stability of this fixed point is determined by considering the eigenvalues of the Jacobian matrix given by

$$J = \begin{pmatrix} w_{11} \frac{\partial}{\partial x} \phi_1(x) & w_{12} \frac{\partial}{\partial y} \phi_2(y) \\ w_{21} \frac{\partial}{\partial x} \phi_1(x) & 0 \end{pmatrix}.$$

The eigenvalues for the fixed point of period one are given by $\lambda_1 = -3.1487$, $\lambda_2 = -0.2550$, and the fixed point is a saddle point. Hence P_{11} is unstable.

The fixed points of period two are found by solving the equations $x_{n+2} = x_n = x$, and $y_{n+2} = y_n = y$, which has two solutions at $P_{21} = (-7.8262, -0.4623)$, and $P_{22} = (0.3107, 2.9976)$. These fixed points are also unstable.

A Bistable Neuromodule. As with many nonlinear dynamical systems, higher-level brain functions can be subject to feedback. The author and Bandar [1] have investigated system (17.7) with the following choice of parameters

(17.10)
$$b_1 = 2, \quad b_2 = 3, \quad w_{11} = 7, \quad w_{21} = 5, \quad w_{12} = -4, \quad \text{and} \quad w_{22} = 0,$$

and using the transfer functions

(17.11) $\quad \phi_1(x) = \tanh(ax) \quad$ and $\quad \phi_2(y) = \tanh(\alpha y)$,

17. Neural Networks

with $a = 1$ and $\alpha = 0.3$. Using numerical techniques, there are three fixed points of period one at $P_{11} = (-2.8331, -1.9655)$, $P_{12} = (0.2371, 4.1638)$, and $P_{13} = (5.0648, 7.9996)$. Using the Jacobian matrix, point P_{11} has eigenvalues $\lambda_1 = 0.0481 + 0.2388i$, $\lambda_2 = 0.0481 - 0.2020i$. The fixed point is stable since $|\lambda_1| < 1$ and $|\lambda_2| < 1$. Points P_{12} and P_{13} have eigenvalues $\lambda_1 = 6.3706$, $\lambda_2 = 0.2502$ and $\lambda_1 = 0.0006 + 0.0055i$, $\lambda_2 = 0.0006 - 0.0055i$, respectively. Therefore, point P_{12} is an unstable saddle point, and point P_{13} is stable, since both eigenvalues have modulus less than one. We conclude that system (17.7), with the parameter values given above (17.10) and the transfer functions defined by equations (17.11), is multistable. That is, there are two stable fixed points for one set of parameter values and the fixed point attained is solely dependent upon the initial conditions chosen.

Now introduce a feedback mechanism. In the first case we vary the parameter α, which determines the gradient of the transfer function $\phi_2(y)$. The other parameters are fixed as above (17.10). The parameter α is increased linearly from $\alpha = -5$ to $\alpha = 5$, and then decreased back down to $\alpha = -5$. Figure 17.14 shows the bifurcation diagrams for the activity of neuron x. Similar bifurcation diagrams may be plotted for the neuron y. The upper figure shows the activity against the number of iterations. The lower figure shows the activity level of neuron x as the parameter α is increased then decreased. As α is increased from -5, the steady state is on the lower branch until $\alpha \approx 1$, where there is a sudden jump to the other steady state. As α increases further, the steady state remains at $x_n \approx 5$. As α is decreased, the steady state remains at $x_n \approx 5$ until $\alpha \approx 0$ where it jumps to $x_n \approx 15$. There is a large bistable region for $-5 < \alpha < 1$, approximately.

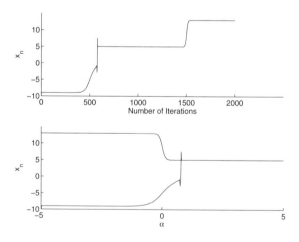

Figure 17.14: Bifurcation diagrams for system (17.7) under conditions (17.10) and (17.11) as α varies. The initial conditions chosen at $\alpha = -5$ were $x_0 = -10$ and $y_0 = -3$.

In the second case, fix the parameters and vary b_1, which is the bias for neuron x. The parameter b_1 is ramped up from $b_1 = -5$ to $b_1 = 5$, and then ramped back down to $b_1 = -5$. There is an isolated counterclockwise bistable region for $-1 < b_1 < 3.5$, approximately. Note the ringing at both ends of the bistable region; see Figure 17.15.

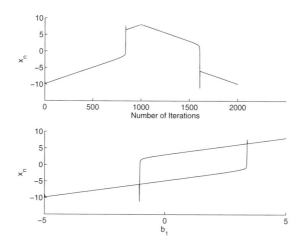

Figure 17.15: Bifurcation diagrams for system (17.7) under conditions (17.10) and (17.11) as b_1 varies. The initial conditions chosen at $b_1 = -5$ were $x_0 = -10$ and $y_0 = -3$.

In the final case, fix the parameters and vary w_{11}, which is the synaptic weight connecting neuron x to itself. The parameter is decreased from $w_{11} = 7$ down to zero and then increased back up to $w_{11} = 7$. The activity of neuron x is on the lower branch until $w_{11} \approx 5.5$, where it jumps to the upper branch. As w_{11} decreases, the system descends into regions of quasiperiodicity and periodicity. As the parameter is increased from zero, the steady state remains on the upper branch, and there is a bistable region for $5.5 < w_{11} < 7$, approximately; see Figure 17.16.

Clearly, the dynamics of this simple two-neuron module are dependent upon the history of the system. The author and his coworkers at Manchester Metropolitan University are currently investigating areas of application for this research.

17.5 MATLAB Commands

```
% Programs_17a - The generalized delta learning rule (see
% Figure 17.7).

function Programs_17a
```

384   17. Neural Networks

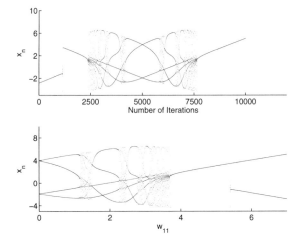

Figure 17.16: Bifurcation diagrams for system (17.7) under conditions (17.10) and (17.11) as w_{11} varies. The initial conditions chosen at $w_{11} = 7$ were $x_0 = -3$ and $y_0 = -2$.

```
% Load Boston housing data.
load housing.txt
X = housing(:,[6 9 13]);
t = housing(:,14);

% Scale to zero mean, unit variance and introduce bias on
% input.
xmean = mean(X);
xstd = std(X);
X = (X-ones(size(X,1),1)*xmean)./(ones(size(X,1),1)*xstd);
X = [ones(size(X,1),1) X];
tmean = (max(t)+min(t))/2;
tstd = (max(t)-min(t))/2;
t = (t-tmean)/tstd;

% Initialise random weight vector.
randn('seed', 123456);
w(:,1) = 0.1*randn(size(X,2),1);
y1 = tanh(X*w(:,1));
e1 = t-y1;
mse(1) = var(e1);


% Do numEpochs iterations.
numEpochs = 10;
```

17.5. MATLAB Commands

```
numPatterns = size(X,1);
eta = 0.001;
k = 1;
for m=1:numEpochs
  for n=1:numPatterns
% Calculate feedforward output, error, and gradient.
yk = tanh(X(n,:)*w(:,k));
err = yk-t(n);
g = X(n,:)'*((1-yk^2)*err);
% Update the weight.
w(:,k+1) = w(:,k) - eta*g;
k = k+1;
  end
end

for m=1:size(w,1)
plot(1:k, w(m,:))
hold on
end

fsize=15;
set(gca,'xtick',[0:2000:6000],'FontSize',fsize)
set(gca,'ytick',[-0.3:0.1:0.3],'FontSize',fsize)
xlabel('Number of Iterations','FontSize',fsize)
ylabel('Weights','FontSize',fsize)
hold off
% End of Programs_17a.
```

```
% Programs_17b - Generalized delta learning rule and
% backpropagation of errors for a network with one
% hidden layer (see Figure 17.8).

function Programs_17b
% Load full Boston housing data.
load housing.txt
X = housing(:,1:13);
t = housing(:,14);

% Scale to zero mean, unit variance and introduce bias on
% input.
xmean = mean(X);
xstd = std(X);
X = (X-ones(size(X,1),1)*xmean)./(ones(size(X,1),1)*xstd);
X = [ones(size(X,1),1) X];
tmean = mean(t);
tstd = std(t);
t = (t-tmean)/tstd;
```

17. Neural Networks

```
% Iterate over a number of hidden nodes
maxHidden = 2;
for numHidden=1:maxHidden

% Initialise random weight vector.
% Wh are hidden weights, wo are output weights.
randn('seed', 123456);
Wh = 0.1*randn(size(X,2),numHidden);
wo = 0.1*randn(numHidden+1,1);

% Do numEpochs iterations of batch error back propagation.
numEpochs = 2000;
numPatterns = size(X,1);
% Set eta.
eta = 0.05/numPatterns;
for i=1:numEpochs
% Calculate outputs, errors, and gradients.
phi = [ones(size(X,1),1) tanh(X*Wh)];
y = phi*wo;
err = y-t;
go = phi'*err;
Gh = X'*((1-phi(:,2:numHidden+1).^2).
    *(err*wo(2:numHidden+1)'));
% Perform gradient descent.
wo = wo - eta*go;
Wh = Wh - eta*Gh;
% Update performance statistics.
mse(i) = var(err);
end

plot(1:numEpochs, mse, '-')
hold on
end

fsize=15;
set(gca,'xtick',[0:500:2000],'FontSize',fsize)
set(gca,'ytick',[0:0.5:1],'FontSize',fsize)
xlabel('Number of Epochs','FontSize',fsize)
ylabel('Mean Squared Error','FontSize',fsize)
hold off
% End of Programs_17b.
```

```
% Programs_17c - The Hopfield network used as an associative
% memory.

function Programs_17c
clear all
```

```
% The 81-dimensional fundamental memories (Figure 17.12).
X = [-1 -1 -1 -1 -1 -1 -1 -1 -1
     -1 -1 -1  1  1  1 -1 -1 -1
     -1 -1  1  1 -1  1  1 -1 -1
 -1 -1  1  1 -1  1  1 -1 -1
 -1 -1  1  1 -1  1  1 -1 -1
 -1 -1  1  1 -1  1  1 -1 -1
 -1 -1  1  1 -1  1  1 -1 -1
 -1 -1 -1  1  1  1 -1 -1 -1
 -1 -1 -1 -1 -1 -1 -1 -1 -1;
     -1 -1 -1  1  1  1 -1 -1 -1
-1 -1 -1  1  1  1 -1 -1 -1
-1 -1 -1  1  1  1 -1 -1 -1
-1 -1 -1  1  1  1 -1 -1 -1
-1 -1 -1  1  1  1 -1 -1 -1
-1 -1 -1  1  1  1 -1 -1 -1
-1 -1 -1  1  1  1 -1 -1 -1
-1 -1 -1  1  1  1 -1 -1 -1
-1 -1 -1  1  1  1 -1 -1 -1;
     -1 -1  1  1  1  1  1 -1 -1
-1  1 -1 -1 -1 -1 -1  1 -1
-1 -1 -1 -1 -1 -1 -1  1 -1
-1 -1 -1 -1 -1 -1  1 -1 -1
-1 -1 -1 -1 -1  1 -1 -1 -1
-1 -1 -1 -1  1 -1 -1 -1 -1
-1 -1 -1  1 -1 -1 -1 -1 -1
-1 -1  1 -1 -1 -1 -1 -1 -1
-1  1  1  1  1  1  1  1 -1;
     -1  1  1 -1 -1 -1  1  1 -1
-1  1  1 -1 -1 -1  1  1 -1
-1  1  1 -1 -1 -1  1  1 -1
-1  1  1 -1 -1 -1  1  1 -1
-1  1  1  1  1  1  1  1 -1
-1 -1 -1 -1 -1 -1  1  1 -1
-1 -1 -1 -1 -1 -1  1  1 -1
-1 -1 -1 -1 -1 -1  1  1 -1
-1 -1 -1 -1 -1 -1  1  1 -1;
      1  1  1  1  1  1 -1 -1 -1
 1  1 -1 -1 -1 -1 -1 -1 -1
 1  1 -1 -1 -1 -1 -1 -1 -1
 1  1  1  1  1  1 -1 -1 -1
 1  1 -1 -1  1  1 -1 -1 -1
 1  1 -1 -1  1  1 -1 -1 -1
 1  1 -1 -1  1  1 -1 -1 -1
 1  1 -1 -1  1  1 -1 -1 -1
 1  1  1  1  1  1 -1 -1 -1];
```

17. Neural Networks

```
% Load data.
X = X([1 2 3 4 5],:);
numPatterns = size(X,1);
numInputs = size(X,2);
% Plot the fundamental memories. They will appear in the
% Figure 1 window.
figure
plotHopfieldData(X)

% STEP 1. Calculate the weight matrix using Hebb's postulate.
W = (X'*X - numPatterns*eye(numInputs))/numInputs;

% STEP 2. Set a probe vector using a predefined noiselevel.
% The probe vector is a distortion of one of the fundamental
% memories.
noiseLevel = 1/3;
patInd = ceil(numPatterns*rand(1,1));
xold = (2*(rand(numInputs,1)> noiseLevel)-1).*X(patInd,:)';

% STEP 3. Asynchronous updates of the elements of the probe
% vector until it converges. To guarantee convergence, the
% algorithm performs at least numPatterns=81 iterations even
% though convergence generally occurs before this.
figure
converged = 0;
x=xold;
while converged==0,
    p=randperm(numInputs);
    for n=1:numInputs
        r = x(p(n));
        x(p(n)) = hsign(W(p(n),:)*x, r);
        plotHopfieldVector(x);
        pause(0.01);
    end
% STEP 4. Check for convergence.
    if (all(x==xold))
        break;
    end
    xold = x;
end

% Step 3. Update the elements asynchronously.
function y = hsign(a, r)
y(a>0) = 1;
y(a==0) = r;
y(a<0) = -1;
```

17.5. MATLAB Commands

```
% Plot the fundamental memories.
function plotHopfieldData(X)
numPatterns = size(X,1);
numRows = ceil(sqrt(numPatterns));
numCols = ceil(numPatterns/numRows);
for i=1:numPatterns
   subplot(numRows, numCols, i);
   axis equal;
   plotHopfieldVector(X(i,:))
end

% Plot the sequence of iterations for the probe vector. The
% sequence is shown in the Figure 2 window. Note that 81
% iterations. You can exit by pressing Ctrl-Shift-C, if you
% wish to interrupt the program.
function plotHopfieldVector(x)
cla;
numInputs = length(x);
numRows = ceil(sqrt(numInputs));
numCols = ceil(numInputs/numRows);
for m=1:numRows
   for n=1:numCols
      xind = numRows*(m-1)+n;
      if xind > numInputs
         break;
      elseif x(xind)==1
         rectangle('Position', [n-1 numRows-m 1 1],
                   'FaceColor', 'k');
      elseif x(xind)==-1
         rectangle('Position', [n-1 numRows-m 1 1],
                   'FaceColor', 'w');
      end
   end
end
set(gca, 'XLim', [0 numCols], 'XTick', []);
set(gca, 'YLim', [0 numRows], 'YTick', []);
% End of Programs_17c.
```

```
% Programs_17d - The minimal chaotic neuromodule.
% A chaotic attractor for a neural network (Figure 17.13).

clear
b1=-2;b2=3;w11=-20;w21=-6;w12=6;a=1;
N=7000;
x(1)=0;
y(1)=2;
for n=1:N
```

17. Neural Networks

```
    x(n+1)=b1+w11*1/(1+exp(-a*x(n)))+w12*1/(1+exp(-a*y(n)));
    y(n+1)=b2+w21*1/(1+exp(-a*x(n)));
end

hold on
axis([-15 5 -5 5])
plot(x(50:N),y(50:N),'.','MarkerSize',1);
fsize=15;
set(gca,'xtick',[-15:5:5],'FontSize',fsize)
set(gca,'ytick',[-5:5:5],'FontSize',fsize)
xlabel('\itx','FontSize',fsize)
ylabel('\ity','FontSize',fsize)
hold off
% End of Programs_17d.
```

```
% Programs_17e - Bifurcation diagram for a bistable
% neuromodule.
% Bifurcation diagram with feedback for a two-neuron module
% (see Figure 17.15).

clear all
format long;
halfN=1000;N=2*halfN+1;N1=1+halfN;Max=10;a=1;alpha=0.3;
b2=3;w11=7;w12=-4;w21=5;start=-5;
x(1)=-10;y(1)=-3;

% Vary bias b1; ramp b1 up
for n=1:halfN
    b1=(start+n*Max/halfN);
    x(n+1)=b1+w11*tanh(a*x(n))+w12*tanh(alpha*y(n));
    y(n+1)=b2+w21*tanh(a*x(n));
end

% Ramp b1 down
for n=N1:N
    b1=(start+2*Max-n*Max/halfN);
    x(n+1)=b1+w11*tanh(a*x(n))+w12*tanh(alpha*y(n));
    y(n+1)=b2+w21*tanh(a*x(n));
end

% Plot the bifurcation diagrams
fsize=14;
subplot(2,1,1)
hold on
set(gca,'xtick',[0:halfN/2:N],'FontSize',fsize);
set(gca,'ytick',[-Max:5:Max],'FontSize',fsize);
plot(x(1:N),'-','MarkerSize',1,'color','k')
```

```
xlabel('Number of Iterations','FontSize',fsize);
ylabel('x_n','FontSize',fsize);
hold off

for n=1:halfN
    x1(n)=x(N+1-n);
    w(n)=start+n*Max/halfN;
end

subplot(2,1,2)
hold on
set(gca,'xtick',[start:5:start+Max],'FontSize',fsize);
set(gca,'ytick',[-Max:5:Max],'FontSize',fsize);
plot(w(1:halfN),x(1:halfN),'-','MarkerSize',1,'color','k');
plot(w(1:halfN),x1(1:halfN),'-','MarkerSize',1,'color','k');
xlabel('b_1','FontSize',fsize);
ylabel('x_n','FontSize',fsize);
hold off

% End of Programs_17e.
```

17.6 Exercises

1. For the following activation functions, show that

   (a) if $\phi(v) = 1/(1 + e^{-av})$, then $\phi'(v) = a\phi(v)(1 - \phi(v))$;

   (b) if $\phi(v) = a\tanh(bv)$, then $\phi'(v) = \frac{b}{a}(a^2 - \phi^2(v))$;

   (c) if $\phi(v) = \frac{1}{2a} \log \frac{\cosh(a(v+1))}{\cosh(a(v-1))}$, then

   $$\phi'(v) = (\tanh(a(v+1)) - \tanh(a(v-1)))/2.$$

2. Prove Theorem 2, showing that when the activation function is nonlinear, say, $y_k = \phi(v_k)$, the generalized delta rule can be formulated as

   $$w_{kj}(n+1) = w_{kj}(n) - \eta g_{kj},$$

   where

   $$g_{kj} = (y_k - t_k)\frac{\partial \phi}{\partial v_k}x_j.$$

3. Edit the M-file Programs 17a to plot a graph of the mean squared error against number of iterations for Example 1.

   (a) Investigate what happens to the mean squared error for varying eta (see Programs 17a) values of your choice.

(b) Edit the M-file Programs 17b to investigate what happens to the mean squared error as the number of hidden neurons increases to five.

4. Use another data set of your choice from the URL

http://www.ics.uci.edu/~mlearn/MLRepository.html

using an edited version of Programs 17a and 17b to carry out your analysis.

5. (a) Prove Theorem 3 regarding Lyapunov functions of continuous Hopfield models.

(b) Consider the recurrent Hopfield network modeled using the differential equations

$$\dot{x} = -x + 7\left(\frac{2}{\pi}\tan^{-1}\left(\frac{\gamma\pi x}{2}\right)\right) + 6\left(\frac{2}{\pi}\tan^{-1}\left(\frac{\gamma\pi y}{2}\right)\right),$$

$$\dot{y} = -y + 6\left(\frac{2}{\pi}\tan^{-1}\left(\frac{\gamma\pi x}{2}\right)\right) - 2\left(\frac{2}{\pi}\tan^{-1}\left(\frac{\gamma\pi y}{2}\right)\right).$$

Plot a vector field portrait and derive a suitable Lyapunov function.

(c) Plot surface plots for the Lyapunov functions for Examples 3 and 4 and Exercise 5(b). Plot the surfaces for $|a_i| \leq 1$, $i = 1, 2$.

6. Consider the discrete Hopfield model investigated in Example 5. Test the vector $\mathbf{x}_7 = (-1, -1, 1, 1, 1)^T$, update in the following orders, and determine to which vector the algorithm converges:

(a) $x_3(1), x_4(1), x_1(1), x_2(1), x_5(1)$;
(b) $x_1(1), x_4(1), x_3(1), x_2(1), x_5(1)$;
(c) $x_5(1), x_3(1), x_2(1), x_1(1), x_4(1)$;
(d) $x_3(1), x_5(1), x_2(1), x_4(1), x_1(1)$.

7. Add suitable characters "3" and "5" to the fundamental memories shown in Figure 17.12. You may need to increase the grids to 10×10 and work with 100 neurons.

8. A simple model of a neuron with self-interaction is described by Pasemann [11]. The difference equation is given by

$$a_{n+1} = \gamma a_n + \theta + w\sigma(a_n), \quad 0 \leq \gamma < 1,$$

where a_n is the activation level of the neuron, θ is a bias, w is a self-weight, γ represents dissipation in a neuron, and the output is given by the sigmoidal transfer function

$$\sigma(x) = \frac{1}{1 + e^{-x}}.$$

(a) Determine an equation for the fixed points of period one and show that the stability condition is given by $|\gamma + w\sigma'(a)| < 1$, where a is a fixed point of period one.

(b) Show that the system is bistable in the region bounded by the parametric equations

$$\theta(a) = (1-\gamma)a - \frac{(1-\gamma)}{(1-\sigma(a))}, \qquad w(a) = \frac{(1-\gamma)}{\sigma'(a)}.$$

(c) Show that the system is unstable in the region bounded by the parametric equations:

$$\theta(a) = (1-\gamma)a + \frac{(1+\gamma)}{(1-\sigma(a))}, \qquad w(a) = -\frac{(1+\gamma)}{\sigma'(a)}.$$

(d) Use the first iterative method to plot a bifurcation diagram when $\theta = 4$ and $w = -16$ for $0 < \gamma < 1$.

(e) Use the second iterative method to plot a bifurcation diagram when $\theta = -2.4$ and $\gamma = 0$ for $3 < w < 7$. Ramp w up and down.

9. Consider the neuromodule defined by the equations

$$x_{n+1} = 2 + 3.5\tanh(x) - 4\tanh(0.3y), \qquad y_{n+1} = 3 + 5\tanh(x).$$

Iterate the system and show that it is quasiperiodic.

10. Use the OGY method to control chaos in the minimal chaotic neuromodule.

Recommended Reading

[1] S. Lynch and Z. G. Bandar, Analysis of a bistable neuromodule, preprint, 2004.

[2] J. A. Rothwell, The word liar, *New Scientist*, March (2003), 51.

[3] Z. G. Bandar, D. A. McLean, J. D. O'Shea, and J. A. Rothwell, *Analysis of the Behaviour of a Subject*, International Publication Number WO 02/087443 A1, 2002.

[4] M. Di Marco, M. Forti, and A. Tesi, Existence and characterization of limit cycles in nearly symmetric neural networks, *IEEE Trans. Circuits Systems 1: Fund. Theory Appl.*, **49** (2002), 979–992.

[5] P. Dayan and L. F. Abbott, *Theoretical Neuroscience: Computational and Mathematical Modeling of Neural Systems*, MIT Press, Cambridge, MA, 2001.

[6] I. W. Sandberg (ed.), J. T. Lo, C. L. Fancourt, J. C. Principe, S. Katagiri, and S. Haykin, *Nonlinear Dynamical Systems: Feedforward Neural Network Perspectives*, Adaptive Learning Systems for Signal Processing, Communications and Control Series, Wiley–Interscience, New York, 2001.

[7] W. J. Freeman, *Neurodynamics: An Exploration in Mesoscopic Brain Dynamics (Perspectives in Neural Computing)*, Springer-Verlag, New York, 2000.

[8] B. Kosko, *Neural Networks and Fuzzy Systems: A Dynamical Systems Approach to Machine Intelligence*, Prentice–Hall, Upper Saddle River, NJ, 1999.

[9] F. Pasemann, Driving neuromodules into synchronous chaos, in J. Mira and J. V. Sanchez-Andres, eds., *Foundations and Tools for Neural Modeling*, Lecture Notes in Computer Science 1606, Springer-Verlag, New York, 1999, 377–384.

[10] S. S. Haykin, *Neural Networks: A Comprehensive Foundation*, 2nd ed. Prentice–Hall, Upper Saddle River, NJ, 1998.

[11] F. Pasemann, A simple chaotic neuron, *Phys. D*, **104** (1997), 205–211.

[12] M. T. Hagan, H. B. Demuth, and M. H. Beale, *Neural Network Design*, Brooks–Cole, Pacific Grove, CA, 1995.

[13] S. J. Schiff, K. Jerger, D. H. Doung, T. Chang, M. L. Spano, and W. L. Ditto, Controlling chaos in the brain, *Nature*, **370** (1994), 615.

[14] D. E. Rumelhart and J. L. McClelland, eds., *Parallel Distributed Processing: Explorations in the Microstructure of Cognition*, Vol. 1, MIT Press, Cambridge, MA, 1986.

[15] J. J. Hopfield and D. W. Tank, Neural computation of decisions in optimization problems, *Biol. Cybernet.*, **52** (1985), 141–154.

[16] J. J. Hopfield, Neurons with graded response have collective computational properties like those of two-state neurons, *Proc. Nat. Acad. Sci.*, **81** (1984), 3088–3092.

[17] J. J. Hopfield, Neural networks and physical systems with emergent collective computational abilities, *Proc. Nat. Acad. Sci.*, **79** (1982), 2554-2558; reprinted in J. A. Anderson and E. Rosenfeld, eds., *Neurocomputing Foundations of Research*, MIT Press, Cambridge, MA, 1988, 448–453.

[18] T. Kohonen, Self-organized formation of topologically correct feature maps, *Biol. Cybernet.*, **43** (1982), 59–69; reprinted in J. A. Anderson and E. Rosenfeld, eds., *Neurocomputing Foundations of Research*, MIT Press, Cambridge, MA, 1988, 554–568.

[19] M. Minsky and S. Papert, *Perceptrons*, MIT Press, Cambridge, MA, 1969.

[20] B. Widrow and M. E. Hoff, Adaptive switching circuits, in 1960 *IRE WESCON Convention Record*, Part 4, IRE, New York, 1960, 96–104.

[21] F. Rosenblatt, The perceptron: A probabalistic model for information storage and organization in the brain, *Psychological Rev.*, **65** (1958), 386–408.

[22] D. O. Hebb, *The Organization of Behavior*, Wiley, New York, 1949.

[23] W. McCulloch and W. Pitts, A logical calculus of the ideas immanent in nervous activity, *Bull. Math. Biophys.*, **5** (1943), 115–133.

18
Simulink

Aims and Objectives

- To model, simulate, and analyze dynamical systems using Simulink.

On completion of this chapter the reader should be able to

- build block diagrams using a graphical user interface (GUI);
- run simulations and see how they are affected by input and parameter changes;
- analyze simulation results.

Simulink is the ideal package to investigate dynamical systems. Unfortunately, it does not come as a standard part of the latest versions of MATLAB. The package is used extensively by engineers and researchers around the world. It is easy to use and is the closest one can get to experimentation without the need for expensive physical apparatus and laboratories.

This chapter is intended to introduce the reader to Simulink using models referred to in earlier sections of the book. Note that in order to use Simulink you must have MATLAB. This chapter is based on Simulink (Version 5^1) of the package, but the programs have been kept as simple as possible and should run on later versions.

---

[1] As this book was in production, The MathWorks, Inc. was preparing to release Simulink 6. Please note that the model files presented in this book will work with the updated version of Simulink.

18.1 Introduction

There is extensive online documentation accompanying the Simulink package, and most of the MATLAB textbooks listed in the references of Chapter 0 contain some information and simple examples. Probably the most popular textbook specializing in Simulink to date is by Dabney and Harman [1]. MathWorks has also published a book that accompanies the package [3].

The author recommends that you download the working Simulink programs before attempting to write your own from scratch. There are examples from electric circuits, mechanics, and nonlinear optics—topics that have all been covered in earlier chapters of this book. There is also an example of chaos synchronization between two Lorenz systems. Most of the results of the simulations have been displayed in earlier chapters of this book using MATLAB and it is easy to compare the results. The main advantages using Simulink are listed below:

- The absence of formal programming.

- It is used extensively in industry.

- You can change parameters easily and quickly rerun the simulation.

- You can change the input to see how it affects the output.

- It encourages you to try new things—like adding a bit of noise.

- It is interactive and fun.

To start Simulink, type **simulink** in the MATLAB command window after the >> prompt. A window of the Simulink Library Browser will appear on your screen. On Microsoft Windows, a list of icons entitled "Continuous," "Discontinuous," "Discrete," etc., appears in the right-hand column of the Simulink Library Browser window. (A similar window opens in UNIX). To open one of these subsystems simply double-click on the relevant icon. For example, the "Integrator" block is the second listed under the "Continuous" subsystem.

To create a Simulink model file, click on **File**, **New**, and **Model** in the toolbars at the top of the Simulink Library Browser menu. All Simulink model files have the extension .mdl, so you could call your first file Simulink1.mdl, for example. Simply click on a block icon, hold the mouse button down, and drag the block into the Simulink1.mdl window and release the mouse button. You can drag other blocks over to the model window and then start to connect them. To connect blocks, simply click on the mouse button at a > port and drag the line to another > port before releasing the mouse button. Once the blocks are connected you can parameterize the model by double-clicking on the relevant blocks. A simulation is run by clicking on **Simulation** and **Start** in the toolbars at the top of the model window. Simulation parameters can also be changed under the **Simulation** toolbar. For example, the

18.1. Introduction   399

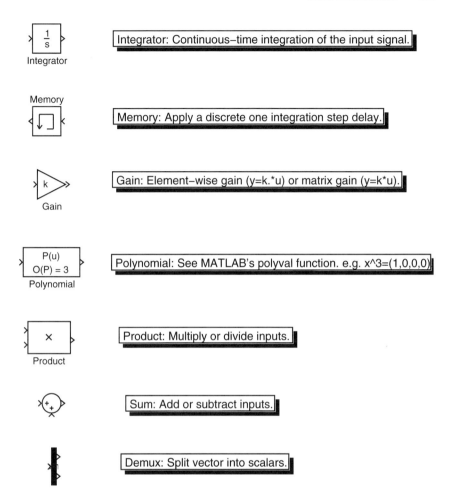

Figure 18.1: Simulink blocks taken from the "Continuous," "Discrete," "Math Operations," and "Signal Routing" subsystems. Note that the blocks may look slightly different in other versions of Simulink.

choice between a continuous and discrete simulation and the length of time that the simulation is to be run can be altered here.

The blocks used in this chapter are listed in Figures 18.1 and 18.2 with a short description of what each one does. Of course, there are many more Simulink blocks not listed here, and the reader should use the extensive online documentation for further information.

As with the MATLAB files, all of the Simulink models can be downloaded

400   18. Simulink

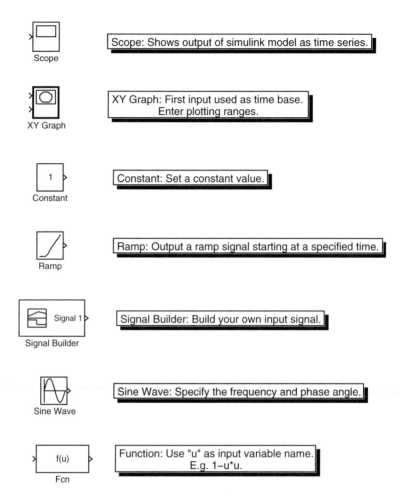

Figure 18.2: Simulink blocks taken from the "Sinks," "Sources," and User-Defined Functions" subsystems.

from the Web at the MathWorks site:

> http://www.mathworks.com/matlabcentral/.

Remember to double-click on the blocks to set continuous or discrete simulation, step-lengths, numerical procedures, tolerances, initial conditions, simulation parameters, and the length of time the simulation is to run.

Continuous systems using differential equations are considered first. Electric circuits, including the van der Pol oscillator, are simulated in Section 18.2. A nonlinear periodically driven pendulum is modeled in Section 18.3.

Discrete systems are considered in the remaining sections. Triangular and Gaussian pulses are input to an SFR resonator and chaos is controlled using a simple feedback of the electric field. Chaos is synchronized between two Lorenz systems in Section 18.5.

18.2 Electric Circuits

Electric circuits were introduced in Chapter 7 when considering differential equations. The first two Simulink models are taken from this chapter. The van der Pol oscillator (Chapter 10) is also modeled here.

Example 1. A series resistor–inductor electrical circuit is modeled using the differential equation

$$\frac{dI}{dt} + 0.2I = 5\sin(t).$$

Create a Simulink model to simulate this simple electric circuit.

Solution. The Simulink model is given in Figure 18.3. Double-click on the **Sine Wave** block and enter **Sine type:** Time based, **Amplitude:** 5, and **Frequency:** 1. Double-click on the **Integrator** block and enter **Initial condition:** 0. Click on **Simulation** and **Start** in the toolbar. Double-click on the **Scope** to see the solution. Click on the **binoculars** in the Scope window to centralize the image. From Example 9 in Chapter 7, the analytic solution is

$$I(t) = \frac{25}{26}\sin(t) - \frac{125}{26}\cos(t) - \frac{125}{126}e^{-\frac{t}{5}}.$$

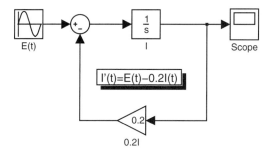

Figure 18.3: Simulink model of a series resistor–inductor circuit.

Example 2. The second-order differential equation used to model a series *RLC* circuit is given by

$$\frac{d^2I}{dt^2} + 5\frac{dI}{dt} + 6I = 10\sin(t).$$

Simulate this circuit using Simulink.

Solution. The Simulink model is given in Figure 18.4. Double-click on the blocks to set up the model as in Example 1. From Example 10 in Chapter 7, the solution is

$$I(t) = 2e^{-2t} - e^{-3t} + \sin(t) - \cos(t).$$

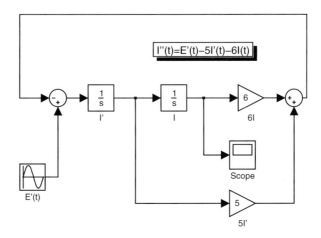

Figure 18.4: Simulink model of a series RLC circuit.

Example 3. The van der Pol equation was introduced in Chapter 10:

$$\ddot{x} + \mu(x^2 - 1)\dot{x} + x = 0,$$

which may be written as a planar system of the form

$$\dot{x} = y, \qquad \dot{y} = -x - \mu(x^2 - 1)y.$$

Use Simulink to produce a phase portrait of this system.

Solution. A Simulink model is shown in Figure 18.5. A phase portrait is shown in Figure 10.2 in Chapter 10.

18.3 A Mechanical System

Example 4. Simulate the motion of a mass suspended from a nonlinear pendulum when it is subject to a periodic force.

Solution. A schematic of the mechanical system is shown in Figure 18.6.

The differential equation used to model this system was introduced in Chapter 14. The Duffing equation is given by

$$\ddot{x} + k\dot{x} + (x^3 - x) = A\cos(\omega t),$$

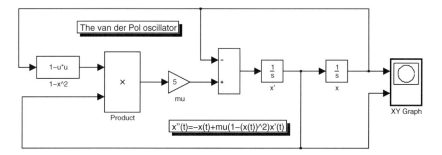

Figure 18.5: Simulink model of a van der Pol oscillator. A phase portrait is plotted in the XY Graph block.

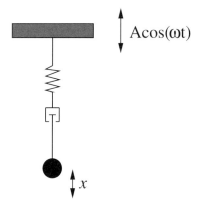

Figure 18.6: A periodically driven pendulum.

where x measures displacement, k is a damping coefficient, ω is a frequency, and A is the amplitude of vibration of the forcing term. A Simulink model of this system is displayed in Figure 18.7.

18.4 Nonlinear Optics

The nonlinear SFR resonator was introduced in Section 4.3. It was shown that the electric field circulating in the loop can be modeled by the discrete complex iterative equation

$$E_{n+1} = A + BE_n \exp[i(|E_n|^2 + \phi_L)],$$

where E_n is the electric field at the nth circulation in the ring, A is the input, B is related to the fiber coupling ratio, and ϕ_L is a linear phase shift due to propagation.

18. Simulink

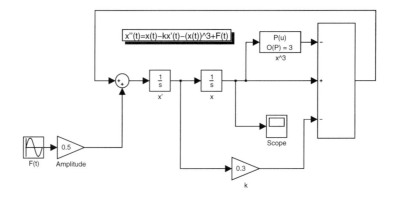

Figure 18.7: Simulink model of the Duffing equation.

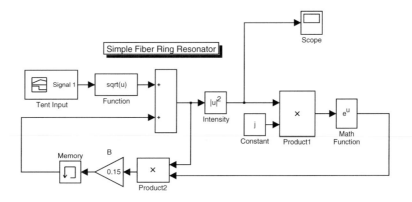

Figure 18.8: An SFR resonator with a triangular input pulse.

Example 5. Model the SFR resonator with a triangular input pulse using 1000 iterations and a maximum input of 9 Wm^{-2}. Set $B = 0.15$ and $\phi_L = 0$ in this simulation.

Solution. A Simulink model is shown in Figure 18.8. Note that the input is square rooted as it is an input power.

For this simulation set **Stop time:** 1000, **Type fixed step:** discrete, and **Fixed step size:** 1.

The input versus output powers may be observed by double-clicking on the **Scope** and then clicking on the binoculars in the scope window. The input pulse can be easily changed to a Gaussian using a function block. A Simulink model can be downloaded from the MathWorks Web site.

Example 6. Use feedback of a proportion of the electric field to control the chaos in an SFR resonator when $B = 0.15$, $A = 2.7$, and $\phi_L = 0$.

Solution. A constant input of $|A|^2 = 7.29$ Wm^{-2} is input to the fiber ring. By changing the gain in the control section of the Simulink model (Figure 18.9), one can obtain different steady states.

Figure 18.9: Chaos control in an SFR resonator.

Note that the output signal in the Scope window is chaotic when the gain in the control section is set to zero.

18.5 The Lorenz Equations and Chaos Synchronization

The Lorenz system of ODEs was introduced in Section 13.4. The Simulink model displayed in Figure 18.10 can be edited to model other continuous three-dimensional systems such as Chua's circuit, the Belousov–Zhabotinski reaction, and the Rössler system. The strange attractor can be viewed by clicking in the XY Graph block. Note that this is a two-dimensional plot of the strange attractor.

Since the publication of the seminal paper of Pecora and Carroll (see [12] in Chapter 6) there has been intense research on the synchronization of chaotic systems. The reader is encouraged to read references [2], [4], and [5] for a detailed introduction to the subject. Chaos synchronization between two Lorenz systems is demonstrated using Simulink in Figure 18.11. The two systems start with different initial conditions but after a short interval of time, they begin to synchronize.

18.6 Exercises

1. Change the input in Example 1 from $E(t) = 5\sin(t)$ to $E(t) = 10\sin(t+\frac{\pi}{3})$ and run the Simulink model in Figure 18.3 when $I(0) = 0$. The file is called Simulink1 at the MathWorks Web site.

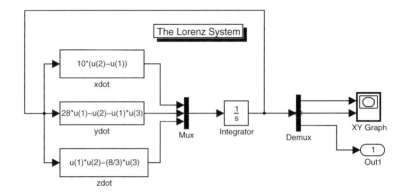

Figure 18.10: A Simulink model of the Lorenz system.

Figure 18.11: Chaos synchronization between two Lorenz systems.

2. Change the input in Example 2 from $E'(t) = 10\sin(t)$ to $E'(t) = \sin(4t)$, $I(0) = \dot{I}(0) = 0$. Run the Simulink model displayed in Figure 18.4 (filename Simulink2).

3. Create a Simulink model to simulate the *RLC* circuit modeled by the differential equation

$$\frac{d^2I}{dt^2} + 3\frac{dI}{dt} + 2I = 5\sin\left(t + \frac{\pi}{4}\right),$$

where $I(0) = \dot{I}(0) = 0$.

4. Use the Simulink model shown in Figure 18.5 to plot a phase portrait for the van der Pol oscillator when

   (i) $\mu = 0.1$;

   (ii) $\mu = 1$;

   (iii) $\mu = 20$.

5. Consider the periodically driven nonlinear pendulum simulated in Figure 18.7. Run the simulation when $k = 0.3$, $\omega = 1.25$, and the driving force is $0.31\cos(\omega t)$. How would you describe this behavior?

6. Run the same simulation as in Exercise 5 but with a driving force of $0.5\cos(\omega t)$. How would you describe this behavior?

7. Consider the Duffing equation in Example 4 with $k = 0.1$, $\omega = 1.25$, and $A = 0.07$. Run the simulation for the initial conditions

   (i) $x(0) = 1.16$ and $\dot{x}(0) = 0.112$;

   (ii) $x(0) = 0.585$ and $\dot{x}(0) = 0.29$.

   What can you say about the system?

8. Consider the simulation of the SFR resonator with a tent input pulse shown in Figure 18.8 (Simulink 5). Change the maximum of the input intensity and see how the image on the scope is affected. Double-click on the Signal 1 block to change the maximum input intensity. It is initially set at 9. Add a small amount of noise to the input and see how it affects the output.

9. Change the input signal in Figure 18.8 from a tent pulse to a Gaussian pulse and run the simulation.

10. Use Simulink to demonstrate synchronization of chaos in Chua's circuit.

Recommended Reading

[1] J. B. Dabney and T. L. Harman, *Mastering Simulink*, Prentice–Hall, Upper Saddle River, NJ, 2003.

[2] S. Boccaletti, J. Kurths, G. Osipov, D. L. Valladares, and C. S. Zhou, The synchronization of chaotic systems, *Phys. Rep.*, **366** (2002), 1–101.

[3] The MathWorks, Inc., *Using Simulink* (*Version* 5), The MathWorks, Inc., Natick, MA, 2002.

[4] E. Mosekilde, Y. Maistrenko, and D. Postnov, *Chaotic Synchronization*, World Scientific, Singapore, 2002.

[5] C. W. Wu, *Synchronization in Coupled Chaotic Circuits and Systems*, World Scientific, Singapore, 2002.

[6] R. H. Bishop, *Modern Control Systems and Design Using MATLAB and Simulink*, Addison–Wesley, Reading, MA, 1997.

[7] L. O. Chua, M. Itoh, L. Kocarev, and K. Eckert, Chaos synchronization in Chua's circuit, *J. Circuits Systems Comput.*, **3** (1993), 93–108.

19
Solutions to Exercises

19.0 Chapter 0

1. (a) 3; (b) 531441; (c) 0.3090; (d) 151; (e) $-\frac{1}{10}$.
2. (a)
$$A + 4BC = \begin{pmatrix} 57 & 38 & 19 \\ 40 & 25 & 16 \\ 35 & 19 & 14 \end{pmatrix}.$$

   (b)
$$A^{-1} = \begin{pmatrix} 0.4 & -0.6 & 0.2 \\ 0 & 1 & 0 \\ -0.6 & 1.4 & 0.2 \end{pmatrix}, \quad B^{-1} = \begin{pmatrix} 0 & 1 & -1 \\ 2 & -2 & -1 \\ -1 & 1 & 1 \end{pmatrix}.$$

   The matrix C is singular.

   (c)
$$A^3 = \begin{pmatrix} -11 & 4 & -4 \\ 0 & 1 & 0 \\ 12 & 20 & -7 \end{pmatrix}.$$

   (d) Determinant of $C = 0$.

   (e) Eigenvalues and corresponding eigenvectors are
   - $\lambda_1 = -0.3772$, $(0.4429, -0.8264, 0.3477)^T$;

19. Solutions to Exercises

- $\lambda_2 = 0.7261, (0.7139, 0.5508, -0.4324)^T$;
- $\lambda_3 = 3.6511, (0.7763, 0.5392, 0.3266)^T$.

3. (a) $-1+3i$; (b) $1-3i$; (c) $1.4687+2.2874i$; (d) $0.3466+0.7854i$; (e) $-1.1752i$.

4. (a) 1; (b) $\frac{1}{2}$; (c) 0; (d) ∞; (e) 0.

5. (a) $9x^2 + 4x$; (b) $\frac{2x^3}{\sqrt{1+x^4}}$; (c) $e^x(\sin(x)\cos(x) + \cos^2(x) - \sin^2(x))$; (d) $1 - \tanh^2 x$; (e) $\frac{2\ln x x^{\ln x}}{x}$.

6. (a) $-\frac{43}{12}$; (b) 1; (c) $\sqrt{\pi}$; (d) 2; (e) divergent.

7. See Section 0.2.

8. (a) $y(x) = \frac{1}{2}\sqrt{2x^2 + 2}$;
   (b) $y(x) = \frac{6}{x}$;
   (c) $y(x) = \frac{(108x^3+81)^{1/4}}{3}$;
   (d) $x(t) = -2e^{-3t} + 3e^{-2t}$;
   (e) $\frac{16}{5}e^{-2t} - \frac{21}{10}e^{-3t} - \frac{1}{10}\cos t + \frac{1}{10}\sin t$.

9.

   (a) When x(0):=0.2,

   ```
   x(91):=0.8779563852
   x(92):=0.4285958836
   x(93):=0.9796058084
   x(94):=0.7991307420e-1
   x(95):=0.2941078991
   x(96):=0.8304337709
   x(97):=0.5632540923
   x(98):=0.9839956791
   x(99):=0.6299273044e-1
   x(100):=0.2360985855
   ```

   (b) When x(0):=0.2001,

   ```
   x(91):=0.6932414820
   x(92):=0.8506309185
   x(93):=0.5082318360
   x(94):=0.9997289475
   x(95):=0.1083916122e-2
   x(96):=0.4330964991e-2
   x(97):=0.1724883093e-1
   x(98):=0.6780523505e-1
   x(99):=0.2528307406
   x(100):=0.7556294285
   ```

10.

```
% Euclids algorithm
function y=gcd(a,b)
while b~=0
    d=rem(a,b);
    a=b;
    b=d;
end
a
disp('is the greatest common divisor.')
% End program.
```

```
>> gcd(12348,14238)

a=126 is the greatest common divisor.
```

19.1 Chapter 1

1. The general solution is $x_n = \pi(4n + cn(n-1))$.
2. (a) $2 \times 3^n - 2^n$;
   (b) $2^{-n}(3n+1)$;
   (c) $2^{\frac{n}{2}}(\cos(\frac{n\pi}{4}) + \sin(\frac{n\pi}{4}))$;
   (d) $F_n = \frac{1}{2^n\sqrt{5}}[(1+\sqrt{5})^n - (1-\sqrt{5})^n]$;
   (e) (i) $x_n = 2^n + 1$; (ii) $x_n = \frac{1}{2}(-1)^n + 2^n + n + \frac{1}{2}$; (iii) $x_n = \frac{1}{3}(-1)^n + \frac{5}{3}2^n - \frac{1}{6}e^n(-1)^n - \frac{1}{3}e^n 2^n + \frac{1}{2}e^n$.
3. The dominant eigenvalue is $\lambda_1 = 1.107$ and

   (a) $$X^{(15)} = \begin{pmatrix} 64932 \\ 52799 \\ 38156 \end{pmatrix};$$

   (b) $$X^{(50)} = \begin{pmatrix} 2.271 \times 10^6 \\ 1.847 \times 10^6 \\ 1.335 \times 10^6 \end{pmatrix};$$

   (c) $$X^{(100)} = \begin{pmatrix} 3.645 \times 10^8 \\ 2.964 \times 10^8 \\ 2.142 \times 10^8 \end{pmatrix}.$$

4. The eigenvalues are $\lambda_1 = 1$ and $\lambda_{2,3} = \frac{-1\pm\sqrt{3}}{2}$. There is no dominant eigenvalue since $|\lambda_1| = |\lambda_2| = |\lambda_3|$. The population stabilizes.
5. The eigenvalues are $0, 0, -0.656 \pm 0.626i$, and $\lambda_1 = 1.313$. Therefore, the population increases by 31.3% every 15 years. The normalized eigenvector is given by

   $$\hat{X} = \begin{pmatrix} 0.415 \\ 0.283 \\ 0.173 \\ 0.092 \\ 0.035 \end{pmatrix}.$$

6. See [4].
7. Before insecticide is applied, $\lambda_1 = 1.465$, which means that the population increases by 46.5% every six months. The normalized eigenvector is

   $$\hat{X} = \begin{pmatrix} 0.764 \\ 0.208 \\ 0.028 \end{pmatrix}.$$

After the insecticide is applied, $\lambda_1 = 1.082$, which means that the population increases by 8.2% every six months. The normalized eigenvector is given by

$$\hat{X} = \begin{pmatrix} 0.695 \\ 0.257 \\ 0.048 \end{pmatrix}.$$

8. For this policy, $d_1 = 0.1$, $d_2 = 0.4$, and $d_3 = 0.6$. The dominant eigenvalue is $\lambda_1 = 1.017$ and the normalized eigenvector is

$$\hat{X} = \begin{pmatrix} 0.797 \\ 0.188 \\ 0.015 \end{pmatrix}.$$

9. Without any harvesting the population would double each year since $\lambda_1 = 2$.

(i) $\quad\quad\quad\quad\quad\quad\quad \lambda_1 = 1; \quad\quad \hat{X} = \begin{pmatrix} 24/29 \\ 4/29 \\ 1/29 \end{pmatrix}.$

(ii) $\quad\quad\quad\quad\quad\quad\quad h_1 = 6/7; \quad\quad \hat{X} = \begin{pmatrix} 2/3 \\ 2/9 \\ 1/9 \end{pmatrix}.$

(iii) $\quad\quad\quad\quad\quad\quad\quad \lambda_1 = 1.558; \quad\quad \hat{X} = \begin{pmatrix} 0.780 \\ 0.167 \\ 0.053 \end{pmatrix}.$

(iv) $\quad\quad h_1 = 0.604, \quad \lambda_1 = 1.433; \quad\quad \hat{X} = \begin{pmatrix} 0.761 \\ 0.177 \\ 0.062 \end{pmatrix}.$

(v) $\quad\quad\quad\quad\quad\quad\quad \lambda_1 = 1.672; \quad\quad \hat{X} = \begin{pmatrix} 0.668 \\ 0.132 \\ 0.199 \end{pmatrix}.$

10. Take $h_2 = h_3 = 1$; then $\lambda_1 = 1$, $\lambda_2 = -1$, and $\lambda_3 = 0$. The population stabilizes.

19.2 Chapter 2

1. The iterates give orbits with periods (i) one, (ii) one, (iii) three, and (iv) nine. There are two points of period one, two points of period two, six points of period three, and twelve points of period four. In general, there are $2^N -$ (sum of points of periods that divide N) points of period N.

2. (a) The functions are given by

$$T^2(x) = \begin{cases} \frac{9}{4}x, & 0 \leq x < \frac{1}{3}, \\ \frac{3}{2} - \frac{9}{4}x, & \frac{1}{3} \leq x < \frac{1}{2}, \\ \frac{9}{4}x - \frac{3}{4}, & \frac{1}{2} \leq x < \frac{2}{3}, \\ \frac{9}{4}(1-x), & \frac{2}{3} \leq x \leq 1, \end{cases}$$

and

$$T^3(x) = \begin{cases} \frac{27}{8}x, & 0 \le x < \frac{2}{9}, \\ \frac{3}{2} - \frac{27}{8}x, & \frac{2}{9} \le x < \frac{1}{3}, \\ \frac{27}{8}x - \frac{3}{4}, & \frac{1}{3} \le x < \frac{4}{9}, \\ \frac{9}{4} - \frac{27}{8}x, & \frac{4}{9} \le x < \frac{1}{2}, \\ \frac{27}{8}x - \frac{9}{8}, & \frac{1}{2} \le x < \frac{5}{9}, \\ \frac{21}{8} - \frac{27}{8}x, & \frac{5}{9} \le x < \frac{2}{3}, \\ \frac{27}{8}x - \frac{15}{8}, & \frac{2}{3} \le x < \frac{7}{9}, \\ \frac{27}{8}(1-x), & \frac{7}{9} \le x < 1. \end{cases}$$

There are two points of period one, two points of period two, and no points of period three.

(b) $x_{1,1} = 0$, $x_{1,2} = \frac{9}{14}$; $x_{2,1} = \frac{45}{106}$, $x_{2,2} = \frac{81}{106}$; $x_{3,1} = \frac{45}{151}$, $x_{3,2} = \frac{81}{151}$, $x_{3,3} = \frac{126}{151}$, $x_{3,4} = \frac{225}{854}$, $x_{3,5} = \frac{405}{854}$, $x_{3,6} = \frac{729}{854}$.

4. Use functions of functions to determine f_μ^N. There are two, two, six, and twelve points of periods one, two, three, and four, respectively.

5. A value consistent with period-two behavior is $\mu = 0.011$. Points of period two satisfy the equation
$$\mu^2 x^2 - 100\mu^2 x - \mu x + 100\mu + 1 = 0.$$

6. Edit Program 2e.

7. Points of period one are $(-\frac{3}{10}, -\frac{3}{10})$ and $(\frac{1}{5}, \frac{1}{5})$. Two points of period two are given by $(\frac{x_1}{2}, \frac{0.1-x_1}{2})$, where x_1 is a root of $5x^2 - x - 1 = 0$. The inverse map is given by
$$x_{n+1} = y_n, \quad y_{n+1} = \frac{10}{9}\left(x_n - \frac{3}{50} + y_n^2\right).$$

8. (a) The eigenvalues are given by $\lambda_{1,2} = -\alpha x \pm \sqrt{\alpha^2 x^2 + \beta}$. A bifurcation occurs when one of the $|\lambda| = 1$. Take the case where $\lambda = -1$.

9. (a) (i) When $a = 0.2$, $c_{1,1} = 0$ is stable, $c_{1,2} = 0.155$ is unstable, and $c_{1,3} = 0.946$ is stable.

   (ii) When $a = 0.3$, $c_{1,1} = 0$ is stable, $c_{1,2} = 0.170$ is unstable, and $c_{1,3} = 0.897$ is unstable.

10. See [1].

19.3 Chapter 3

1. (i) The orbit remains bounded forever, $z_{500} \approx -0.3829 + 0.1700i$.
   (ii) The orbit is unbounded, $z_{10} \approx -0.6674 \times 10^{197} + 0.2396 \times 10^{197}$.

2. Fixed points of period one are given by

$$z_{1,1} = \frac{1}{2} + \frac{1}{4}\sqrt{10 + 2\sqrt{41}} - \frac{i}{4}\sqrt{2\sqrt{41} - 10},$$

$$z_{1,2} = \frac{1}{2} - \frac{1}{4}\sqrt{10 + 2\sqrt{41}} + \frac{i}{4}\sqrt{2\sqrt{41} - 10}.$$

Fixed points of period two are given by

$$z_{2,1} = -\frac{1}{2} + \frac{1}{4}\sqrt{2 + 2\sqrt{17}} - \frac{i}{4}\sqrt{2\sqrt{17} - 2},$$

$$z_{2,2} = -\frac{1}{2} - \frac{1}{4}\sqrt{2 + 2\sqrt{17}} + \frac{i}{4}\sqrt{2\sqrt{17} - 2}.$$

3. Use the MATLAB program given in Section 3.3; $J(0, 0)$ is a circle and $J(-2, 0)$ is a line segment.

4. There is one fixed point located approximately at $z_{1,1} = 1.8202 - 0.0284i$.

5. See the example in the text. The curves are again a cardioid and a circle but the locations are different in this case.

7. Fixed points of period one are given by

$$z_{1,1} = \frac{3 + \sqrt{9 - 4c}}{2}, \quad z_{1,2} = \frac{3 - \sqrt{9 - 4c}}{2}.$$

Fixed points of period two are given by

$$z_{2,1} = \frac{1 + \sqrt{5 - 4c}}{2}, \quad z_{2,2} = \frac{1 - \sqrt{5 - 4c}}{2}.$$

9. (i) Period four and (ii) period three.

19.4 Chapter 4

1. There are 11 points of period one.

3. Find an expression for E_n in terms of E_{n+1}.

5. See the paper of Li and Ogusu [4].

6. (a) Bistable: $4.765 - 4.766$ Wm^{-2}. Unstable: $6.377 - 10.612$ Wm^{-2}.

   (b) Bistable: $3.936 - 5.208$ Wm^{-2}. Unstable: $4.74 - 13.262$ Wm^{-2}.

   (c) Bistable: $3.482 - 5.561$ Wm^{-2}. Unstable: $1.903 - 3.995$ Wm^{-2}.

8. Use the function $G(x) = ae^{-bx^2}$ to generate the Gaussian pulse. The parameter b controls the width of the pulse.

19.5 Chapter 5

1. (a) The length remaining at stage k is given by

$$L = 1 - \frac{2}{5} - \frac{2 \times 3}{5^2} - \cdots - \frac{2 \times 3^{k-1}}{5^k}.$$

   The dimension is $D_f = \frac{\ln 3}{\ln 5} \approx 0.6826$.

   (b) See reference [5], for example; $D_f = \frac{\ln 2}{\ln \sqrt{2}} = 2$. If the fractal were constructed to infinity there would be no holes and the object would have the same dimension as a plane. Thus this mathematical object is not a fractal.

2. See Figure 5.8 as a guide.

3. The figure is similar to the stage 3 construction of the Sierpiński triangle. In fact, this gives yet another method for constructing this fractal as Pascal's triangle is extended to infinity.

4. The dimension is $D_f = \frac{\ln 8}{\ln 3} \approx 1.8928$.

6. Note that $\sum_{i=1}^{N} p_i(l) = 1$ and show that

$$\frac{d}{dq} \ln\left(\sum_{i=1}^{N} p_i^q\right) = \frac{\left(\sum_{i=1}^{N} p_i^q \ln p_i\right)}{\left(\sum_{i=1}^{N} p_i^q\right)},$$

   and take the limit $q \to 1$.

7. (i) The fractal is homogeneous;

   (ii) $\alpha_{max} \approx 1.26$ and $\alpha_{min} \approx 0.26$;

   (iii) $\alpha_{max} \approx 0.83$ and $\alpha_{min} \approx 0.46$. Take $k = 500$ in the plot commands.

8. Using the same methods as in Example 1:

$$D_0 = \frac{\ln 4}{\ln 3}, \quad \alpha_s = \frac{s \ln p_1 + (k-s) \ln p_2}{-k \ln 3}, \quad \text{and} \quad -f_s = \frac{\ln\left(2^k \binom{k}{s}\right)}{-k \ln 3}.$$

9. At the kth stage, there are 5^k segments of length 3^{-k}. A number

$$N_s = 3^{k-s} 2^s \binom{k}{s}$$

   of these have weight $p_1^{k-s} p_2^s$. Use the same methods as in Example 1.

10. Using multinomials,

$$\alpha_s = \frac{n_1 \ln p_1 + n_2 \ln p_2 + n_3 \ln p_3 + n_4 \ln p_4}{\ln 3^{-k}} \quad \text{and} \quad -f_s = \frac{\ln \frac{4!}{n_1! n_2! n_3! n_4!}}{\ln 3^{-k}},$$

    where $n_1 + n_2 + n_3 + n_4 = k$.

19.6 Chapter 6

1. Take the transformations $x_n = \frac{1}{a}u_n$ and $y_n = \frac{b}{a}v_n$.
2. There are one control range when $p = 1$, three control ranges when $p = 2$, seven control ranges when $p = 3$, and twelve control ranges when $p = 4$.
3. Points of period one are located at approximately $(-1.521, -1.521)$ and $(0.921, 0.921)$. Points of period two are located near $(-0.763, 1.363)$ and $(1.363, -0.763)$.
4. See the paper of Chau [4].
5. See Section 6.3.
6. The two-dimensional mapping is given by

$$x_{n+1} = A + B(x_n \cos(x_n^2 + y_n^2) - y_n \sin(x_n^2 + y_n^2)),$$
$$y_{n+1} = B(x_n \sin(x_n^2 + y_n^2) + y_n \cos(x_n^2 + y_n^2)).$$

7. The one point of period one is located near $(2.731, 0.413)$.
8. (i) There are three points of period one.
   (ii) There are nine points of period one.
9. See our research paper [2].
10. The control region is very small and targeting is needed in this case. The chaotic transients are very long. Targeting is not required in Exercise 9, where the control region is much larger. Although there is greater flexibility (nine points of period one) with this system, the controllability is reduced.

19.7 Chapter 7

1. (a) $y = \frac{C}{x}$; (b) $y = Cx^2$; (c) $y = C\sqrt{x}$; (d) $\frac{1}{y} = \ln(\frac{C}{x})$;
   (e) $\frac{y^4}{4} + \frac{x^2 y^2}{2} = C$; (f) $y = Ce^{-\frac{1}{x}}$.
2. The fossil is 8.03×10^6 years old.
3. (a) $\dot{d} = k_f(a_0 - d)(b_0 - d)(c_0 - d) - k_r(d_0 + d)$;
   (b) $\dot{x} = k_f(a_0 - 3x)^3 - k_r x$, where $a = [A]$, $x = [A_3]$, $b = [B]$, $c = [C]$, and $d = [D]$.
4. (a) The current is $I = 0.733$ amps.
   (b) The charge is $Q(t) = 50(1 - \exp(-10t - t^2))$ coulombs.
5. Time 1.18 hours.
6. The concentration of glucose is

$$g(t) = \frac{G}{100kV} - Ce^{-kt}.$$

7. The differential equations are

$$\dot{A} = -\alpha A, \qquad \dot{B} = \alpha A - \beta B, \qquad \dot{C} = \beta B.$$

9. The differential equations are

$$\dot{H} = -aH + bI, \qquad \dot{I} = aH - (b+c)I, \qquad \dot{D} = cI.$$

The number of dead is given by

$$D(t) = acN\left(\frac{\alpha - \beta + \beta e^{\alpha t} - \alpha e^{\beta t}}{\alpha\beta(\alpha - \beta)}\right),$$

where α and β are the roots of $\lambda^2 + (a+b+c)\lambda + ac = 0$. This is not realistic as the whole population eventually dies. In reality people recover and some are immune.

10. (a) (i) The solution is $x^3 = 1/(1 - 3t)$, with maximal interval (MI) $-\infty < t < \frac{1}{3}$;
   (ii) $x(t) = (e^t + 3)/(3 - e^t)$, with MI $-\infty < t < \ln 3$;
   (iii) $x(t) = 6/(3 - e^{2t})$, with MI $-\infty < t < \ln\sqrt{3}$;

   (b) The solution is $x(t) = (t + x_0^{1/2} - t_0)^2$, with MI $t_0 - x_0^{1/2} < t < \infty$.

19.8 Chapter 8

1. (a) Eigenvalues and eigenvectors are $\lambda_1 = -10$, $(-2, 1)^T$; $\lambda_2 = -3$, $(\frac{3}{2}, 1)^T$. The origin is a stable node.

   (b) Eigenvalues and eigenvectors are $\lambda_1 = -4$, $(1, 0)^T$; $\lambda_2 = 2$, $(-\frac{4}{3}, 1)^T$. The origin is a saddle point.

2. (a) All trajectories are vertical and there are an infinite number of critical points on the line $y = -\frac{x}{2}$.

   (b) All trajectories are horizontal and there are an infinite number of critical points on the line $y = -\frac{x}{2}$.

   (c) Eigenvalues and eigenvectors are $\lambda_1 = 5$, $(2, 1)^T$; $\lambda_2 = -5$, $(1, -2)^T$. The origin is a saddle point.

   (d) Eigenvalues are $\lambda_1 = 3 + i$, $\lambda_2 = 3 - i$, and the origin is an unstable focus.

   (e) There are two repeated eigenvalues and one linearly independent eigenvector: $\lambda_1 = -1$, $(-1, 1)^T$. The origin is a stable degenerate node.

   (f) This is a nonsimple fixed point. There are an infinite number of critical points on the line $y = x$.

3. (a) $\dot{x} = y$, $\dot{y} = -25x - \mu y$;

   (b) (i) unstable focus, (ii) center, (iii) stable focus, (iv) stable node;

   (c) (i) oscillations grow, (ii) periodic oscillations, (iii) damping, (iv) *critical damping*.

   The constant μ is called the *damping coefficient*.

19. Solutions to Exercises

4. (a) There is one critical point at the origin, which is a col. Plot the isoclines. The eigenvalues are $\lambda = \frac{-1 \pm \sqrt{5}}{2}$ with eigenvectors

$$\begin{pmatrix} 1 \\ \lambda_1 \end{pmatrix} \text{ and } \begin{pmatrix} 1 \\ \lambda_2 \end{pmatrix}.$$

   (b) There are two critical points at $A = (0, 2)$ and $B = (1, 0)$. A is a stable focus and B is a col with eigenvalues and corresponding eigenvectors given by

$$\lambda_1 = 1, \quad \begin{pmatrix} 1 \\ -3 \end{pmatrix} \text{ and } \lambda_2 = -2, \quad \begin{pmatrix} 1 \\ 0 \end{pmatrix}.$$

   (c) There are two critical points at $A = (1, 1)$ and $B = (1, -1)$. A is an unstable focus and B is a stable focus. Plot the isoclines where $\dot{x} = 0$ and $\dot{y} = 0$.

   (d) There are three critical points at $A = (2, 0)$, $B = (1, 1)$, and $C = (1, -1)$; A is a col and B and C are both stable foci.

   (e) There is one nonhyperbolic critical point at the origin. The solution curves are given by $y^3 = x^3 + C$. The line $y = x$ is invariant, the flow is horizontal on $\dot{y} = x^2 = 0$, and the flow is vertical on the line $\dot{x} = y^2 = 0$. The slope of the trajectories is given by $\frac{dy}{dx} = \frac{x^2}{y^2}$.

   (f) There is one nonhyperbolic critical point at the origin. The solution curves are given by $y = \frac{x}{1+Cx}$. The line $y = x$ is invariant.

   (g) There is one nonhyperbolic critical point at the origin. The solution curves are given by $2y^2 = x^4 + C$. The slope of the orbits is given by $\frac{dy}{dx} = \frac{x^3}{y}$.

   (h) When $\mu < 0$ there are no critical points. When $\mu = 0$, the solution curves are given by $|x| = Ce^{\frac{1}{y}}$. When $\mu > 0$, there are two critical points at $A = (0, \sqrt{\mu})$ and $B = (0, -\sqrt{\mu})$; A is a col and B is an unstable node.

5. One possible system is

$$\dot{x} = y^2 - x^2, \qquad \dot{y} = x^2 + y^2 - 2,$$

   for example.

6. There are three critical points at $O = (0, 0)$, $A = (1, 0)$, and $B = (-1, 0)$. If $a_0 > 0$, since $\det J_O > 0$ and trace $J_O < 0$, the origin is stable and A and B are cols because $\det J < 0$ for these points. If $a_0 < 0$, the origin is unstable and A and B are still cols. Therefore, if $a_0 > 0$, the current in the circuit eventually dies away to zero with increasing time. If $a_0 < 0$, the current increases indefinitely, which is physically impossible.

7. There are three critical points at $O = (0, 0)$, $A = (\frac{a}{b}, 0)$, and $B = (\frac{c+a}{b}, \frac{c(c+a)}{b})$. The origin is an unstable node and A is a col. The critical point at B is stable since $\det J_B > 0$ and trace $J_b < 0$. Therefore, the population and birth rate stabilize to the values given by B in the long term.

8. When $\alpha\beta > 1$, there is one stable critical point at $A = (0, \frac{1}{\beta})$. When $\alpha\beta < 1$, A becomes a col and $B = (\sqrt{1-\alpha\beta}, \alpha)$ and $C = (-\sqrt{1-\alpha\beta}, \alpha)$ are both stable. When $\alpha\beta > 1$, the power goes to zero and the velocity of the wheel tends to $\frac{1}{\beta}$ and when $\alpha\beta < 1$, the power and velocity stabilize to the point B.

9. (a) There is one critical point at $(\frac{KG_0}{K-C}, \frac{G_0}{K-C})$, which is in the first quadrant if $K > C$. When $C = 1$, the critical point is nonhyberbolic. The system can be solved and there are closed trajectories around the critical point. The economy oscillates (as long as $I(t), S(t) > 0$). If $C \neq 1$, then the critical point is unstable if $0 < C < 1$ and stable if $C > 1$.

   (b) The critical point is stable and the trajectory tends to this point. The choice of initial condition is important to avoid $I(t)$ or $S(t)$ from going negative, where the model is no longer valid.

10. Note that $\frac{d\eta}{d\tau} = e^t$ and $\frac{d^2\eta}{d\tau^2} = \frac{d\eta}{d\tau}\frac{dt}{d\tau}$. There are four critical points: $O = (0, 0)$, an unstable node; $A = (-1, 0)$, a col; $B = (0, 2)$, a col; and $C = (-\frac{3}{2}, \frac{1}{2})$, a stable focus.

19.9 Chapter 9

1. This is a competing species model. There are four critical points in the first quadrant at $O = (0, 0), P = (0, 3), Q = (2, 0)$, and $R = (1, 1)$. The point O is an unstable node, P and Q are both stable nodes, and R is a saddle point. There is mutual exclusion and one of the species will become extinct depending on the initial populations.

2. This is a Lotka–Volterra model with critical points at $O = (0, 0)$ and $A = (3, 2)$. The system is structurally unstable. The populations oscillate but the cycles are dependent on the initial values of x and y.

3. This is a predator–prey model. There are three critical points in the first quadrant at $O = (0, 0), F = (2, 0)$, and $G = (\frac{3}{2}, \frac{1}{2})$. The points O and F are saddle points and G is a stable focus. In terms of species behavior, the two species coexist and approach constant population values.

4. Consider the three cases separately.

   (i) If $0 < \mu < \frac{1}{2}$, then there are four critical points at $O = (0, 0), L = (2, 0), M = (0, \mu)$, and $N = (\frac{\mu-2}{\mu^2-1}, \frac{\mu(2\mu-1)}{\mu^2-1})$. The point O is an unstable node, L and M are saddle points, and N is a stable point. To classify the critical points, consider det J and trace J. The two species coexist.

   (ii) If $\frac{1}{2} < \mu < 2$, then there are three critical points in the first quadrant, all of which lie on the axes. The point O is an unstable node, L is a stable node, and M is a saddle point. Species y becomes extinct.

   (iii) If $\mu > 2$, then there are four critical points in the first quadrant. The point O is an unstable node, L and M are stable nodes, and N is a saddle point. One species becomes extinct.

5. (a) A predator–prey model. There is coexistence; the populations stabilize to the point $(\frac{5}{4}, \frac{11}{4})$.

(b) A competing species model. There is mutual exclusion; one species becomes extinct.

6. There are three critical points in the first quadrant if $0 \leq \epsilon < 1$: at $O = (0, 0)$, $A = (\frac{1}{\epsilon}, 0)$ and $B = (\frac{1+\epsilon}{1+\epsilon^2}, \frac{1-\epsilon}{1+\epsilon^2})$. There are two when $\epsilon \geq 1$. The origin is always a col. When $\epsilon = 0$, the system is Lotka–Volterra, and trajectories lie on closed curves away from the axes. If $0 < \epsilon < 1$, A is a col, and B is stable since the trace of the Jacobian is negative and the determinant is positive. When $\epsilon \geq 1$, A is stable.

7. There are three critical points at $O = (0, 0)$, $P = (1, 0)$, and $Q = (0.6, 0.24)$. Points O and P are cols and Q is stable. There is coexistence.

8. There is a limit cycle enclosing the critical point at $(0.48, 0.2496)$. The populations vary periodically and coexist.

9. One example would be the following. X and Y prey on each other; Y has cannibalistic tendencies and also preys on Z. A diagram depicting this behavior is plotted in Figure 19.1.

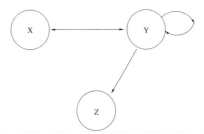

Figure 19.1: One possible interaction between three interacting insect species.

10. Let species X, Y, and Z have populations $x(t)$, $y(t)$, and $z(t)$, respectively. The interactions are as follows: X preys on Y; Z preys on X; Y and Z are in competition.

19.10 Chapter 10

1. Convert to polar coordinates to get

$$\dot{r} = r\left(1 - r^2 - \frac{1}{2}\cos^2\theta\right), \quad \dot{\theta} = -1 + \frac{1}{2}\cos\theta\sin\theta.$$

Since $\dot{\theta} < 0$, the origin is the only critical point. On $r = \frac{1}{2}$, $\dot{r} > 0$, and on $r = 2$, $\dot{r} < 0$. Therefore, there exists a limit cycle by the corollary to the Poincaré–Bendixson theorem.

3. Plot the graph of $y = x - x^3 \cos^3(\pi x)$ to prove that the origin is the only critical point inside the square. Linearize to show that the origin is an unstable focus. Consider the flow on the sides of the rectangle, for example, on $x = 1$, with $-1 \leq y \leq 1$, $\dot{x} = -y + \cos\pi \leq 0$. Hence the flow is from right to left on this line. Show that the rectangle is invariant and use the corollary to the Poincaré–Bendixson theorem.

5. Plot the graph of $y = x^8 - 3x^6 + 3x^4 - 2x^2 + 2$ to prove that the origin is a unique critical point. Convert to polar coordinates to get

$$\dot{r} = r(1 - r^2(\cos^4\theta + \sin^4\theta)), \qquad \dot{\theta} = 1 - r^2\cos\theta\sin\theta(\sin^2\theta - \cos^2\theta).$$

Now div(**X**) $= 2 - 3r^2$ and so div(**X**) is nonzero in the annulus $A = \{1 < r < 2\}$. On the circle $r = 1 - \epsilon$, $\dot{r} > 0$, and on the circle $r = 2 + \epsilon$, $\dot{r} < 0$. Therefore, there is a unique limit cycle contained in the annulus by Dulac's criteria.

7. Consider the isocline curves as in Figures 10.4(b) and 10.5(b). If the straight line intersects the parabola to the right of the maximum, then there is no limit cycle. If the straight line intersects the parabola to the left of the maximum, then there exists a limit cycle.

8. (a) The limit cycle is circular.

   (b) The limit cycle has fast and slow branches.

9. It will help if you draw rough diagrams.

   (a) Now div(**X**) $= -(1 + x^2 + x^4) < 0$. Hence there are no limit cycles by Bendixson's criteria.

   (b) Now div(**X**) $= 2 - x$. There are four critical points at $(0, 0)$, $(1, 0)$, $(-1, 1)$, and $(-1, -1)$. The x-axis is invariant. On $x = 0$, $\dot{x} = 2y^2 \geq 0$. Hence there are no limit cycles in the plane.

   (c) Now div(**X**) $= -6 - 2x^2 < 0$. Hence there are no limit cycles by Bendixson's criteria.

   (d) Now div(**X**) $= -3 - x^2 < 0$. Hence there are no limit cycles by Bendixson's criteria.

   (e) Now div(**X**) $= 3x - 2$, and div(**X**) $= 0$ on the line $x = \frac{2}{3}$. There are three critical points at $(1, 0)$, $(-1, 0)$, and $(2, 3)$. The x-axis is invariant, and $\dot{x} < 0$ for $y > 0$ on the line $x = \frac{2}{3}$. Hence there are no limit cycles by Bendixson's criteria.

   (f) Now div(**X**) $= -3x^2y^2$. Therefore, there are no limit cycles lying entirely in one of the quadrants. However, $\dot{x} = -y^2$ on the line $x = 0$ and $\dot{y} = x^5$ on the line $y = 0$. Hence there are no limit cycles by Bendixson's criteria.

   (g) Now div(**X**) $= (x - 2)^2$. On the line $x = 2$, $\dot{x} = -y^2$, and so no limit cycle can cross this line. Hence there are no limit cycles by Bendixson's criteria.

10. (a) The axes are invariant. Now div$(\psi\mathbf{X}) = \frac{1}{xy^2}(2 - 2x)$ and so div$(\psi\mathbf{X}) = 0$ when $x = 1$. There are four critical points and only one, (-16,38), lying wholly in one of the quadrants. Since the divergence is nonzero in this quadrant, there are no limit cycles.

    (b) Now div$(\psi\mathbf{X}) = -\frac{\delta}{y} - \frac{d}{x}$ and so div$(\psi\mathbf{X}) = 0$ when $y = -\frac{\delta x}{d}$. Since $\delta > 0$ and $d > 0$, there are no limit cycles contained in the first quadrant.

19.11 Chapter 11

1. The Hamiltonian is $H(x, y) = \frac{y^2}{2} - \frac{x^2}{2} + \frac{x^4}{4}$. There are three critical points: $(0, 0)$, which is a saddle point and $(1, 0)$ and $(-1, 0)$, which are both centers.

2. There are three critical points: $(0, 0)$, which is a center, and $(1, 0)$ and $(-1, 0)$, which are both saddle points.

3. The critical points occur at $(n\pi, 0)$, where n is an integer. When n is odd, the critical points are saddle points, and when n is even the critical points are stable foci. The system is now damped and the pendulum swings less and less, eventually coming to rest at $\theta = 2n\pi$ degrees. The saddle points represent the unstable equilibria when $\theta = (2n + 1)\pi$ degrees.

4. The Hamiltonian is $H(x, y) = \frac{y^4}{4} - \frac{y^2}{2} - \frac{x^2}{2} + \frac{x^4}{4}$. There are nine critical points.

5. (a) The origin is asymptotically stable.

    (b) The origin is asymptotically stable if $x < \alpha$ and $y < \beta$.

    (c) The origin is unstable.

6. The origin is asymptotically stable. The positive limit sets are either the origin or the ellipse $4x^2 + y^2 = 1$, depending on the value of p.

7. The function $V(x, y)$ is a Lyapunov function if $a > \frac{1}{4}$.

8. The basin of attraction of the origin is the circle $x^2 + y^2 < 4$.

10. Now $\dot{V} = -8(x^4 + 3y^6)(x^4 + 2y^2 - 10)^2$. The origin is unstable, and the curve $x^4 + 2y^2 = 10$ is an attractor.

19.12 Chapter 12

1. (a) There is one critical point when $\mu \leq 0$, and there are two critical points when $\mu > 0$. This is a saddle-node bifurcation.

    (b) When $\mu < 0$, there are two critical points and the origin is stable. When $\mu > 0$, there is one critical point at the origin which is unstable. The origin undergoes a transcritical bifurcation.

    (c) There is one critical point at the origin when $\mu \leq 0$, and there are three critical points—two are unstable—when $\mu > 0$. This is called a *subcritical pitchfork bifurcation*.

3. Possible examples include

    (i) $\dot{x} = \mu x(\mu^2 - x^2)$;

    (ii) $\dot{x} = x^4 - \mu^2$; and

    (iii) $\dot{x} = x(\mu^2 + x^2 - 1)$.

4. The critical points are given by $O = (0, 0)$, $A = \frac{12+\sqrt{169-125h}}{5}$, and $B = \frac{12-\sqrt{169-125h}}{5}$. There are two critical points if $h \leq 0$, the origin is unstable, and A is stable (but negative harvesting is discounted). There are three critical points if

$0 < h < 1.352$, the origin and A are stable, and B is unstable. There is one stable critical point at the origin if $h \geq 1.352$.

The term $x(1 - \frac{x}{5})$ represents the usual logistic growth when there is no harvesting. The term $\frac{hx}{0.2+x}$ represents harvesting from h is zero up to a maximum of h, no matter how large x becomes (plot the graph).

When $h = 0$, the population stabilizes to 5×10^5; when $0 < h < 1.352$, the population stabilizes to $A \times 10^5$; and when $h > 1.352$, the population decreases to zero. Use the animate command in MATLAB to plot $\dot{x}$ as h varies from zero to eight. The harvesting is *sustainable* if $0 < h < 1.352$, where the fish persist, and it is *unsustainable* if $h > 1.352$, when the fish become extinct from the lake.

5. (a) No critical points if $\mu < 0$. There is one nonhyperbolic critical point at $O = (0, 0)$ if $\mu = 0$, and there are two critical points at $A = (0, \sqrt[4]{\mu})$ and $B = (0, -\sqrt[4]{\mu})$. Both A and B are unstable.

   (b) There are two critical points at $O = (0, 0)$ and $A = (\mu^2, 0)$ if $\mu \neq 0$ (symmetry). O is stable and A is unstable. There is one nonhyperbolic critical point at $O = (0, 0)$ if $\mu = 0$.

   (c) There are no critical points if $\mu < 0$. There is one nonhyperbolic critical point at $O = (0, 0)$ if $\mu = 0$, and there are four critical points at $A = (2\sqrt{\mu}, 0)$, $B = (-2\sqrt{\mu}, 0)$, $C = (\sqrt{\mu}, 0)$, and $D = (-\sqrt{\mu}, 0)$ if $\mu > 0$. The points A and D are stable, while B and C are unstable.

6. (a) If $\mu < 0$, there is a stable critical point at the origin and an unstable limit cycle of radius $r = -\mu$. If $\mu = 0$, the origin is a center, and if $\mu > 0$, the origin becomes unstable. The flow is counterclockwise.

   (b) If $\mu \leq 0$, the origin is an unstable focus. If $\mu > 0$, the origin is unstable, and there is a stable limit cycle of radius $r = \frac{\mu}{2}$ and an unstable limit cycle of radius $r = \mu$.

   (c) If $\mu \neq 0$, the origin is unstable and there is a stable limit cycle of radius $|r| = \mu$. If $\mu = 0$, the origin is stable.

7. There is one critical point at the origin and there are at most two stable limit cycles. As μ increases through zero there is a Hopf bifurcation at the origin. Next there is a saddle-node bifurcation to a large-amplitude limit cycle. If μ is then decreased back through zero, there is another saddle-node bifurcation back to the steady state at the origin.

8. If $\mu < 0$, the origin is unstable, and if $\mu = 0$, $\dot{r} > 0$ if $r \neq 0$ the origin is unstable and there is a semistable limit cycle at $r = 1$. If $\mu > 0$, the origin is unstable, there is a stable limit cycle of radius $r = \frac{2+\mu-\sqrt{\mu^2+4\mu}}{2}$ and an unstable limit cycle of radius $r = \frac{2+\mu+\sqrt{\mu^2+4\mu}}{2}$. It is known as a fold bifurcation because a fold in the graph of $y = (r-1)^2 - \mu r$ crosses the r-axis at $\mu = 0$.

9. If $\mu < 0$, the origin is a stable focus and as μ passes through zero, the origin changes from a stable to an unstable spiral. If $\mu > 0$, convert to polars. The origin is unstable and a stable limit cycle bifurcates.

424   19. Solutions to Exercises

10. The critical points occur at $A = (0, -\frac{\alpha}{\beta})$ and $B = (\alpha + \beta, 1)$. Thus there are two critical points everywhere in the (α, β) plane apart from along the line $\alpha = -\beta$ where there is only one. The eigenvalues for the matrix J_A are $\lambda_1 = \beta$ and $\lambda_2 = -\frac{(\alpha+\beta)}{\beta}$. The eigenvalues for the matrix J_B are $\lambda = \frac{-\alpha \pm \sqrt{\alpha^2 - 4(\alpha+\beta)}}{2}$. There is a codimension-2 bifurcation along the line $\alpha = -\beta$ and it is a transcritical bifurcation.

19.13 Chapter 13

1. Eigenvalues and eigenvectors given by $[3, (-2, -2, 1)^T]$, $[-3, (-2, 1, -2)^T]$, and $[9, (1, -2, -2)^T]$. The origin is unstable; there is a col in two planes and an unstable node in the other.

2. Eigenvalues are $\lambda_{1,2} = 1 \pm i\sqrt{6}, \lambda_3 = 1$. The origin is unstable and the flow is rotating. Plot solution curves using MATLAB.

3. There are two critical points at $O = (0, 0, 0)$ and $P = (-1, -1, -1)$. The critical points are both hyperbolic and unstable. The eigenvalues for O are $[1, 1, -1]$ and those for P are $[1, -1, -1]$.

4. Consider the flow on $x = 0$ with $y \geq 0$ and $z \geq 0$, etc. The first quadrant is positively invariant. The plane $x + y + 2z = k$ is invariant since $\dot{x} + \dot{y} + 2\dot{z} = 0$. Hence if a trajectory starts on this plane, then it remains there forever. The critical points are given by $(\frac{\lambda y}{1+y}, y, \frac{y}{2})$. Now on the plane $x + y + 2z = k$, the critical point satisfies the equation $\frac{\lambda y}{1+y} + y + y = k$, which has solutions $y = \frac{(2-\lambda) \pm \sqrt{(2-\lambda)^2 + 32}}{4}$. Since the first quadrant is invariant, $\lambda^+(p)$ must tend to this critical point.

5. (a) Take $V = x^2 + y^2 + z^2$. Then $\dot{V} = -2(x^2 + y^4 + (y - z^2)^2 + (z - x^2)^2) \leq 0$. Now $\dot{V} = 0$ if and only if $x = y = z = 0$; hence the origin is globally asymptotically stable.

   (b) Consider $V = ax^2 + by^2 + cz^2$. Now $\dot{V} = -2(a^2x^2 + b^2y^2 + c^2z^2) + 2xyz(ax + by + cz)$. Hence $\dot{V} < \frac{V^2}{c} - 2cV$ and $\dot{V} < 0$ in the set $V < 2c^2$. Therefore, the origin is asymptotically stable in the ellipsoid $V < 2c^2$.

7. There are eight critical points at $(0, 0, 0)$, $(0, 0, \frac{1}{2})$, $(0, \frac{1}{2}, 0)$, $(0, 1, -1)$, $(\frac{1}{2}, 0, 0)$, $(-\frac{1}{3}, 0, \frac{1}{3})$, $(\frac{1}{3}, -\frac{1}{3}, 0)$, and $(\frac{1}{14}, \frac{3}{14}, \frac{3}{14})$. The plane $x + y + z = \frac{1}{2}$ is a solution plane since $\dot{x} + \dot{y} + \dot{z} = (x + y + z) - 2(x + y + z)^2 = 0$ on this plane. There are closed curves on the plane representing periodic behavior. The three species coexist and the populations oscillate in phase. The system is structurally unstable.

8. (i) The populations settle on to a period-two cycle.

   (ii) The populations settle on to a period-four cycle.

10. A Jordan curve lying wholly in the first quadrant exists, similar to the limit cycle for the Liénard system when a parameter is large (see Chapter 16). The choice of q and C are important; see [7].

19.14 Chapter 14

1. Starting with $r_0 = 4$, the returns are $r_1 = 1.13854$, $r_2 = 0.66373$, ..., $r_{10} = 0.15307$, to five decimal places.

2. The Poincaré map is given by
$$r_{n+1} = \mathbf{P}(r_n) = \frac{\mu r_n}{r_n + e^{-2\mu\pi}(\mu - r_n)}.$$

3. Now $\frac{d\mathbf{P}}{dr}|_\mu = e^{-2\mu\pi}$. Therefore, the limit cycle at $r = \mu$ is hyperbolic stable if $\mu > 0$ and hyperbolic unstable if $\mu < 0$. What happens when $\mu = 0$?

4. The Poincaré map is given by
$$r_{n+1} = \mathbf{P}(r_n) = \left(\frac{r_n^2}{r_n^2 + e^{-4\pi}(1 - r_n^2)}\right)^{\frac{1}{2}}.$$

5. The limit cycle at $r = 1$ is stable since $\frac{d\mathbf{P}}{dr}|_{r=1} = e^{-4\pi}$.

6. The Poincaré section in the $p_1 q_1$ plane is crossed 14 times.

7. The trajectory is quasiperiodic.

9. A chaotic attractor is formed.

10. (a) See Figure 19.2(a).

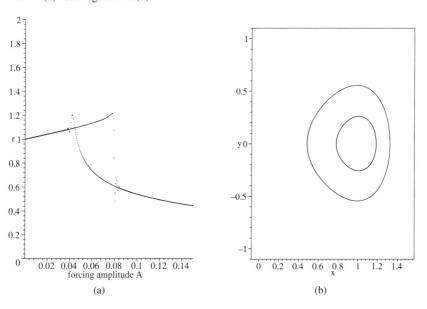

Figure 19.2: (a) Bifurcation diagram. (b) Multistable behavior.

(b) Take $\Gamma = 0.07$. For example, choose initial conditions (i) $x_0 = 1.16$, $y_0 = 0.112$ and (ii) $x_0 = 0.585$, $y_0 = 0.29$.

19.15 Chapter 15

1. Differentiate to obtain $2u\dot{u} = G'(x)\dot{x}$ and find $\frac{dy}{du}$.
2. The Lyapunov quantities are given by $L(i) = a_{2i+1}$, where $i = 0$ to 6.
3. The Lyapunov quantities are given by $L(0) = -a_1$, $L(1) = -3b_{03} - b_{21}$, $L(2) = -3b_{30}b_{03} - b_{41}$, and $L(3) = b_{03}^3$.
8. The homoclinic loop lies on the curve $y^2 = x^2 + \frac{2}{3}x^3$.
10. There are three limit cycles when $\lambda = -0.9$.

19.16 Chapter 16

1. There is one critical point in the finite plane at the origin which is a stable node. The eigenvalues and eigenvectors are given by $\lambda_1 = -1$, $(1, -1)^T$ and $\lambda_2 = -4$, $(1, -4)^T$, respectively. The function $g_2(\theta)$ is defined as

$$g_2(\theta) = -4\cos^2\theta - 5\cos\theta\sin\theta - \sin^2\theta.$$

There are four critical points at infinity at $\theta_1 = \tan^{-1}(-1)$, $\theta_2 = \tan^{-1}(-1) + \pi$, $\theta_3 = \tan^{-1}(-4)$, and $\theta_4 = \tan^{-1}(-4) + \pi$. The flow in a neighborhood of a critical point at infinity is qualitatively equivalent to the flow on $X = 1$ given by

$$\dot{y} = -y^2 - 5y - 4, \qquad \dot{z} = -yz.$$

There are two critical points at $(-1, 0)$, which is a col, and $(-4, 0)$, which is an unstable node. Since n is odd, antinodal points are qualitatively equivalent.

2. There is one critical point in the finite plane at the origin, which is a col. The eigenvalues and eigenvectors are given by $\lambda_1 = 1, (1, 1)^T$ and $\lambda_2 = -1, (2, 1)^T$, respectively. The function $g_2(\theta)$ is defined as

$$g_2(\theta) = -2\cos^2\theta + 6\cos\theta\sin\theta - 4\sin^2\theta.$$

There are four critical points at infinity at $\theta_1 = \tan^{-1}(1)$, $\theta_2 = \tan^{-1}(1) + \pi$, $\theta_3 = \tan^{-1}(\frac{1}{2})$, and $\theta_4 = \tan^{-1}(\frac{1}{2}) + \pi$. The flow in a neighborhood of a critical point at infinity is qualitatively equivalent to the flow on $X = 1$ given by

$$\dot{y} = -4y^2 + 6y - 2, \qquad \dot{z} = 3z - 4yz.$$

There are two critical points at $(1, 0)$, which is a stable node, and $(\frac{1}{2}, 0)$, which is an unstable node. Since n is odd, antinodal points are qualitatively equivalent.

3. There are no critical points in the finite plane. The function $g_3(\theta)$ is given by

$$g_3(\theta) = 4\cos^2\theta\sin\theta - \sin^3\theta.$$

The function has six roots in the interval $[0, 2\pi)$, at $\theta_1 = 0$, $\theta_2 = 1.10715$, $\theta_3 = 2.03444$, $\theta_4 = 3.14159$, $\theta_5 = 4.24874$, and $\theta_5 = 5.1764$. All of the angles are measured in radians. The behavior on the plane $X = 1$ is determined from the system

$$\dot{y} = 4y - 5z^2 - y^3 + yz^2, \qquad \dot{z} = -z - zy^2 + z^3.$$

There are three critical points at $O = (0, 0)$, $A = (2, 0)$, and $B = (-2, 0)$. Points A and B are stable nodes and O is a col. Since n is even, antinodal points are qualitatively equivalent, but the flow is reversed.

All of the positive and negative limit sets for this system are made up of the critical points at infinity.

4. There is one critical point at the origin in the finite plane which is a stable focus. The critical points at infinity occur at $\theta_1 = 0$ radians, $\theta_2 = \frac{\pi}{2}$ radians, $\theta_3 = -\frac{\pi}{2}$ radians, and $\theta_4 = \pi$ radians. Two of the points at infinity are cols and the other two are unstable nodes.

5. There is a unique critical point in the finite plane at the origin which is an unstable node. The critical points at infinity occur at $\theta_1 = 0$ radians, $\theta_2 = \frac{\pi}{2}$ radians, $\theta_3 = -\frac{\pi}{2}$ radians, and $\theta_4 = \pi$ radians. Two of the points at infinity are cols and the other two are unstable nodes. There is at least one limit cycle surrounding the origin by the corollary to the Poincaré–Bendixson theorem.

7. If $a_1 a_3 > 0$, then the system has no limit cycles. If $a_1 a_3 < 0$, there is a unique hyperbolic limit cycle. If $a_1 = 0$ and $a_3 \neq 0$, then there are no limit cycles. If $a_3 = 0$ and $a_1 \neq 0$, then there are no limit cycles. If $a_1 = a_3 = 0$, then the origin is a center by the classical symmetry argument.

8. When ϵ is small one may apply the Melnikov theory of Chapter 15 to establish where the limit cycles occur. The limit cycles are asymptotic to circles centered at the origin. If the degree of F is $2m + 1$ or $2m + 2$, there can be no more than m limit cycles.

When ϵ is large, if a limit cycle exists, it shoots across in the horizontal direction to meet a branch of the curve $y = F(x)$, where the trajectory slows down and remains near the branch until it shoots back across to another branch of $F(x)$ where it slows down again. The trajectory follows this pattern forever. Once more there can be no more than m limit cycles.

9. Use a similar argument to that used in the proof of Theorem 4. See Liénard's paper [5, Chapter 10].

10. The function F has to satisfy the conditions $a_1 > 0$, $a_3 < 0$, and $a_3^2 > 4a_1$, for example. This guarantees that there are five roots for $F(x)$. If there is a local maximum of $F(x)$ at, say, $(\alpha_1, 0)$, a root at $(\alpha_2, 0)$, and a local minimum at $(\alpha_3, 0)$, then it is possible to prove that there is a unique hyperbolic limit cycle crossing $F(x)$ in the interval (α_1, α_2) and a second hyperbolic limit cycle crossing $F(x)$ in the interval (α_3, ∞). Use similar arguments to those used in the proof of Theorem 4.

See [19] for a proof.

19.17 Chapter 17

2. Use the chain rule.

5. (a) Show that

$$\frac{d\mathbf{V}(\mathbf{a})}{dt} = -\sum_{i=1}^{n} \left(\frac{d}{da_i} \left(\phi^{-1}(a_i) \right) \right) \left(\frac{da_i}{dt} \right)^2.$$

(b)

$$V(\mathbf{a}) = -\frac{1}{2}\left(7a_1^2 + 12a_1a_2 - 2a_2^2\right) - \frac{4}{\gamma\pi^2}\left(\log\left(\cos(\pi a_1/2)\right) + \log\left(\cos(\pi a_2/2)\right)\right).$$

There are two stable critical points, one at $(12.98, 3.99)$ and the other at $(-12.98, -3.99)$.

6. The algorithm converges to (a) $\mathbf{x}_2$; (b) $\mathbf{x}_1$; (c) $\mathbf{x}_3$; (d) $-\mathbf{x}_1$.
8. (a) Fixed points of period one satisfy the equation $a = \gamma a + \theta + w\sigma(a)$.
   (b)–(d) See Pasemann's paper ([2] in Chapter 2).
   (e) There is a bistable region for $4.5 < w < 5.5$, approximately.
9. Iterate $10,000$ times. A closed loop starts to form, indicating that the system is quasiperiodic.

19.18 Chapter 18

1. The analytic solution is $I(t) = -3.14\cos(t) + 9.29\sin(t) + 3.14e^{-2t}$.
2. The analytic solution is $I(t) = (-2\cos(4t) - \sin(4t) + 10e^{-2t} - 8e^{-3t})/50$.
3. The analytic solution is $I(t) = (2\sin(t) - \cos(t) + e^{-2t})/\sqrt{2}$.
4. As $\mu \to 0$, the limit cycle becomes a circle.
5. Period four.
6. Chaotic behavior.
7. There are two distinct limit cycles. The system is multistable.
10. See Strogatz's book ([3] in Chapter 12) and Chua's paper [7].

References

Textbooks

P. S. Addison, *Fractals and Chaos: An Illustrated Course*, Institute of Physics, Bristol, PA, 1997.

G. P. Agrawal, *Applications in Nonlinear Fiber Optics*, Academic Press, New York, 2001.

G. P. Agrawal, *Nonlinear Fiber Optics*, 3rd ed., Academic Press, New York, 2001.

N. Arnold, *Chemical Chaos*, Hippo, London, 1997.

D. K. Arrowsmith and C. M. Place, *Dynamical Systems*, Chapman and Hall, London, 1992.

S. Barnet, *Discrete Mathematics*, Addison–Wesley, Reading, MA, 1998.

R. H. Bishop, *Modern Control Systems and Design Using MATLAB and Simulink*, Addison–Wesley, Reading, MA, 1997.

F. Brauer and C. Castillo-Chavez, *Mathematical Models in Population Biology and Epidemiology*, Springer-Verlag, New York, 2001.

C-ODE-E (Consortium for ODE Experiments), *ODE Architect: The Ultimate ODE Power Tool*, Wiley, New York, 1999.

References

J. Cronin, *Differential Equations: Introduction and Qualitative Theory*, 2nd ed., Marcel Dekker, New York, 1994.

R. M. Crownover, *Introduction to Fractals and Chaos*, Jones and Bartlett, Sudbury, MA, 1995.

J. B. Dabney and T. L. Harman, *Mastering Simulink*, Prentice–Hall, Upper Saddle River, NJ, 2003.

B. Daku, *Learn MATLAB Fast: The M-Tutor Interactive Tutorial*, Wiley, New York, 2002.

P. W. Davis, *Differential Equations: Modeling with MATLAB*, Prentice–Hall, Upper Saddle River, NJ, 1999.

P. Dayan and L. F. Abbott, *Theoretical Neuroscience: Computational and Mathematical Modeling of Neural Systems*, MIT Press, Cambridge, UK, 2001.

W. R. Derrick and S. I. Grossman, *Elementary Differential Equations*, Fourth ed., Addison–Wesley, Reading, MA, 1997.

M. Demazure and D. Chillingworth (translator), *Bifurcations and Catastrophes: Geometry of Solutions to Nonlinear Problems*, Springer-Verlag, New York, 2000.

R. L. Devaney, *The Mandelbrot and Julia Sets: A Tool Kit of Dynamics Activities*, Key Curriculum Press, Emeryville, CA, 2002.

K. J. Falconer, *Techniques in Fractal Geometry*, Wiley, New York, 1997.

R. Field and M. Burger, eds., *Oscillations and Travelling Waves in Chemical Systems*, Wiley, New York, 1985.

G. W. Flake, *The Computational Beauty of Nature: Computer Explorations of Fractals*, MIT Press, Cambridge, MA, 1998.

W. J. Freeman, *Neurodynamics: An Exploration in Mesoscopic Brain Dynamics (Perspectives in Neural Computing)*, Springer-Verlag, New York, 2000.

H. M. Gibbs, *Optical Bistability: Controlling Light with Light*, Academic Press, New York, 1985.

S. I. Grossman, *Applications for Elementary Linear Algebra*, Wadsworth, Belmont, CA, 1980.

J. Guckenheimer and P. Holmes, *Nonlinear Oscillations, Dynamical Systems, and Bifurcations of Vector Fields*, 3rd ed., Springer-Verlag, New York, 1990.

M. T. Hagan, H. B. Demuth, and M. H. Beale, *Neural Network Design*, Brooks–Cole, Pacific Grove, CA, 1995.

D. Hanselman and B.R. Littlefield, *Mastering MATLAB* 6, Prentice–Hall, Upper Saddle River, NJ, 2000.

P. Hartman, *Ordinary Differential Equations*, Wiley, New York, 1964.

S. S. Haykin, *Neural Networks: A Comprehensive Foundation*, 2nd ed., Prentice–Hall, Upper Saddle River, NJ, 1998.

D. O. Hebb, *The Organization of Behavior*, Wiley, New York, 1949.

R. C. Hilborn, *Chaos and Nonlinear Dynamics: An Introduction for Scientists and Engineers*, 2nd ed., Oxford University Press, London, 2000.

M. W. Hirsch and S. Smale, *Differential Equations, Dynamical Systems, and Linear Algebra*, Academic Press, New York, 1974.

R. A. Holmgrem, *A First Course in Discrete Dynamical Systems*, Springer-Verlag, New York, 1996.

B. R. Hunt, R. L. Lipsman, J. Rosenberg, and K. R. Coombes, *A Guide to MATLAB*, Cambridge University Press, Cambridge, UK, 2001.

K. Ikeda and M. Murota, *Imperfect Bifurcations in Structures and Materials*, Springer-Verlag, New York, 2002.

Yu. S. Il'yashenko, *Finiteness Theorems for Limit Cycles*, Translations of Mathematical Monographs 94, American Mathematical Society, Providence, 1991.

G. Iooss and D. D. Joseph, *Elementary Stability and Bifurcation Theory*, Springer-Verlag, New York, 1997.

D. W. Jordan and P. Smith, *Nonlinear Ordinary Differential Equations*, 3rd ed., Oxford University Press, London, 1999.

T. Kapitaniak, *Controlling Chaos: Theoretical and Practical Methods in Non-Linear Dynamics*, Academic Press, New York, 1996.

T. Kapitaniak, *Chaos for Engineers: Theory, Applications and Control*, 2nd ed., Springer-Verlag, New York, 2000.

D. Kaplan and L. Glass, *Understanding Nonlinear Dynamics*, Springer-Verlag, New York, 1995.

B. Kosko, *Neural Betworks and Fuzzy Systems: A Dynamical Systems Approach to Machine Intelligence*, Prentice–Hall, Upper Saddle River, NJ, 1999.

E. J. Kostelich and D. Armbruster, *Introductory Differential Equations*, Addison–Wesley, Reading, MA, 1997.

M. R. S Kulenovic and O. Merino, *Discrete Dynamical Systems and Differential Equations with Mathematica*, Chapman and Hall, London, 2002.

H. A. Lauwerier, *Fractals: Images of Chaos*, Penguin, New York, 1991.

J. H. Liu, *A First Course in the Qualitative Theory of Differential Equations*, Prentice–Hall, Upper Saddle River, NJ, 2002.

A. J. Lotka, *Elements of Physical Biology*, William and Wilkins, Baltimore, 1925.

S. Lynch, *Dynmical Systems with Applications Using Maple*, Birkhaüser, Boston, 2001.

R. N. Madan, *Chua's Circuit: A Paradigm for Chaos*, Singapore, World Scientific, 1993.

P. Mandel, *Theoretical Problems in Cavity Nonlinear Optics*, Cambridge University Press, Cambridge, UK, 1997.

B. B. Mandelbrot, *The Fractal Geometry of Nature*, W. H. Freeman, New York, 1983.

The MathWorks, Inc., *Using Simulink (Version 5)*, The MathWorks, Inc., Natick, MA, 2002.

R. M. May, *Stability and Complexity in Model Ecosystems*, Princeton University Press, Princeton, NJ, 1974.

M. Minsky and S. Papert, *Perceptrons*, MIT Press, 1Cambridge, MA, 969.

E. Mosekilde, Y. Maistrenko and D. Postnov, *Chaotic Synchronization*, World Scientific, Singapore, 2002.

H. Nagashima and Y. Baba, *Introduction to Chaos: Physics and Mathematics of Chaotic Phenomena*, Institute of Physics, Bristol, PA, 1998.

V. V. Nemitskii and V. V. Stepanov, *Qualitative Theory of Differential Equations*, Princeton University Press, Princeton, NJ, 1960.

W. Palm III, *Introduction to MATLAB 7 for Engineers*, McGraw–Hill, New York, 2004.

H.-O. Peitgen, H. Jürgens, and D. Saupe, *Chaos and Fractals: New Frontiers of Science*, Springer-Verlag, New York, 1992.

H.-O. Peitgen, H. Jürgens, D. Saupe, and C. Zahlten, *Fractals: An Animated Discussion*, Spektrum Akademischer Verlag, Heidelberg, 1989; W. H. Freeman, New York, 1990.

H.-O. Peitgen (ed.), E. M. Maletsky, H. Jürgens, T. Perciante, D. Saupe, and L. Yunker, *Fractals for the Classroom: Strategic Activities*, Vol. 1, Springer-Verlag, New York, 1991.

H.-O. Peitgen (ed.), E. M. Maletsky, H. Jürgens, T. Perciante, D. Saupe, and L. Yunker, *Fractals for the Classroom: Strategic Activities*, Vol. 2, Springer-Verlag, New York, 1992.

H.-O. Peitgen and P. H. Richter, *The Beauty of Fractals*, Springer-Verlag, New York, 1986.

L. Perko, *Differential Equations and Dynamical Systems*, 3rd ed., Springer-Verlag, New York, 2001.

E. C. Pielou, *Mathematical Ecology*, Wiley, New York, 1977.

R. Pratab, *Getting Started with MATLAB: Version 6: A Quick Introduction for Scientists and Engineers*, Oxford University Press, London, 2001.

Ye Yan Qian, *Theory of Limit Cycles*, Translations of Mathematical Monographs 66, American Mathematical Society, Providence, 1986.

J. Rayleigh, *The Theory of Sound*, Dover, New York, 1945.

D. E. Rumelhart and J. L. McClelland, eds., *Parallel Distributed Processing: Explorations in the Microstructure of Cognition*, Vol. 1, MIT Press, Cambridge, MA, 1986.

I. W. Sandberg (ed.), J. T. Lo, C. L. Fancourt, J. C. Principe, S. Katagiri, and S. Haykin, *Nonlinear Dynamical Systems: Feedforward Neural Network Perspectives*, Adaptive Learning Systems for Signal Processing, Communications and Control Series, Wiley–Interscience, New York, 2001.

S. K. Scott, *Oscillations, Waves, and Chaos in Chemical Kinetics*, Oxford Science Publications, Oxford, UK, 1994.

R. Seydel, *Practical Bifurcation and Stability Analysis: From Equilibrium to Chaos*, Springer-Verlag, New York, 1994.

C. Sparrow, *The Lorenz Equations: Bifurcations, Chaos, and Strange Attractors*, Springer-Verlag, New York, 1982.

M. R. Spiegel, *Schaum's Outline of Laplace Transforms*, Mc-Graw–Hill, New York, 1965.

S. H. Strogatz, *Nonlinear Dynamics and Chaos with Applications to Physics, Biology, Chemistry and Engineering*, Perseus Books, New York, 2001.

J. M. T. Thompson and H. B. Stewart, *Nonlinear Dynamics and Chaos*, 2nd ed., Wiley, New York, 2002.

G. Turrell, *Mathematics for Chemistry and Physics*, Academic Press, New York, 2001.

B. West, S. Strogatz, J. M. McDill, J. Cantwell, and H. Holn, *Interactive Differential Equations*, Version 2.0, Addison–Wesley, Reading, MA, 1997.

S. Wiggins, *Introduction to Applied Nonlinear Dynamical Systems and Chaos*, 2nd ed., Texts in Applied Mathematics 2, Springer-Verlag, New York, 2003.

R. Williams, *Introduction to Differential Equations and Dynamical Systems*, McGraw–Hill, New York, 1997.

C. W. Wu, *Synchronization in Coupled Chaotic Circuits and Systems*, World Scientific, Singapore, 2002.

J. A. Yorke (contributor), K. Alligood (ed.), and T. Sauer (ed.), *Chaos: An Introduction to Dynamical Systems*, Springer-Verlag, New York, 1996.

Zhang Zhifen, Ding Tongren, Huang Wenzao, and Dong Zhenxi, *Qualitative Theory of Differential Equations*, Translation of Mathematical Monographs 102, American Mathematical Society, Providence, 1992.

G. M. Zaslavsky, *Physics of Chaos in Hamiltonian Systems*, World Scientific, Singapore, 1998.

Research Papers

E. Ahmed, A. El-Misiery, and H. N. Agiza, On controlling chaos in an inflation-unemployment dynamical system, *Chaos Solitons Fractals*, **10**-9 (1999), 1567–1570.

M. Alber and J. Peinke, Improved multifractal box-counting algorithm, virtual phase transitions, and negative dimensions, *Phys. Rev.* E, **57**-5 (1998), 5489–5493.

Z. G. Bandar, D. A. McLean, J. D. O'Shea, and J. A. Rothwell, *Analysis of the Behaviour of a Subject*, International Publication Number WO 02/087443 A1, 2002.

N. Bautin, On the number of limit cycles which appear with the variation of the coefficients from an equilibrium point of focus or center type, *Amer. Math. Soc. Trans.*, **5** (1962), 396–414.

T. Bischofberger and Y. R. Shen, Theoretical and experimental study of the dynamic behavior of a nonlinear Fabry-Perot interferometer, *Phys. Rev.* A, **19** (1979), 1169–1176.

S. Blacher, F. Brouers, R. Fayt, and P. Teyssié, Multifractal analysis: A new method for the characterization of the morphology of multicomponent polymer systems, *J. Polymer Sci. B*, **31** (1993), 655–662.

T. R. Blows and N. G. Lloyd, The number of small-amplitude limit cycles of Liénard equations, *Math. Proc. Cambridge Philos. Soc.*, **95** (1984), 359–366.

T. R. Blows and L. M. Perko, Bifurcation of limit cycles from centers and separatrix cycles of planar analytic systems, *SIAM Rev.*, **36** (1994), 341–376.

T. R. Blows and C. Rousseau, Bifurcation at infinity in polynomial vector fields, *J. Differential Equations*, **104** (1993), 215–242.

S. Boccaletti, J. Kurths, G. Osipov, D. L. Valladares, and C. S. Zhou, The synchronization of chaotic systems, *Phys. Rep.*, **366** (2002), 1–101.

J. Borresen and S. Lynch, Further investigation of hysteresis in Chua's circuit, *Internat. J. Bifurcation Chaos*, **12** (2002), 129–134.

M. Buchanan, Fascinating rhythm, *New Scientist*, 3 January (1998), 20–25.

N. P. Chau, Controlling chaos by periodic proportional pulses, *Phys. Lett.* A, **234** (1997), 193–197.

E. S. Cheb-Terrab and H. P. de Oliveira, Poincaré sections of Hamiltonian systems, *Comput. Phys. Comm.*, **95** (1996), 171.

L. A. Cherkas, Conditions for a Liénard equation to have a center, *Differentsial'nye Uravneniya*, **12** (1976), 201–206.

A. B. Chhabra, C. Meneveau, R. V. Jensen, and K. R. Sreenivasan, Direct determination of the $f(\alpha)$ singularity spectrum and its application to fully developed turbulence, *Phys. Rev.* A, **40**-9 (1989), 5284–5294.

C. J. Christopher and N. G. Lloyd, Polynomial systems: A lower bound for the Hilbert numbers, *Proc. Roy. Soc. London Ser.* A, **450** (1995), 219–224.

C. J. Christopher and S. Lynch, Small-amplitude limit cycle bifurcations for Liénard systems with quadratic or cubic damping or restoring forces, *Nonlinearity*, **12** (1999), 1099–1112.

L. O. Chua, M. Itoh, L. Kocarev, and K. Eckert, Chaos synchronization in Chua's circuit, *J. Circuits Systems Comput.*, **3** (1993), 93–108.

W. A. Coppel, Some quadratic systems with at most one limit cycle, in U. Kirchgraber and H. O. Walther, eds., *Dynamics Reported*, Vol. 2, Wiley/Teubner, New York/Stuttgart, 1988, 61–68.

R. H. Day, Irregular growth cycles, *Amer. Econ. Rev.*, **72** (1982), 406–414.

M. Di Marco, M. Forti, and A. Tesi, Existence and characterization of limit cycles in nearly symmetric neural networks, *IEEE Trans. Circuits Systems* 1: *Fund. Theory Appl.*, **49** (2002), 979–992.

W. L. Ditto, S. N. Rausseo, and M. L. Spano, Experimental control of chaos, *Phys. Rev. Lett.*, **65** (1990), 3211–3214.

N. J. Doran and D. Wood, Nonlinear-optical loop mirror, *Optics Lett.*, **13** (1988), 56–58.

F. Dumortier and L. Chengzhi, On the uniqueness of limit cycles surrounding one or more singularities for Liénard equations, *Nonlinearity*, **9** (1996), 1489–1500.

F. Dumortier and L. Chengzhi, Quadratic Liénard equations with quadratic damping, *J. Differential Equations*, **139** (1997), 41–59.

J. Ecalle, J. Martinet, J. Moussu, and J. P. Ramis, Non-accumulation des cycles-limites I, *C. R. Acad. Sci. Paris Sér. I Math.*, **304** (1987), 375–377.

J. P. Eckmann, S. O. Kamphorst, D. Ruelle, and S. Ciliberto, Lyapunov exponents from time series, *Phys. Rev. A*, **34**-6 (1986), 4971–4979.

K. J. Falconer and B. Lammering, Fractal properties of generalized Sierpiński triangles, *Fractals*, **6**-1 (1998), 31–41.

F. S. Felber and J. H. Marburger, Theory of nonresonant multistable optical devices, *Appl. Phys. Lett.*, **28** (1976), 731.

W. J. Firth, Stability of nonlinear Fabry-Perot resonators, *Optics Comm.*, **39**-5 (1981), 343–346.

A. Garfinkel, M. L. Spano, W. L. Ditto, and J. N. Weiss, Controlling cardiac chaos, *Science*, **257** (1992), 1230–1235.

K. Geist, U. Parlitz, and W. Lauterborn, Comparison of different methods for computing Lyapunov exponents, *Progr. Theoret. Phys.*, **83** (1990), 875–893.

H. Giacomini and S. Neukirch, Improving a method for the study of limit cycles of the Liénard equation, *Phys. Rev. E*, **57** (1998), 6573–6576.

G. A. Gottwald and I. Melbourne, A new test for chaos in deterministic systems, *Proc. Roy. Soc. London Ser. A*, **460**-2042 (2004), 603–611.

T. C. Halsey, M. H. Jensen, L. P. Kadanoff, I. Procaccia, and B. I. Shraiman, Fractal measures and their singularities, *Phys. Rev. A*, **33** (1986), 1141.

S. M. Hammel, C. K. R. T. Jones, and J. V. Maloney, Global dynamical behavior of the optical field in a ring cavity, *J. Optim. Soc. Amer.* B, **2** (1985), 552–564.

M. Hénon, Numerical study of quadratic area-preserving mappings, *Quart. Appl. Math.*, **27** (1969), 291–311.

J. J. Hopfield, Neural networks and physical systems with emergent collective computational abilities, *Proc. Nat. Acad. Sci.*, **79** (1982), 2554-2558; reprinted in J. A. Anderson and E. Rosenfeld, eds., *Neurocomputing Foundations of Research*, MIT Press, Cambridge, MA, 1988, 448–453.

J. J. Hopfield and D. W. Tank, Neural computation of decisions in optimization problems, *Biol. Cybernet.*, **52** (1985), 141–154.

J. J. Hopfield, Neurons with graded response have collective computational properties like those of two-state neurons, *Proc. Nat. Acad. Sci.*, **81** (1984), 3088–3092.

Li Hua, D. Ze-jun, and Wu Ziqin, Multifractal analysis of the spatial distribution of secondary-electron emission sites, *Phys. Rev.* B, **53**-24 (1996), 16631–16636.

E. R. Hunt, Stabilizing high-period orbits in a chaotic system: The diode resonator, *Phys. Rev. Lett.*, **67** (1991), 1953–1955.

Y. Il'yashenko, Centennial history of Hilbert's 16th problem, *Bull. Amer. Math. Soc*, **39** (2002), 301–354.

K. Ikeda, H. Daido, and O. Akimoto, Optical turbulence: Chaotic behavior of transmitted light from a ring cavity, *Phys. Rev. Lett.*, **45**-9 (1980), 709–712.

Y. H. Ja, Multiple bistability in an optical-fiber double-ring resonator utilizing the Kerr effect, *IEEE J. Quantum Electron.*, **30**-2 (1994), 329–333.

S. Jal and T. G. Hallam, Effects of delay, truncations and density dependence in reproduction schedules on stability of nonlinear Leslie matrix models, *J. Math. Biology*, **31**-4 (1993), 367–395.

Li Jibin and Li Chunfu, Global bifurcation of planar disturbed Hamiltonian systems and distributions of limit cycles of cubic systems, *Acta Math. Sinica*, **28** (1985), 509–521.

T. Kohonen, Self-organized formation of topologically correct feature maps, *Biol. Cybernet.*, **43** (1982), 59–69; reprinted in J. A. Anderson and E. Rosenfeld, eds., *Neurocomputing Foundations of Research*, MIT Press, Cambridge, MA, 1988, 554–568.

R. E. Kooij and A. Zegeling, Qualitative properties of two-dimensional predator-prey systems, *Nonlinear Anal. Theory Meth. Appl.*, **29** (1997), 693–715.

A. Lasota, Ergodic problems in biology, *Astérisque*, **50** (1977), 239–250.

Y. Lenbury S. Rattanamongkonkul, N. Tumrasvin, and S. Amornsamankul, Predator-prey interaction coupled by parasitic infection: Limit cycles and chaotic behavior, *Math. Comput. Model.*, **30**-9/10 (1999), 131–146.

H. Li and K. Ogusu, Analysis of optical instability in a double-coupler nonlinear fiber ring resonator, *Optics Comm.*, **157** (1998), 27–32.

T. Y. Li and J. A. Yorke, Period three implies chaos, *Amer. Math. Monthly*, **82** (1975), 985–992.

A. Liénard, Étude des oscillations entrenues, *Rev. Gén. Électricité*, **23** (1928), 946–954.

A. Lins, W. de Melo, and C. Pugh, On Liénards equation with linear damping, in J. Palis and M. do Carno, eds., *Geometry and Topology* III, Lecture Notes in Mathematics 597, Springer-Verlag, Berlin, 1977, 335–357.

N. G. Lloyd, Limit cycles of polynomial systems, in T. Bedford and J. Swift, eds., *New Directions in Dynamical Systems*, London Mathematical Society Lecture Notes Series 127, Cambridge University Press, Cambridge, UK, 1988, 192–234.

N. G. Lloyd and S. Lynch, Small-amplitude limit cycles of certain Liénard systems, *Proc. Roy. Soc. London Ser.* A, **418** (1988), 199–208.

E. N. Lorenz, Deterministic non-periodic flow, *J. Atmos. Sci.*, **20** (1963), 130–141.

S. Lynch and Z. G. Bandar, Analysis of a bistable neuromodule, preprint, 2004.

S. Lynch and A. L. Steele, Controlling chaos in nonlinear bistable optical resonators, *Chaos Solitons Fractals*, **11**-5 (2000), 721–728.

S. Lynch and C. J. Christopher, Limit cycles in highly nonlinear differential equations, *J. Sound Vibration*, **224**-3 (1999), 505–517.

S. Lynch, A. L. Steele, and J. E. Hoad, Stability analysis of nonlinear optical resonators, *Chaos Solitons Fractals*, **9**-6 (1998), 935–946.

J. Mach, F. Mas, and F. Sagués, Two representations in multifractal analysis, *J. Phys.* A, **28** (1995), 5607–5622.

J. H. Marburger and F. S. Felber, Theory of a lossless nonlinear Fabry-Perot interferometer, *Phys. Rev.* A, **17** (1978), 335–342.

R. Matthews, Catch the wave, *New Scientist*, **162**-2189 (1999), 27–32.

W. McCulloch and W. Pitts, A logical calculus of the ideas immanent in nervous activity, *Bull. Math. Biophys.*, **5** (1943), 115–133.

S. L. Mills, G. Lees, C. Liauw, and S. Lynch, Dispersion assessment of flame retardent filler/polymer systems using a combination of X-ray mapping and multifractal analysis, *Polymer Testing*, **21** (2002), 941–948.

H. N. Moreira, Liénard-type equations and the epidemiology of malaria, *Ecol. Model.*, **60** (1992), 139–150.

J. Moser, Recent developments in the theory of Hamiltonian systems, *SIAM Rev.*, **28**-4 (1986), 459–485.

J. Muller, O. K. Huseby, and A. Saucier, Influence of multifractal scaling of pore geometry on permeabilies of sedimentary rocks, *Chaos Solitons Fractals*, **5**-8 (1995), 1485–1492.

H. Natsuka, S. Asaka, H. Itoh, K. Ikeda, and M. Matouka, Observation of bifurcation to chaos in an all-optical bistable system, *Phys. Rev. Lett.*, **50** (1983), 109–112.

K. Ogusu, A. L. Steele, J. E. Hoad, and S. Lynch, Corrections to and comments on "Dynamic behaviour of reflection optical bistability in a nonlinear fibre ring resonator," *IEEE J. Quantum Electron.*, **33** (1997), 2128–2129.

E. Ott, C. Grebogi, and J. A. Yorke, Controlling chaos, *Phys. Rev. Lett.*, **64** (1990), 1196–1199.

F. Pasemann, Driving neuromodules into synchronous chaos, in J. Mira and J. V. Sanchez-Andres, eds., *Foundations and Tools for Neural Modeling*, Lecture Notes in Computer Science 1606, Springer-Verlag, New York, 1999, 377–384.

F. Pasemann and N. Stollenwerk, Attractor switching by neural control of chaotic neurodynamics, *Comput. Neural Systems*, **9** (1998), 549–561.

L. M. Pecora and T. L. Carroll, Synchronization in chaotic systems, *Phys. Rev. Lett.*, **64** (1990), 821–824.

H. Poincaré, Mémoire sur les courbes définies par une equation différentielle, *J. Math.*, **7** (1881), 375–422; oeuvre, Gauthier-Villars, Paris, 1890.

F. Rosenblatt, The perceptron: A probabalistic model for information storage and organization in the brain, *Psychological Rev.*, **65** (1958), 386–408.

O. E. Rössler, An equation for continuous chaos, *Phys. Lett.*, **57A** (1976), 397–398.

J. A. Rothwell, The Word Liar, *New Scientist*, March (2003), 51.

R. Roy, T. W. Murphy, T. D. Maier, Z. Gills, and E. R. Hunt, Dynamical control of a chaotic laser: Experimental stabilization of a globally coupled system, *Phys. Rev. Lett.*, **68** (1992), 1259–1262.

G. S. Rychkov, The maximum number of limit cycles of the system $\dot{x} = y - a_0 x - a_1 x^3 - a_2 x^5$, $\dot{y} = -x$ is two, *Differentsial'nye Uravneniya*, **11** (1973), 380–391.

N. Sarkar and B. B. Chaudhuri, Multifractal and generalized dimensions of gray-tone digital images, *Signal Process.*, **42** (1995), 181–190.

S. J. Schiff, K. Jerger, D. H. Doung, T. Chang, M. L. Spano, and W. L. Ditto, Controlling chaos in the brain, *Nature*, **370** (1994), 615.

Chao-Xiang Shi, Nonlinear fiber loop mirror with optical feedback, *Optics Comm.*, **107** (1994), 276–280.

L. P. Shil'nikov, Chua's circuit: Rigorous results and future problems, *Intermat. J. Bifurcation Chaos*, **4** (1994), 489–519.

T. Shinbrot, C. Grebogi, E. Ott, and J. A. Yorke, Using chaos to direct trajectories to targets, *Phys. Rev. Lett.*, **65** (1990), 3215–3218.

K. S. Sibirskii, The number of limit cycles in the neighbourhood of a critical point, *Differential Equations*, **1** (1965), 36–47.

V. Silberschmidt, Fractal and multifractal characteristics of propagating cracks, *J. Phys. IV*, **6** (1996), 287–294.

J. Singer, Y.-Z. Wang, and H. H. Bau, Controlling a chaotic system, *Phys. Rev. Lett.*, **66** (1991), 1123–1125.

S. Smale, Differentiable dynamical systems, *Bull. Amer. Math. Soc.*, **73** (1967), 747–817.

P. W. Smith and E. H. Turner, A bistable Fabry-Perot resonator, *Appl. Phys. Lett.*, **30** (1977), 280–281.

S. D. Smith, Towards the optical computer, *Nature*, **307** (1984), 315–316.

Shi Songling, A concrete example of the existence of four limit cycles for plane quadratic systems, *Sci. Sinica* A, **23** (1980), 153–158.

A. L. Steele, S. Lynch, and J. E. Hoad, Analysis of optical instabilities and bistability in a nonlinear optical fibre loop mirror with feedback, *Optics Comm.*, **137** (1997), 136–142.

A. Szöke, V. Daneu, J. Goldhar, and N. A. Kurnit, Bistable optical element and its applications, *Appl. Phys. Lett.*, **15** (1969), 376.

B. van der Pol, On relaxation oscillations, *Philos. Magazine*, **7** (1926), 901–912, 946–954.

V. Volterra, Variazioni e fluttuazioni del numero d'individui in specie animali conviventi, *Mem. R. Accad. Naz. Lincei*, **2**-3 (1926), 30–113.

B. Widrow and M. E. Hoff, Adaptive switching circuits, in 1960 *IRE WESCON Convention Record*, Part 4, IRE, New York, 1960, 96–104.

H. Zoladek, Eleven small limit cycles in a cubic vector field, *Nonlinearity*, **8** (1995), 843–860.

M. Zoltowski, An adaptive reconstruction of chaotic attractors out of their single trajectories, *Signal Process.*, **80**-6 (2000), 1099–1113.

MATLAB Program File Index

These files can be downloaded at the MathWorks file exchange on the Web.

Program 1a — Solving Recurrence Relations

Program 1b — The Leslie Matrix

Program 2a — Graphical Iteration of the Tent Map

Program 2b — Graphical Iteration of the Logistic Map

Program 2c — Lyapunov Exponent

Program 2d — Bifurcation Diagram of the Logistic Map

Program 2e — Bifurcation Diagram of the Gaussian Map

Program 2f — Iteration of the Hénon Map

Program 3a — A Julia Set

Program 3b — The Mandelbrot Set

Program 3c — Color Mandelbrot Set (on the Web).

Program 4a — Plotting Chaotic Attractors for Ikeda Map

Program 4b — Bifurcation Diagrams for the SFR

Program 5a — The Koch Curve

Program 5b — The Sierpiński Triangle

Program 5c — Barnsley's Fern

Program 5d — τ Curve

Program 5e — D_q Curve

Program 5f — $f(\alpha)$ Curve

Program 6a — Chaos Control in the Logistic Map

Program 6b — Chaos Control in the Hénon Map

Program 7a — Solving Simple Differential Equations Symbolically

Program 7b — Solution Curves, Initial Value Problem

Program 7c — Solving Simple Differential Equations

Program 8a — The Vector Field (Save M-file as vectorfield.m)

Program 8b — Phase Portrait of a Linear System

Program 8c — Phase Portrait of a Nonlinear System

Program 9a — The Holling–Tanner Model

Program 9b — The Holling–Tanner Model (Time Series)

Programs 10a — Vector Field Program (Save M-file as vectorfield.m)

Programs 10b — Limit Cycle of a Van der Pol System

Programs 10c — Nonconvex Limit Cycle of a Liénard System

Programs 11a — Lyapunov Functions

Programs 12a — Animation of a Simple Curve

Programs 12b — Finding Critical Points

Programs 13a — 3D Phase Portrait

Programs 13b — 3D Phase Portrait: Chua's Chaotic Attractor

Programs 13c — The Chapman Cycle

Programs 14a — Solving an Initial Value Problem

Programs 14b — Phase Portraits for Hamiltonian Systems with Two Degrees of Freedom

Programs 14c — Phase Portrait for the Duffing System

Programs 14d — Poincaré Section for the Duffing System

Programs 14e — Periodic Behavior

Programs 14f — Needed for Programs 14g

Programs 14g — Bifurcation Diagram of the Duffing Equation

Programs 15a — Computing the Coefficients of V3 and V4 Lyapunov Quantities for the Liénard System

Programs 17a — The Generalized Delta Learning Rule

Programs 17b — Backpropagation of Errors for a Network with One Hidden Layer

Programs 17c — The Hopfield Network Used as an Associative Mapping

Programs 17d — The Minimal Chaotic Neuromodule

Programs 17e — Bifurcation Diagram for a Bistable Neuromodule

Simulink Model File Index

Simulink1.mdl — Series Resistor Inductor Circuit

Simulink2.mdl — Electric Circuit

Simulink3.mdl — The van der Pol Circuit

Simulink4.mdl — A Periodically Forced Pendulum

Simulink5.mdl — SFR Resonator

Simulink6.mdl — Chaos Control in a SFR Resonator

Simulink7.mdl — The Lorenz Attractor

Simulink8.mdl — Chaos Synchronization for Two Lorenz Systems

Simulink9.mdl — SFR with Gaussian Input

Index

absorptive nonlinearity, 85
AbsTol, 208
activation
    function, 361, 391
    level, 371
    potential, 362
ADALINE network, 363
affine linear transformation, 115
age class, 21, 224
algebraicity of limit cycles, 346
ampere, 172
Ampere's law, 83
anemia, 59
angular frequency of the wave, 84
animate command, 268
ants and termites, 215
aperiodic, 47, 285
    behavior, 282
applying a damper, 145
arrhythmic, 146
Artificial Intelligence Group, 365
artificial neural networks, 360
associative memory, 364, 370

asymptotically stable critical point, 249
asynchronous updating, 375
attractor, 282
attributes, 368
autocatalysis, 289
autonomous
    differential equation, 176
    system, 272
average Lyapunov exponent, 54
axon, 361

backpropagation, 366
    algorithm, 368
backward training iterative process, 72
bandwidth, 84
Barnsley's fern, 116
basin of attraction, 59, 72, 255, 282
batch data, 370
Belousov–Zhabotinski reaction, 287, 294
Bendixson's criteria, 237

450 **Index**

bias, 361
bifurcating limit cycles from a center, 328
bifurcation
   at infinity, 337
   curve, 260
   diagram, 260
      Duffing equation, 313
      for the Hénon map, 57
      for the SFR resonator, 99
      neuromodule, 382
      of the logistic map, 52
   point, 50
   value, 258
bifurction
   diagram
      for the CR resonator, 87
biology, 59
bipolar activation function, 362
bistability, 55, 98, 264, 265, 379
bistable, 57, 85, 265, 290
   cycle, 287
   device, 84, 98
   neuromodule, 381
   optical resonator, 146
   region, 87, 92, 314, 321, 383
   solution, 266
block icon, 398
blowflies, 48
Boston housing data, 368
boundaries of periodic orbits, 74
box-counting dimension, 118, 123
brain functions, 360
butterfly effect, 284

canonical form, 186, 187, 272
Cantor
   multifractal, 128
   set, 110, 316
capacitance, 173
cardioid, 75
cardiology, 143
carrying capacity, 165

cavity ring (CR) resonator, 86
cavity round-trip time, 86
center, 189, 245, 324
   manifold, 277
   theorem, 277
changing the system parameters, 144
chaologist, 144
chaos, 38, 40, 87, 280, 282
   control, 379, 393
      Hénon map, 151
      in an SFR, 405
      periodic proportional pulses, 157
   game, 114
   synchronization, 398, 405
chaotic
   attractor, 93, 282, 310
      Hénon map, 152
      neuromodule, 381
      Sierpiński, 115
   dynamics, 283
   phenomena, 36
Chapman cycle, 293
characteristic
   equation, 18
   exponent, 346
   multiplier, 301
charge density, 83
chemical, 181
   kinetics, 169, 205, 232, 287
   law of mass action, 169
   reaction, 180
   signals, 361
Chua's circuit, 145, 286, 407
circle map, 302
circuit, 231
circular frequency of light, 86
classical symmetry argument, 325
classification of critical points, 192
clipping problem, 126
clockwise
   bistable cycle, 425
   hysteresis, 88

loop, 321
cluster, 364
coarse Hölder exponent, 125
codimension-1 bifurcation, 269
codimension-2 bifurcation, 269
coexistence, 216
col, 188
Command Window, 2
competing species, 215, 227, 242
completely integrable, 304
complex
  eigenvalues, 18, 188
  iterative equation, 94
compound interest, 17
concentration, 169
conductivity, 83
conformal mapping, 70
conservation
  of energy, 244
  of mass, 206
conservative, 244
contact rate, 294
content-addressable memory, 371
control
  curves, 150
  engineering, 365
  parameter, 147
  region, 147
controlling chaos in the logistic map, 148
conversational agents, 366
convex closed curve, 237
core area of the fiber, 90
corollary to Poincaré–Bendixson theorem, 234
correlation dimension, 124
coulomb, 172
Coulomb's law, 173
counterclockwise hysteresis, 88
coupler, 89
critical point, 177, 192, 272
  at infinity, 340
culling, 225

policy, 25
current, 172
  density, 83
cusp, 205
cylindrical polar coordinates, 278

damping, 231
  coefficient, 345
dangerous bifurcation, 264
data mining, 360
databases, 368
defibrillator, 146
degenerate
  critical point, 245
  node, 190
degree, 336
deleted neighborhood, 235
delta learning rule, 363, 366
dendrites, 361
derivative of the Poincaré map test, 301
desired vector, 366
deterministic, 282
  chaos, 144, 282
  system, 360
D_f, 116
dielectric, 84
difference equation, 16, 392
differential amplifiers, 84
diffusion limited aggregates (DLA), 128
dimension, 123
direction
  field, 186
  vector, 186
dispersive nonlinearity, 85
displacement function, 325
divergence test, 325
domain of stability, 72, 217, 282
double-coupler fiber ring resonator, 87, 105
double-scroll attractor, 287
double-well potential, 249

D_q, 123
Duffing
    equation, 309, 404
    system, 145
Dulac's
    criteria, 236
    theorem, 336

E_C, 277
economic model, 20
economics, 61, 67, 212
eigenvector, 190
electric
    circuit, 172, 211, 232, 286, 371, 398, 401
    displacement, 83
        vector, 83
    field, 89, 94, 146
        strength, 82
    flux density, 83
electromotive force (EMF), 174
energy level, 306
enrichment of prey, 224
environmental effects, 224
epidemic, 182, 204, 224, 231, 293
epoch, 364
equilibrium point, 177
ergodicity, 54, 147
error
    backpropagation rule, 368
    function, 366
erythrocytes, 59
E_S, 192, 277
E_U, 192, 277
Euclidean dimension, 123
exact, 165
    differential equation, 165
excitory, 361
existence and uniqueness
    of limit cycle, 232
    theorem, 176
extinct, 216

Fabry–Perot
    interferometer, 85
    resonator, 85
$f(\alpha)$ spectrum, 123
Faraday's law, 173
    of induction, 82
feedback, 84, 97, 265
    mechanism, 97, 382
feedforward single layer network, 363
Feigenbaum constant, 53
fiber parameters, 99
Fibonacci sequence, 7, 31
fine focus, 324
first
    integral, 244
    iterative method, 97, 98, 393
    return map, 298
first-order difference equation, 16
fish population, 32, 164, 268
fixed point, 40, 177
    of period m, 302
    of period N, 45
    of period one, 91, 148, 298
    of period two, 149
fixed-size box-counting algorithm, 126
fixed-weight box-counting algorithm, 126
flow, 232
focal values, 324
fold bifurcation, 269
forced system, 308
forward rate constant, 170
fossil dating, 180
fractal, 110, 117
    attractor, 116, 282
    dimenion
        Cantor set, 117
    dimension, 116
        Koch,
            curve, 117
            square, 117
        Sierpiński triangle, 118
    geometry, 110
    structure, 70, 282, 285

fragmentation ratios, 124
function approximators, 365
fundamental memory, 374
fuzzy discs, 126

gain, 405
Gauss–Newton method, 367
Gauss's law
    for electricity, 83
    for magnetism, 83
Gaussian
    input pulse, 98
    map, 55
    pulse, 404, 414
generalized
    delta rule, 367
    fractal dimensions, 123, 124
    mixed Rayleigh–Liénard equations, 333
global bifurcations, 330, 337
globally asymptotically stable, 279, 293
glucose in blood, 181
gradient, 186
    vector, 367
graph block, 403
graphic, 234
graphical method, 38, 95
Green's theorem, 236
grey scale, 132
Gross National Product (GNP), 61

Hamiltonian, 244
    systems with two degrees of freedom, 304
handcrafted patterns, 379
hard bifurction, 264
Hartman's theorem, 198
harvesting, 225, 268
    policy, 25
Hausdorff
    –Besicovich dimension, 125
    dimension, 123

index, 116
Heaviside function, 362
Hebb's
    learning law, 363
    postulate of learning, 375
Hénon
    –Heiles Hamiltonian, 305
    map, 57, 119, 126
heteroclinic
    bifurcation, 314
    orbit, 234, 248, 314, 338
    tangle, 314
heterogeneous, 121
hidden layer, 364, 368
Hilbert numbers, 336
history of the system, 266
Holling–Tanner model, 221, 241, 258
homoclinic
    bifurcation, 285, 314, 330
    loop, 330, 333
    orbit, 248, 314, 338
    tangle, 314
homogeneous, 121
    differential equation, 167
Hopf bifurcation, 262, 269
Hopfield
    model
        continuous, 371
        discrete, 374
    network, 370, 392
    neural, 365
horseshoe dynamics, 316
housing.txt, 368
human population, 31, 205
hyperbolic
    attracting, 346
    critical point, 198
    fixed point, 58, 301
    iterated function system, 116
    repelling, 346
    stable limit cycle, 301
    unstable limit cycle, 301
hysteresis, 57, 98, 265

Ikeda map, 61, 91, 104, 158
image
    analysis, 127
    compression, 110
incident, 85
index, 239
inductance, 173
infectives, 205
inflation unemployment model, 67
information dimension, 124
inhibitory, 361
initial value problem, 164
input vector, 361
insect population, 32, 227
instability, 98
instant physician, 365
integrable, 304
integrating factor, 168
integrator block, 398
intensity, 89
interacting species, 215, 420
intermittency, 52, 287, 294
    route to chaos, 53
invariant, 92, 232, 285
inverted Koch square, 114
isoclines, 186
isolated periodic solution, 229
isothermal, 205
iterated function system (IFS), 115
iteration, 16

Jacobian matrix, 58, 152, 198, 277, 290
Jordan curve, 235, 349
jth point of period i, 45
Julia set, 70, 73, 110

KAM
    theorem, 307
    tori, 307
kernel machines, 365
Kerr
    effect, 85, 90
    type, 90
kinetic energy, 244
Kirchhoff's
    current law, 173
    laws, 371
    voltage law, 174
Koch
    curve, 111
    snowflake, 137
    square, 112

ladybirds and aphids, 218
laminarize, 146
Laplace transform, 174
large-amplitude limit cycle, 265
    bifurcation, 265
laser, 61, 87, 146, 269
learning
    process, 361
    rate, 367
least mean squared (LMS) algorithm, 363
Legendre transformation, 125
Leslie
    matrix, 22
    model, 21
lie detector, 366
Liénard
    equation, 326
    plane, 345
    system, 231, 237, 241, 330, 345
        large parameter, 349
        local results, 352
    theorem, 355
limit cycle, 229, 233, 280, 290
    hyperbolic, 329
linear
    differential equation, 168
    phase shift, 90, 99
    stability analysis, 94, 178
    transformation, 273
linearization, 197
linearized system, 198

Index   455

Lipschitz
    condition, 176
    continuous, 176
local bifurcation, 337
log–log plot, 121
logic gates, 84
logistic
    equation, 164
    growth, 221
    map, 48, 148
Lorenz
    attractor, 286
    equations, 146, 284, 405
loss in the fiber, 90
Lotka–Volterra model, 218, 258, 293
low-gain saturation function, 362
Lyapunov
    domain of stability, 253
    exponent, 54, 282
    function, 249, 251, 279, 324, 347, 392
        Hopfield network, 371
    quantity, 325, 353
    stability, 370
        theorem, 250
lynx and snowshoe hares, 218

magnetic
    field vector, 82
    flux, 82
magnetostrictive ribbon, 146
Mandelbrot, 118
    set, 72, 74, 110
manifold, 190
MATLAB Central File Exchange, 2
maximal interval (MI) of existence, 176, 182, 232
Maxwell
    –Bloch equations, 84
    –Debye equations, 84
Maxwell's electromagnetic equations, 82
McCulloch–Pitts neuron, 363

mean
    infectivity period, 294
    latency period, 294
    squared error, 391
mechanical
    oscillator, 211
    system, 231, 266, 398, 402
Melnikov
    function, 329
    integral, 328
memory devices, 84
meteorology, 284
method of steepest descent, 366
minimal chaotic neuromodule, 380
mixed fundamental memories, 376
mixing, 40
mortgage assessment, 365
motif, 110
multifractal, 121
    formalism, 121
    Hénon map, 133
    Sierpiński triangle, 133
    spectra, 123
multistability, 264
multistable, 249, 265, 290, 321, 382, 428
murder, 181
mutual exclusion, 216

national income, 20
negative
    limit set, 232
    semiorbit, 232
negatively invariant, 232
net reproduction rate, 27
network architecture, 361
neural network, 61, 360
neurodynamics, 379
neuromodule, 379
neuron, 361, 392
    module, 61
neuronal model, 361
Newton's

law
  of cooling, 181
  of motion, 244
  method, 367
noise, 147, 398, 407
nonautonomous system, 307
nonconvex closed curve, 238
nondegenerate
  critical point, 245, 324
nondeterministic
  chaos, 144, 282
  system, 360
nonexistence of limit cycles, 237
nonhyperbolic
  critical point, 198, 249
  fixed point, 58
nonlinear
  center, 325
  optical loop mirror (NOLM) with feedback, 88
  optics, 61, 398, 403
  phase shift, 90
  refractive index coefficient, 90
nonlinearity, 84, 265
nonperiodic behavior, 282
nonsimple canonical system, 187
normal form, 258
normalized eigenvector, 28
not robust, 221
numerical solutions, 6

occasional proportional feedback (OPF), 145
OGY method, 145
Ohm's law, 172
optical
  bistability, 84
  computer, 84
  fiber, 87
    double-ring, 88
  memories, 84
  resonator, 84, 266
optimal sustainable, 28

orbit, 186, 232
ordinary differential equation (ODE), 162
oscillation of a violin string, 230
Ott, Grobogi, and Yorke (OGY) method, 146
output vector, 361
ozone production, 293

parasitic infection, 228
partial differential equation, 162
partition function, 123
Pascal's triangle, 137
passive circuit, 175
Peixoto's theorem in the plane, 258
pendulum, 244, 253, 309, 321, 402
perceptron, 363
period, 329
  bubblings, 55
  doubling, 287
    bifurcations to chaos, 52
  of limit cycle, 223, 233, 290
  undoublings, 55
period-one
  behavior, 37
  cycle, 280
period-three behavior, 38
period-two, 281
  behavior, 37
periodic
  behavior, 230
  orbit, 329
  windows, 52
periodicity, 36, 40
permittivity of free space, 83
phase
  portrait, 186, 402
  shift, 89
piecewise linear function, 362
pitchfork bifurcation, 261
pixels, 126
planar manifold, 273
plastics, 127

Poincaré
    −Bendixson theorem, 234, 307, 344, 349
    compactification, 338
    map, 57, 146, 233, 284, 298, 329
    section, 298
Poisson brackets, 304
polar coordinates, 189, 339
pole placement technique, 147
pollution, 225
polymer, 127
population, 211
    of rabbits, 206
positive
    limit set, 232
    semiorbit, 232
positively invariant, 232
potato man, 76
potential
    difference, 172
    energy, 244, 249
    function, 249
power, 89
    law, 119
    of a waterwheel, 212
    splitting ratio, 90
Prandtl number, 284
preallocate, 10
predation rate, 221
predator–prey, 231
    models, 218
    system, 227
probe vector, 375
propagation, 90
psychological profiling, 365
Pyragas's method, 145

qth moment, 123
qualitative behavior, 186
qualitatively equivalent, 193
quasiperiodic, 393
    route to chaos, 287
quasiperiodicity, 304

random behavior, 282
rate
    constant, 170
    -determining steps, 170
rationally independent, 303
Rayleigh
    number, 284
    system, 230
reaction rate equation, 170
real distinct eigenvalues, 187
recurrence relation, 16
recurrent neural network, 364, 370
red and grey squirrels, 215
red blood cells, 59
reflected, 85
refractive
    index, 85
    nonlinearity, 85
refuge, 225
regulator poles, 147
relative
    permeabilities, 83
    permittivities, 83
RelTol, 208
repeated real eigenvalues, 189
resistance, 172
restoring
    coefficient, 345
    force, 231
return map, 325
reverse rate constant, 170
reversed fundamental memories, 376
ringing, 321
RLC circuit, 175, 309, 401
robust, 223
Rössler
    attractor, 280
    system, 280
rubbers, 127

saddle point, 188, 245
saddle-node bifurcation, 258
safe bifurcation, 264

scaling, 119, 123
scope, 401
sea lions and penguins, 215
seasonal effects, 225
seasonality, 294
second
    iterative method, 97, 99, 393
    part of Hilbert's sixteenth problem, 336
second-order
    differential equation, 174
    linear difference equation, 18
sedimentary rocks, 128
self-similar, 123
    fractal, 116
self-similarity, 110
semistable
    critical point, 179
    limit cycle, 233, 301
sensitivity to initial conditions, 36, 40, 144, 282
separable differential equation, 162
separation of variables, 162
separatrix, 248
    cycle, 331
sharks and fish, 218
Sierpiński triangle, 114
sigmoid function, 362
signal processing, 128
simple
    canonical system, 187
    fiber ring (SFR) resonator, 86, 89
    nonlinear pendulum, 244
simply connected domain, 237
simulation, 400
singlet, 293
singular node, 189
Smale
    –Birkhoff theorem, 316
    horseshoe map, 59, 314
small
    -amplitude limit cycle, 324

perturbation, 94, 178
soft bifurcation, 264
solar system, 144
solution curves, 163
soma, 361
spatial vector, 82
speed of light, 84
spin-glass states, 376
spirals, 42
spurious steady state, 376
stability, 249, 276
    diagram, 97
stable
    critical point, 177, 249
    fixed point, 49, 72
    focus, 189
    limit cycle, 223, 233
    manifold, 147, 190, 192, 198, 272
    node, 188
staircases, 42
stationary point, 177
steady state, 175, 223
stiff system, 171, 293
stiffness, 231
stochastic methods, 360
stock market analysis, 128
stoichiometric equations, 170
Stokes's theorem, 83
strange attractor, 282, 405
stretching and folding, 36
strictly dominant, 24
structurally
    stable, 223, 258
    unstable, 221, 258
subcritical Hopf bifurcation, 264, 285
subharmonic oscillations, 310
summing junction, 361
supercritical Hopf bifurcation, 264
supervised learning, 364
susceptibles, 205
sustainable, 26
switches, 84

synaptic weights, 361
synchronization, 145, 379
synchronous updating, 376

target vector, 364, 366
targeting, 148
$\tau(q)$, 123
Taylor series expansion, 58, 94, 178, 198
tent map, 36
three-dimensional system, 272
threshold value, 205
time series, 51
    plot, 283
Toda Hamiltonian, 321
tolerance, 400
topological dimension, 123
topologically equivalent, 193
torus, 309
totally
    connected, 72
    disconnected, 72
trajectory, 186, 232
transcritical bifurcation, 260
transfer function, 361, 392
transient, 175
transmitted, 85
transversal, 324
transversely, 298
traveling salesman problem, 370
triangular
    input pulse, 404
    pulse, 98
trivial fixed point, 44
turbulence, 127, 143
two-neuron module, 372

unconstrained optimization problem, 366
uncoupled, 273
uniform harvesting, 28
unipolar activation function, 362

universality, 53
Unix, 1
unstable
    critical point, 177, 250
    fixed point, 49, 72
    focus, 189
    limit cycle, 233
    manifold, 190, 192, 198, 272
    node, 188
unsupervised learning, 364

vacuum, 83
value of homes in Boston, 368
van der Pol
    equation, 402
    system, 230, 329
vector field, 186, 208, 290
    plot, 373
vectorize, 10
velocity of light, 90
Verhulst's equation, 164, 216
viscosity, 284
viscous fingering, 110
volt, 172
voltage drop, 172

wave
    equations, 82
    vector, 84
wavelength, 84
    of light, 90
W_C, 277
Windows, 1
W_S, 198, 272
W_U, 198, 272

XOR gate, 363
X-ray spectroscopy, 127

youngest class harvesting, 27

Z_q, 123